MONOGRAPHS ON THE PHYSICS AND CHEMISTRY OF MATERIALS

General Editors

RICHARD J. BROOK   ANTHONY CHEETHAM
ARTHUR HEUER   SIR PETER HIRSCH
TOBIN J. MARKS   DAVID G. PETTIFOR
MANFRED RUHLE   JOHN SILCOX
ADRIAN P. SUTTON   MATTHEW V. TIRRELL
VACLAV VITEK

MONOGRAPHS ON THE PHYSICS AND CHEMISTRY OF MATERIALS

SMART STRUCTURES

Blurring the Distinction Between the Living and the Nonliving

VINOD K. WADHAWAN

Raja Ramanna Fellow,
Bhabha Atomic Research Centre, Mumbai, India

T A
418.9
S.62 W33
2007

WEB

OXFORD
UNIVERSITY PRESS

OXFORD
UNIVERSITY PRESS

Great Clarendon Street, Oxford OX2 6DP

Oxford University Press is a department of the University of Oxford.
It furthers the University's objective of excellence in research, scholarship,
and education by publishing worldwide in

Oxford New York

Auckland Cape Town Dar es Salaam Hong Kong Karachi
Kuala Lumpur Madrid Melbourne Mexico City Nairobi
New Delhi Shanghai Taipei Toronto

With offices in

Argentina Austria Brazil Chile Czech Republic France Greece
Guatemala Hungary Italy Japan Poland Portugal Singapore
South Korea Switzerland Thailand Turkey Ukraine Vietnam

Oxford is a registered trade mark of Oxford University Press
in the UK and in certain other countries

Published in the United States
by Oxford University Press Inc., New York

© Vinod K. Wadhawan 2007

The moral rights of the author have been asserted
Database right Oxford University Press (maker)

First published 2007

All rights reserved. No part of this publication may be reproduced,
stored in a retrieval system, or transmitted, in any form or by any means,
without the prior permission in writing of Oxford University Press,
or as expressly permitted by law, or under terms agreed with the appropriate
reprographics rights organization. Enquiries concerning reproduction
outside the scope of the above should be sent to the Rights Department,
Oxford University Press, at the address above

You must not circulate this book in any other binding or cover
and you must impose the same condition on any acquirer

British Library Cataloguing in Publication Data

Data available

Library of Congress Cataloging in Publication Data
Wadhawan, Vinod K.
  Smart structures : blurring the distinction between the living and the nonliving /
Vinod K. Wadhawan.
    p. cm.
  Includes bibliographical references.
  ISBN 978-0-19-922917-8
1. Smart materials. 2. Nanoelectromechanical systems. I. Title.
TA418.9.S62W33 2007
624.1–dc22   2007025749

Typeset by Newgen Imaging Systems (P) Ltd., Chennai, India
Printed in Great Britain
on acid-free paper by
Biddles Ltd., King's Lynn, Norfolk

ISBN 978–0–19–922917–8

1 3 5 7 9 10 8 6 4 2

Technology, hailed as the means of bringing nature
under the control of our intelligence,
is enabling nature to exercise intelligence over us.
– George Dyson, *Darwin Among the Machines*

Dedicated to the memory to my younger brother Ajay,
who left this world so unjustly soon.

CONTENTS

PREFACE

Smart or adaptronic structures are defined as structures with an ability to respond in a pre-designed useful and efficient manner to changing environmental conditions, including any changes in their own condition.

Spillman had this to say in 1992: '... it is reasonable to assume that highly integrated smart structures very similar to the biological model will be technologically possible in the near future. These smart structures will need to interface with human beings. Given the level of their anticipated sophistication and adaptive abilities, they will appear as living conscious entities to the majority of those people interacting with them. This will be the case in spite of the fact that they will not meet the formal requirements of either life or consciousness'. Spillman's expectations are being fulfilled sooner than was anticipated at that time. One reason for this is the exponentially increasing computing power at our disposal. Another is the rapid progress being made in the fields of multiferroics, soft matter, and nanostructured functional materials. A third factor is our increasing understanding of how the human brain functions, and the possibility that we may be able to build machines with human-like (i.e. cortical-like) intelligence.

Truly exciting times are ahead in science and engineering. The full potential of smart structures will be realized when the field of nanostructured functional materials comes of age. At length-scales of a few nanometres the traditional boundaries between the physical, chemical, biological and materials sciences begin to lose their meaning. This means, among other things, that the training packages for the research scientists and engineers of the future will have to be rather different from what we have at present.

While there are several good monographs and review articles on smart structures, there is no single book providing a *comprehensive* and updated overview of this ever-growing field, with due emphasis on what has come to be called *computational intelligence*. The present book will hopefully meet that need. I have tried to create a text that is comprehensive without being voluminous, and that maps out the field of smart structures across diverse disciplines. Whereas all the existing books on smart structures have been written by engineers, the present one is perhaps the first to be written by a scientist, with the attendant difference in perspective.

This book is about a subject that involves a large number of cutting-edge and 'advanced' concepts in science and engineering. How to make such a subject accessible to persons who may not have the necessary background for understanding them easily? I have tried to tackle this problem by writing a number of appendices at an introductory level. A fairly large glossary of technical terms is also included. The glossary is an integral part of the book. Sometimes, for the sake of continuity of narrative, a technical term may not be explained in the main text, but is explained in the glossary. I hope that the appendices and the glossary, along with the main text, would help in making the book accessible to a large readership. Unexplained technical jargon has been kept to the minimum.

To understand my choice of contents and the level of narrative and discussion, let us first consider the two extreme scenarios: The first would be that in which each of the highly diverse and large number of topics is treated in substantial depth and detail, leading to the creation of an encyclopaedic text, running into thousands of pages. At the other extreme, one could think of doing a rather perfunctory job of writing a small book at an elementary, semipopular level. I have adopted a middle course, so that portions of the book can be read at all levels, whether as a student or as a research scientist or engineer. While the book should be useful to seasoned researchers, it will also serve as an introduction for newcomers to the field, especially as a guide to the expansive literature.

It is my hope that the availability of this book will induce at least some teachers to design courses on smart structures with contents somewhat on the lines of this book, and substantially different from what is now taught in many universities. Needless to say, this book alone will not suffice as course material for that purpose, and other, more specialized and elaborate, books will also have to be referred to. The books/articles listed in *Further Reading* are my favourites, and will together provide a wealth of material (along with the present book) to the teacher for designing a course of his/her choice.

One of my objectives in writing this book was to put forth my viewpoint about how the subject of smart structures *should* be approached and taught. As the contents of the book demonstrate, the emphasis *has* to shift towards what Kevin Kelly calls *a new biology of machines*. I have tried to present a fairly comprehensive overview of the subject.

The book can also be used as self-study material by a wide variety of readers with some formal exposure to science, preferably up to the graduate or senior undergraduate level (physics, chemistry, mathematics, biology, computer science). To such readers my advice is: If you find some portions a bit too technical for your taste or comfort, just ignore them and move along! On a second reading you may want to explore the rather extensive original references cited by me.

The future of smart structures is linked, among other things, to developments in computer science and brain science. If a truly intelligent artificial brain can indeed be successfully developed, several smart structures will become synonymous with *artificial life* (outside the computer). Such a development will have mind-boggling consequences. It is likely that 'fact' will turn out to be stranger than even recent 'fiction'. Isaac Asimov, the great science-fiction writer, imagined the evolution of humanoid robots a millennium into the future. He died at the age of seventy-two in 1992. That makes him almost a contemporary of the eminent robotics expert Hans Moravec, who wrote two scientific books (not science fiction): *Mind Children: The Future of Robot and Human Intelligence* (1988), and *Robot: Mere Machine to Transcendent Mind* (1999). So far as projections into the near-future go, Asimov was far more conservative than Moravec! The latter predicts that by 2050 (if not earlier) machine intelligence will overtake human intelligence. His central tenet is that the development of low-cost advanced computers (and therefore ultra-smart structures) is inevitable, in spite of the slow progress so far. One of the reasons cited for the slow progress in the past is the lack of adequate economic incentive. But things are changing rapidly. We are living in a technologically explosive age, in which nations will be forced to develop sophisticated smart robots for the sake of economic competitiveness.

Moravec has also made a case that proteins are very fragile material for supporting life and intelligence in harsh conditions, particularly in outer space. They can survive only in a narrow temperature and pressure range, and are very sensitive to radiation. It is therefore inevitable that, aided by human beings, an empire of *inorganic life* will also evolve, just as biological or organic life has evolved. We are about to enter a *post-biological world*, in which machine intelligence, once it has crossed a certain threshold, will not only undergo Darwinian evolution on its own, but will do so millions of times faster than the biological evolution we are familiar with. The result will be smart or intelligent structures with a composite, i.e. organic–inorganic or man–machine intelligence. An important factor responsible for this rapid evolution will be the *distributed* nature of this networked global intelligence.

As a balancing act, we should also take note of the fact that some scientific luminaries have indeed argued *against* the idea of machine intelligence. The subject of ultra-smart structures is bound to raise a serious debate on several ethical, economical, and philosophical aspects of life, consciousness, and existence. For instance, will autonomous robots of the future have a consciousness and an inner life? What will be the nature of the relationship between humans and intelligent robots? The least we can do at present is to make the public at large adequately aware of what smart structures are all about, and about what may be in store for humanity in the near and distant future. This book does its bit by addressing scientists and engineers, at a level of presentation that is not too elementary.

As far as pedagogical considerations go, the subject of smart structures divides itself fairly neatly into *present* or possible smart structures, and *future* or probable smart structures. This book attempts to cover both types, though very briefly. For the next few decades, smart structures will still have use for inorganic single crystals, polymers, ferroic materials (particularly some of the multiferroics), MEMS/NEMS, and other nanostructured functional materials. Progress in soft-matter research, and its marriage with research on nanostructured materials, will throw up some unique and unprecedented types of materials for use in smart structures. Such assemblies will have to make do with whatever computational-intelligence capabilities are available at the time a smart or intelligent device is fabricated.

Opinion is strongly divided on *future* smart structures. For one thing, we need to develop much more computing power for creating thinking robots, or for designing an artificial superbrain (envisaged, for example, by Isaac Asimov in 1950 in his science-fiction book *I, Robot*). After Moore's law has run its course, computing based on three-dimensional semiconductor design, and other kinds like DNA computing, quantum computing, and approaches enabling computing at the level of atomic particles, will hopefully provide the necessary breakthroughs for achieving the computing power (speed and memory capacity) needed for creating truly intelligent artificial structures. The course of progress in brain science is still a question mark. Shall we ever be able to properly understand how the human brain functions, so that we can build something similar, artificially?

While this debate goes on, the *desirability* of developing better and better smart structures cannot be doubted. Stephen Hawking has commented that the present century will be the century of complexity. As we shall see in this book, complexity and evolution have a strong linkage. Evolution of really smart artificial structures will go hand in hand with our improving understanding of complexity in Nature. Apart from the economic and military power such developments will afford to nations which acquire a lead in these aspects of science and technology, there will also be a very welcome fallout for the earth as a whole. The unprecedented levels of efficiency and economy achievable for manufacturing goods of all kinds with the help of (or entirely by) intelligent robots will not only bring prosperity for all, but will also have highly salutary effects on the ecology of our planet. What more can one ask for?

Acknowledgements

There is little or no original material in this book. What is perhaps new is the comprehensive approach to the subject of smart structures, and the choice of material included in the book, with a substantial tilt towards biomimetics. The material itself has come from the work and writings of other people. I have tried to acknowledge my sources, but my apologies if I have not done it adequately enough at any place.

Portions of the book were written during my tenures at the Centre for Advanced Technology, Indore, the Abdus Salam International Centre for Theoretical Physics, Trieste, and the Bhabha Atomic Research Centre, Mumbai. I am happy to acknowledge the support and the facilities I got at these Centres. I am particularly thankful to the Board of Research in Nuclear Sciences (BRNS) of the Department of Atomic Energy, Government of India, for the award of a Raja Ramanna Fellowship, during the tenure of which the book got its revisions and finishing touches.

One trait that will continue to set us apart from humanoid robots for a long time to come is that we humans make friends. I acknowledge Kahlil Gibran for wording the sentiments so aptly on my behalf:

> *And in the sweetness of friendship let there be laughter,*
> *and sharing of pleasures.*
> *For in the dew of little things the heart finds its morning,*
> *and is refreshed.*

A large number of friends and professional colleagues have helped me in various ways. I am particularly grateful to Dr. Anil Kakodkar, Dr. S. Banerjee, Dr. V. C. Sahni, and Dr. R. B. Grover for their help and support. Special mention must also be made of Prof. V. E. Kravtsov, Prof. Subodh Shenoy, Dr. Avadh Saxena,

Prof. A. M. Glazer, Dr. Ashish Ray, Dr. S. L. Chaplot, Dr. Ashwani Karnal, Ms. Pragya Pandit, Mr. Srinibas Satapathy, Dr. P. K. Pal, Dr. Vinay Kumar, and Dr. A. K. Rajarajan for their help. Some of them read portions of the manuscript, and gave their reactions and suggestions for improvement. I also want to thank Rajni, Namrata, Girish, Twishi, and Anupam for the moral support and affection.

It was indeed a matter of pleasure to work with the Oxford University Press team, particularly Mrs. Lynsey Livingston. Their high degree of professionalism contributed substantially to the final product in your hands.

V. K. Wadhawan

LIST OF ABBREVIATIONS

A	Adenine
ACO	Ant colony optimization
A-F	Antiferroelectric–ferroelectric
AFM	Atomic force microscopy
AI	Artificial intelligence
AL	Artificial life
ALR	Anisotropic long range
AMD	Autonomous mental development
AMR	Anisotropic magnetoresistance
ANN	Artificial neural network
ASIC	Application-specific integrated circuit
ASM	Actively smart material
BBD	Brain-based device
BCC	Body-centred cubic
BMR	Ballistic magnetoresistance
BZ reaction	Belousov–Zhabotinsky reaction
C	Cytosine
C_{12}PyCl	n-dodecyl pyridinium chloride
CA	Cellular automata
CAD	Computer-assisted design
CAS	Complex adaptive system
CI	Computational intelligence
CMC	Critical micelle concentration
CMR	Colossal magnetoresistance
CNT	Carbon nanotube
CPU	Central processing unit
DAE	Domain-average engineering
DCC	Dynamical combinatorial chemistry
DCL	Dynamical combinatorial library
DI	Distributed intelligence
DLP	Digital light projector
DMKD	Data mining and knowledge discovery
DNA	Deoxyribonucleic acid
DPT	Diffuse phase transition

DRAM	Dynamic random-access memory
DSE	Domain shape engineering
DWE	Domain wall engineering
EAP	Electroactive polymer
EC	Evolutionary computing
EMR	Enormous magnetoresistance
ENIAC	Electronic numerical integrator and computer
EP	Evolutionary programming
ER	Electrorheological
F-A	Ferroelectric–antiferroelectric
FC	Field cooling
FCC	Face-centred cubic
FL	Fuzzy logic
FPGA	Field-programmable gate array
FPT	Ferroic phase transition
FRBS	Fuzzy-rule-based system
G	Guanine
GA	Genetic algorithm
GL	Ginzburg–Landau
GMR	Giant magnetoresistance
GP	Genetic programming
HEL	Hot-embossing lithography
HI	Humanistic intelligence
IC	Integrated circuit
IR	Infrared
I-V	Current–voltage
ISP	Intelligent signal processing
KDP	Potassium dihydrogen phosphate
KTP	Potassium titanyl phosphate
LBL	Layer by layer
LED	Light-emitting diode
LIGA	German acronym for lithographie, galvanoformung, abformung (lithography, galvanoforming, moulding)
LUT	Look-up table
MB	Machine brain
MBB	Molecular building block
MBE	Molecular-beam epitaxy
MD	Molecular dynamics
MEMS	Microelectromechanical system
MI	Machine intelligence
MIPS	Million instructions per second
MPB	Morphotropic phase boundary
MR	Magnetorheological
mRNA	Messenger RNA

MSME	Magnetic shape-memory effect
MST	Microsystems technology
MT	Machine translation
μCP	Micro-contact printing
NBT	Sodium bismuth titanate
NEMS	Nanoelectromechanical system
NFC	Negative factor counting
NIL	Nanoimprint lithography
NLO	Nonlinear optics, or nonlinear-optical
NITINOL	Nickel titanium (alloy) (developed at) Naval Ordnance Laboratory
NM	Network motif
NN	Neural network
OGY	Ott, Grebogi and Yorke
OOP	Object-oriented program
PA	Polyacetylene
PANI	Polyaniline
PAMPS	Poly(2-acrylamido-2-methyl propane) sulphonic acid
PDI	Periodic domain inversion
PDMS	Poly-(dimethylsiloxane)
PDP	Parallel distributed processing
PE	Polyethylene
PE	Processing element
PLZT	Lead lanthanum zirconate titanate
PMMA	Poly(methyl methacrylate)
PMN	Lead magnesium niobate
PMN-PT	Lead magnesium niobate lead titanate
PP	Polypropylene
PPy	Polypyrrole
PSM	Passively smart material
PT	Lead titanate
PT	Polythiophene
PTFE	Polytetrafluoroethylene, better known as 'Teflon'
PVDF or PVF$_2$	Poly(vinylidene fluoride)
PVF$_2$EF$_3$	Copolymer of PVF$_2$ with trifluoroethylene
P(VDF-TrFE)	Copolymer of PVF$_2$ with trifluoroethylene
PZ	Lead zirconate
PZN-PT	Lead zinc niobate lead titanate
PZT	Lead zirconate titanate
PZTS	Lead zirconate titanate stannate
QD	Quantum dot
QPM	Quasi-phase matching
R-brain	Reptilian brain
RC	Reconfigurable computer

RCC	Reinforced cement concrete
RF	Radio frequency
RKKY interaction	Ruderman–Kittel–Kasuya–Yosida interaction
RMS	Root mean square
RNA	Ribonucleic acid
RP	Rapid prototyping
R-phase	Rhombohedral phase
RSB	Replica symmetry breaking
RTD	Resonant tunnelling device
SAM	Self-assembled monolayer
SAW	Surface acoustic wave
SC	Simple cubic
SET	Single-electron transistor
SMA	Shape-memory alloy
SME	Shape-memory effect
SMFM model	Statistical-multifragmentation-of-nuclei model
SMP	Shape-memory polymer
SNM	Sensory network motif
SOC	Self-organized criticality
SPM	Scanning probe microscope or microscopy
SRAM	Static random-access memory
STM	Scanning tunnelling microscope or microscopy
STN	Sensory transcription network
SWOT	Strengths, Weaknesses, Opportunities, Threats
T	Thymine
TERFENOL	Terbium iron (alloy) (developed at) Naval Ordnance Laboratory
TERFENOL-D	TERFENOL doped with dysprosium
TMO	Transition-metal oxide
TMR	Tunnelling magnetoresistance
TN	Transcription network
T-phase	Tetragonal phase
TSP	Travelling-salesman problem
TWSME	Two-way SME
UV	Ultraviolet
UVM	UV moulding
VANTs	Virtual ants
VCL	Virtual combinatorial library
ZFC	Zero-field cooling

1

INTRODUCTION AND OVERVIEW

There is a single light of science,
and to brighten it anywhere
is to brighten it everywhere.
— Isaac Asimov

1.1 Smart structures

A structure is an assembly that serves an engineering function. A *smart* structure is that which has the ability to respond adaptively in a pre-designed useful and efficient manner to changes in environmental conditions, as also any changes in its own condition.

In conventionally engineered structures, there has been a tendency towards *over-designing*, usually for meeting safety requirements. Ideally, one would like to have 'passive design for purpose, and adaptive design for crises'. By contrast, the usual conventional approach has been to ensure that the passive structure is adequate for crises also, thus entailing higher costs due to over-designing. Some examples wherein, ideally, adaptive action should come into play only for crises or special situations are: buildings in earthquake zones, aircraft during take-off and landing, and vehicles in crashes. A smart configuration would be that in which normal loads are taken care of in normal conditions, and suitable actuation systems are activated to tackle abnormal loads.

Even for normal loads, corrosion and other ageing effects can render the original passive design unsuitable (even unsafe) with the passage of time. If continuous monitoring can be built into the design through distributed, embedded, *smart sensors*, timely repairs can be taken up, thus saving costs and ensuring a higher degree of safety.

Smart bridges are an example of the need for building smart structures. Bridges involve an enormous amount of investment on construction, maintenance, repair, upgrade, and finally replacement. Possible vehicular accidents, earthquakes, and terrorism are additional problems to contend with. Embedding optical fibres as distributed sensors at the construction stage itself is not a very costly proposition (Amato 1992). On-line monitoring and processing of the vast amount of sensor data is again not a difficult thing to do by present-day standards. And the overall advantages in terms of lower maintenance costs, higher safety and security, and avoidance of inconvenience caused by closure for repair work can be enormous, not to mention the prevention of disasters like bridge collapses.

Prehistoric man used materials as they existed, rather than designed them. The next stage in the evolution of what is now called materials science and technology was the deliberate design of materials (alloys, ceramics, composites) with certain desired properties. Bronze, steel, fired clay, and plywood are examples of this type of materials. This trend continues to this day, and the motivation always is to develop the best material for a given task, at a reasonable cost.

There is, however, a limit to the best possible performance one can obtain from a single material. The case of RCC (reinforced cement concrete) illustrates the point. Neither cement nor concrete nor steel can individually match the performance of RCC for the construction of large beams, columns, and floor slabs subject to bending loads. Such loads involve tension on one side of the structural element and compression on the other. Steel has a much better tensile strength than unreinforced concrete, so the structure is designed to transfer the tensile load to the steel in RCC. This results in better performance characteristics for RCC, compared to those of any of its constituents.

RCC is an example of a *composite* material (cf. the appendix on composites). A composite is made from two or more single-phase materials. Conventional composites, as well as other materials (crystals, polycrystals, polymers), ordinarily have *fixed* properties, and a given force or stimulus generates a response *linearly proportional* to the force. There is nothing smart about that. Smartness requires that the response of the material or structure be *changeable*, rather than fixed. Then only can there be an adaptability to changing environmental conditions, with the possibility of pre-designed useful or smart behaviour. This requires *nonlinear* response, among other things. We shall presently see what that means.

One must distinguish between smart *materials* and smart *structures*. We shall take up the still-debatable question of defining smart materials later in this chapter. There is reasonable consensus, however, about the definition of smart *structures*:

Smart or adaptronic structures are structures with an ability to respond in a pre-designed useful and efficient manner to changing environmental conditions, including any changes in their own condition.

As this definition indicates, apart from the need to use the right materials, there is also great scope for smart engineering, and for incorporating expert systems and other tools from the fields of artificial intelligence and neural networks, etc. (Rogers and Giurgiutiu 1999). In fact, as Culshaw (1999) put it, *smart structures are nothing other than a synonym for good engineering*.

Rogers and Giurgiutiu (1999) gave an example of what a well-designed and well-engineered smart structure should be: 'Adaptronic structures must, as their basic premise, be designed for a given purpose; and, by the transduction of energy, must be able to modify their behaviour to create an envelope of utility. As an example, a ladder that is overloaded could use electrical energy to stiffen or strengthen it while alerting the user that the ladder is overloaded. The overload response should also be based upon the actual 'life experience' of the ladder to account for ageing or a damaged rung; therefore, the ladder must determine its current state of health and use this information as the metric for when the ladder has been overloaded. At some point in time, the ladder will then announce its retirement, as it can no longer perform even minimal tasks'.

As far as structures go, human beings are the smartest of them all (even the dumbest among them!). And there is a lot we can learn from how Nature has gone about creating (or rather *evolving*) its smart structures (Thompson, Gandhi and Kasiviswanathan 1992). Let us take a look at this all-important aspect of the subject of smart structures.

1.2 Biomimetics

A smart structure should respond in a pre-designed useful manner to changes in itself or in the environment. This is what biological structures or systems are doing all the time (Phillips and Quake 2006). Figure 1.1 shows the basic essential configuration of a biological system. There are sensors (which we may crudely label as just nerves), actuators (muscles), a control centre (the brain), and the host structure (the body, with or without bones). A source of energy is also needed.

The living system senses the changes in the environment, and the information from the sensors is sent to the brain. The organism has a purpose or objective (for example, the need to survive and sustain itself). In keeping with this objective, the brain sends command signals to the muscles or other actuators to take appropriate action. For example, if the sensors sense excessive heat in the environment, the actuators take the system away from the hot region. All along, there is continuous interaction (*feedback*) among the sensors, the actuators and the decision-making centre(s).

Artificial smart structures are designed to mimic biological systems to a small or large extent. Figure 1.2 describes the basic design philosophy evolved by Nature for all levels of hierarchy of a living system.

One can, in fact, give an alternative definition for smart structures as follows (Takahashi 1992):

Smart structures are those which possess characteristics close to, and, if possible, exceeding, those found in biological structures.

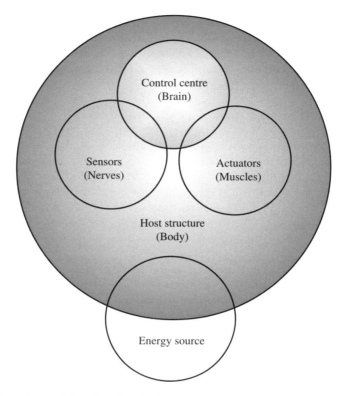

Fig. 1.1 The basic essential configuration of a living system.

Biological structures have several characteristics: sensing; actuation; adaptability; sustainability and survival; selectivity; stability, including multistability; self-diagnosis; self-repair; multifunctionality; self-replication or reproduction; standby properties; switchability; memory; recognition; discrimination; etc. Designers of smart structures strive to achieve as many of these features for the structure as possible, at minimal cost. *Very smart structures* can be defined as those which have several of the biological features listed here; it is a matter of *degree* of smartness, which can vary from case to case.

The conventional approach to design and engineering has been to imagine a worst-case situation and build in enough redundancies and a large safety margin. This can result in over-designing and high costs. The smart-structure approach, by contrast, learns from biological systems and introduces *distributed* and *on-line* sensors, actuators, and microprocessors. The idea is to design a structure or system which senses the changing environment at all times, and takes corrective or preventive action globally or locally, before it is too late. Such a smart structure can be designed for self-diagnosis and self-repair, just like living systems. One tries to replace one-time design and fabrication by on-line or lifelong adaptability (Rogers 1999).

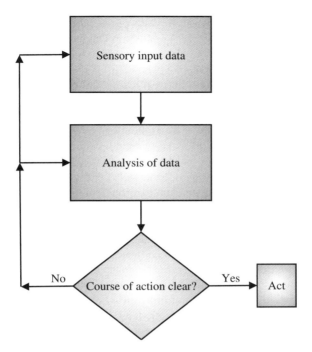

Fig. 1.2   Basic action plan of a living smart structure.

1.3   Bio-inspired computing

If smart structures are to graduate to *intelligent* structures, there has to be a provision for learning and decision-making. This requires inputs from computational science for putting together at least the crude equivalent of the animal brain. There is a deep connection between life processes and computation (or information processing).

The classical *artificial intelligence* (AI) approach was born with the digital computer. Alan Turing was one of the inventors of the idea of the general-purpose computer. He demonstrated the concept of universal computation, viz.: all computers are fundamentally equivalent, regardless of the details of how they are built. This led to the notion of *the Universal Turing Machine*. Turing proved that if you choose the right set of rules for the central processing unit (CPU), and give it an infinitely long tape (memory device) to work with, it can perform any definable set of operations in the universe. In other words, it is all 1s and 0s underneath, and any Turing machine can be programmed to handle it: All digital computers are logically equivalent.

The famous *Turing test* was postulated for defining machine intelligence: 'If a computer can fool a human interrogator into thinking that it too is a person, then by definition the computer must be intelligent'; cf. Moravec (1999a) for an in-depth discussion of Turing's stand on this issue.

This definition of intelligence was *behaviour-based*. The Turing test and the Turing machine helped launch the field of AI. The Central Dogma of AI is: The brain is just another kind of computer. It does not matter how you design an artificially intelligent system, it just has to produce humanlike behaviour.

This approach got a boost from the work of McCulloch and Pitts (1943), who postulated how neurons perform digital functions, viz. by acting as logic gates (a CPU in a computer is just a collection of logic gates).

Such an approach has been viewed as a fallout of *behaviourism*, a dominant trend in psychology during the first half of the twentieth century. Its basic tenets can be summarized as follows (Hawkins and Blakeslee 2004): 'It is not possible to know what goes on inside the brain. It is an impenetrable black box. But one could observe and measure an animal's environment and its behaviour – what it senses and what it does, its inputs and outputs. The brain contains reflex mechanisms that can be used to condition an animal into adopting new behaviours through reward and punishment. There is no need to study the brain, especially messy subjective feelings such as hunger, fear, or what it means to understand something.'

Some notable failures of classical AI are language translation and machine vision. Classical AI has not been a great success because conventional computers and animal brains are built on completely different principles. One is programmed, the other is self-learning. One has to be perfect to work at all, the other is naturally flexible and tolerant of errors. One is fragile, the other robust. One has a central processor, the other has no centralized control.

In the conventional AI approach, one represents problem states by symbols, and defines a set of IF–THEN rules to describe transitions in the problem states. A knowledge base is provided, so that a large number of rules can be defined. The states of the problem are compared against the IF part of the rules. If the matching is successful, the rule involved is *fired*, i.e. the THEN part of the rule is applied for causing a transition from the existing state to a new state. Such an approach, to be sufficiently successful, requires the availability of a very large knowledge base. This degrades the efficiency and efficacy of the AI approach.

One way out of this problem is to have a knowledge base with fewer rules, and to allow *partial matching* of the problem states with the IF part of the rules. This is done by using the logic of *fuzzy sets* (Zadeh 1975).

Another approach is that of working with *artificial neural networks* (ANNs). These are based, albeit only to some extent, on models of the brain and its behaviour. They are a set of 'neurons' working concurrently (Amit 1989; Konar 2005). They can learn system dynamics without requiring *a priori* information regarding the system structure. ANNs are circuits consisting of a large number of *processing elements* (PEs), and designed in such a way as to significantly exploit aspects of *collective behaviour*, rather than rely on the precise behaviour of each PE. ANNs are a distinct conceptual approach to computation, depending in an essential way on statistical-physics concepts. A typical application for ANNs is to help in making decisions based on a large number of input data having comparable *a priori* importance: For instance, reconstructing the characters on a license plate (a few bits of information) from the millions of pixels of a noisy and somewhat blurred and distorted camera image. We shall discuss ANNs in more detail in Chapter 2, along with several other types of bio-inspired computing.

There is a viewpoint that fields like conventional AI and even ANNs have not gone very far because they are not based very substantially on how the human brain really works. ANNs are indeed a genuine improvement over AI. Their architecture is based, though very loosely, on real nervous systems. An ANN is unlike a conventional computer in that it has no CPU, and does not store information in a centralized memory. The ANN's knowledge and memories are distributed throughout its connectivity – just like in a real brain.

According to Hawkins and Blakeslee (2004), even the ANNs leave much to be desired. There is no adequate inclusion of *time* in the ANN function. Real (human) brains possess and process rapidly changing streams of information. Moreover, there is not enough emphasis on *feedback* in ANNs. The real brain is saturated with feedback. It is also necessary that any theory or model of the brain should account for *the physical architecture of the brain*. The neocortex is not a simple structure: it is organized as a repeating *hierarchy*. The ANNs are too simplistic by comparison.

As emphasized by Hawkins and Blakeslee (2004), the trouble with AI and ANNs is that both focus on behaviour. Both assume that intelligence lies in the behaviour that a program or ANN produces after processing a given input. In these approaches, correct or desired output is the most important attribute. As inspired by Turing, 'intelligence equals behaviour', whereas the fact is that one can be intelligent just lying in the dark, thinking and understanding!

Hawkins has put forward his *memory-prediction framework* as a model for under- standing human intelligence. This work is fairly representative of the current thinking in brain science. We shall describe it later in the book (Chapter 7). An important point that emerges from it is this: *It should be possible to build truly intelligent machines on the lines of the human neocortex.* This bodes well for the subject matter of the present book (cf. Moravec 1999b).

There have been approaches other than Hawkins', on somewhat similar lines. 'Consciousness', for example, is not easy to define or understand (Horgan 1994; Velmans 2006; Nunn 2006). Some researchers take the view that if we can model it, howsoever crudely, the very act of modelling it would lead to at least some incre- mental understanding (Aleksander 2003; Mareschal and Thomas 2006). The so-called *global workspace theory of consciousness* represents consciousness as a phenomenon that emerges when a number of sensory inputs, such as images or sounds, activate competing mechanisms in the brain, such as memory or basic feelings like fear or pleasure. These momentarily activated mechanisms then compete with one another to determine the most relevant action.

Action plans are believed to be the basis of conscious thought in this model. Many constituent parts play a role in determining the action plans. One of the designs for a conscious machine starts by assuming that there is *a neural 'depiction' in the brain* that exactly matches every scrap of our inner sensations.

In order to form 'consciousness', these depictions have to have at least five major qualities: (i) a sense of place; (ii) awareness of the past, and imagination; (iii) the ability to focus on what one wants to be conscious about; (iv) the ability to predict and plan; and (v) decision making and emotions.

The artificial smart structures we build will have to draw parallels from the human body and brain for best results, although the task looks quite daunting at present. Several aspects of bio-inspired computing (involving both software and hardware) will be taken up in Chapter 2. We have to build artificial sensors, artificial actuators, artificial brains, and integrate them into working smart structures that supplant and surpass our own capabilities.

1.4   Nonlinear and tuneable response

Field-tuneability of properties of primary sensor and actuator materials is an important requirement for building smart actuators. As stated earlier, this requires nonlinear response.

Suppose a force X is applied on a material, and there is a response Y (the force can be anything: mechanical, electrical, magnetic). Naturally, Y will depend on X:

$$Y = A_1 X \tag{1.1}$$

Here A_1 is a property of the material that is a measure of how strongly or weakly Y depends on X. For example, X can be a compressive force applied to the material along a specified direction, and Y is the change of length produced in the material along that direction. Then, if the material is such and the experimental conditions are such that the familiar Hooke's law is obeyed, A_1 is a *constant* factor that is independent of X. In other words, if we double the value of X, the compression Y is also doubled. If we were to plot the value of Y as a function of X, we would get a straight line for such a system, so we speak of *linear* behaviour or response.

There are many real-life situations in which the proportionality factor is not independent of the value of X. In such cases,

$$Y = A_2(X)X \tag{1.2}$$

$A_2(X)$ may have, for example, the following dependence on X:

$$A_2(X) = A_1 + C_1 X + C_2 X^2 \tag{1.3}$$

If we were to now plot Y as a function of X, we would not get a straight line, but rather a curved line. This is a simple example of nonlinear response.

Knowledge of the exact detailed form of eqn 1.2, or of its variants, is of central importance in the design of smart materials and structures (Banks, Smith and Wang 1996). Nonlinear response implies that the relevant property of the material is not fixed; rather it is field-dependent. This dependence on an external field offers a direct

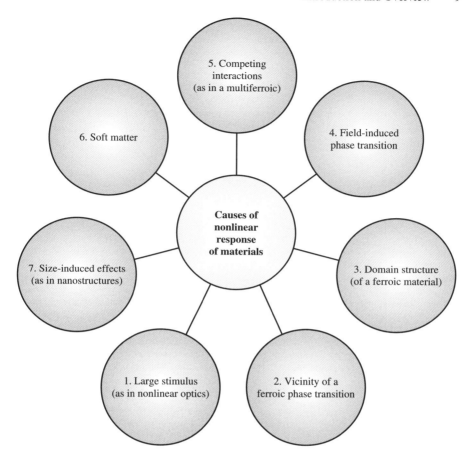

Fig. 1.3 Causes of nonlinear response in materials. Details of the seven causes depicted here are discussed at appropriate places in the book.

way to achieve adaptability for fulfilling pre-designed requirements in certain smart-structure applications (Newnham 1991).

Field-dependence of a property can also be used indirectly for achieving *field-tuneability* through a negative-feedback mechanism. This requires the use of a variable biasing field, and therefore entails the need for a power source.

There are several ways of achieving nonlinear response in materials (Fig. 1.3).

One way is to have a large force field X, so that, although the material normally exhibits linear behaviour (at small X, as shown by Curve 1 in Fig. 1.4), it enters the nonlinear regime because X is large (Curve 2 in Fig. 1.4). This is the central theme of the subject of nonlinear optics. The electric fields associated with laser beams are so high that the nonlinear component of the optical response of a material interacting with a laser beam becomes very significant.

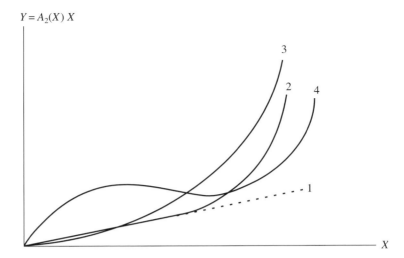

$Y = A_2(X)\, X$

Fig. 1.4   Nonlinear response situations in practice, as embodied in eqn 1.2. The initial portion of Curve 1, a straight line, depicts linear response (for small X), which becomes nonlinear at large X (Curve 2). A ferroic material near a phase transition is inherently nonlinear, even for small X (Curve 3). Such a material may exhibit nonlinear response even when there is no phase transition nearby, because of the domain structure. Curve 4 indicates arbitrarily the strong and complicated response of soft materials to a given stimulus. Curve 4 may also symbolize the behaviour of nanomaterials because of the nonextensive thermostatistics obeyed by them (cf. the appendix on nonextensive thermostatistics), and also because of the strong influence of surrounding materials on their response functions and other properties.

A second way is to work with a material that undergoes certain types of phase transitions (*ferroic* phase transitions), so that its response is inherently nonlinear (even for small X, as indicated by Curve 3 in Fig. 1.4) (Newnham 1998; Wadhawan 2000). We discuss ferroic materials and their phase transitions in Chapter 3.

A third way is to exploit the properties of what has come to be called *soft matter* (Kleman and Lavrentovich 2003). Such materials exhibit a complicated, highly non-linear, dependence of properties on the applied external field (Curve 4 in Fig. 1.4). Some basic features of soft matter are described in Chapters 4 and 5.

Fourthly, one could work with nanostructured materials, which offer exciting possibilities for achieving nonlinear response, particularly *size-dependent* response (Koch 2002; Di Ventra, Evoy and Heflin 2004). In Chapter 6 we consider some relevant properties of nanostructured materials.

In the context of smart structures, it is necessary to enlarge the scope of field-tuneability of properties by bringing in *chemical potential* as another controlling field. A large number of materials used in smart structures have mixed-valence elements in their molecular structure, and their properties can be varied over a wide range by fine-tuning the chemical composition. Change of composition in such materials can have a drastic effect on the oxidation states and band structure, with an attendant change of physical properties (Wang and Kang 1998).

The nonlinearity feature is present, not only in smart materials and structures, but also in smart systems like *adaptive controllers*. An adaptive controller is defined as a controller with adjustable parameters and a mechanism for adjusting the parameters (Astrom and Wittenmark 1995). The parameter-adjustment feature makes such a controller nonlinear. By contrast, a constant-gain feedback system is not an adaptive system.

The very simple description of nonlinear behaviour given here is sufficient for many purposes. We discuss nonlinear dynamical systems in some more detail in an appendix. *Chaos* is a particularly important manifestation of nonlinearity that can be exploited for an effective control of the dynamics of smart structures, and it is discussed in another appendix. Nonlinear systems are a fascinating and challenging field of study. In a general sense, this term encompasses systems with nonlinear interactions among the degrees of freedom (Abarbanel 1994).

1.5   Smart systems, structures, and materials

A *system* is an assembly of components working together to serve one or more *purposes* (Drexler 1992). A *smart system* is a device (or a set of devices) that can sense changes in the environment and then respond in a pre-designed useful or optimal manner, either by changing material properties and geometry, or by changing the mechanical or electromagnetic responsivity, etc. (Varadan, Jiang and Varadan 2001). A conventional temperature-controller module is an example of a smart system. Information about the temperature to be kept constant is fed into the system. Any deviation from this set temperature generates an 'error' signal, and a negative-feedback mechanism determines the amount of corrective action for restoring the temperature to the set value. There is complicated electronic circuitry, and the system comprises of a number of distinct and separate subsystems. The configuration is *not that of an integrated or embedded structure*. If the system were to be split into, say, two parts arbitrarily, it would stop functioning.

In the older literature, a thermistor has been described as an example of a smart *material*. Its resistivity is a pre-designed useful function of temperature. What is more, if it is cut into two or more parts arbitrarily, each part is still a thermistor, which can act as a smart material, adjusting its electrical resistance autonomously against variations in temperature.

Thermistors have been described as *passively* smart materials (PSMs). Many other examples of PSMs can be found: varistors, photochromic glasses, optical limiters, etc. (Davidson 1992). The response of PSMs can be likened to that of the spinal cord of a vertebrate: There is an 'involuntary' reflex action, without involving an overt signal processing.

Let us discuss a varistor to illustrate two features of such materials, namely nonlinear response, and standby phenomena. The electrical resistivity of a ceramic varistor decreases rapidly and highly *nonlinearly* when the applied voltage becomes very high (cf. eqn 1.2, with X standing for voltage, and $A_2(X)$ for resistivity). ZnO

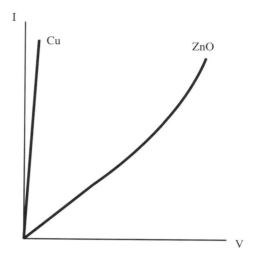

Fig. 1.5   Zinc oxide ceramic as a passively smart material. Compared to copper, it has a high resistivity at low applied voltages (which can be a useful feature in certain applications), but the resistivity varies highly nonlinearly at high voltages and becomes very small. The I–V curves are only schematic, and not to scale.

polycrystalline ceramic is an example of such a material, and it can be used for protection of electronic circuitry against surges in voltage. Its resistance at low voltages is very high, but when a high voltage is applied, the resistance falls rapidly to a low value (Fig. 1.5). The I–V characteristic in this case is an example of a *standby property*, which comes into operation only when needed, i.e. when a high-voltage surge is encountered.

Smart *structures* are not very different from smart *systems*; a highly integrated system may be termed a structure. But more on this later.

Actively smart structures (ASMs) is another term used in the older literature. They not only have pre-designed sensor and actuator features, but, unlike PSMs, they also involve external biasing or feedback. There is thus an externally aided, nonlinear, self-tuneability in their design with respect to one or more properties of the material(s) used (cf. eqn 1.2).

There is a viewpoint that, by and large, a *material* cannot be smart; only a *structure* or *system* designed by a suitable choice and configuration of materials can possibly be smart (Culshaw 1996; Srinivasan and McFarland 2001). Culshaw (1996) has introduced the so-called *information-reduction criterion* in this context:

Culshaw takes the view that there is nothing smart about a material (or a structure) making a *fixed* (linear or nonlinear) response to a stimulus. What is required for smartness is that the response be according to some pre-designed functional dependence on the input parameter, *irrespective of other inputs impinging on the material (or structure)*.

Although the point made by Culshaw is certainly valid (also see Duerig 1995, and Melton 1998), his information-reduction criterion is not quoted widely in the literature. There is, nevertheless, a certain amount of caution now exercised in the use of the phrase 'smart materials'; the expression used more often is: *materials for smart systems* (George *et al.* 1995, 1997; Wun-Fogle *et al.* 2000).

In several publications, smart materials (as also smart structures) are defined as those which have a purpose and which have sensing, actuation and control capabilities. It is reasonable to expect that if a material (or structure) indeed has all these capabilities together, it would, in all probability, also be able to pass Culshaw's test!

A possible distinction between 'smart systems' and 'smart-material systems' is worth noting. The conventional temperature controller mentioned above is an example of the former. Its intended smart action is achieved through a negative-feedback strategy, and by using electronic components which do not involve the use of smart materials. A smart-material system, on the other hand, is a system which involves the conscious use of at least one smart material.

Finally, one should make a distinction between smart structures and *intelligent* structures, reserving the latter term only for applications incorporating cognitive capabilities and adaptive learning in the design, generally entailing the use of fast, real-time, information processing with tools of computational intelligence (Grossman *et al.* 1989).

A distinction is sometimes made between smart systems and smart *adaptive* systems. According to EUNITE (European Network on Intelligent Technologies), 'smart adaptive systems are systems developed with the aid of intelligent technologies including neural networks, fuzzy systems, methods from machine learning, and evolutionary computing and aim to develop adaptive behaviour that converge faster, more effectively and in a more appropriate way than standard adaptive systems. These improved characteristics are due to either learning and/or reasoning capabilities or to the intervention of and/or interaction with smart human controllers or decision makers' (cf. the website http:///www.eunite.org/eunite.index.htm).

1.6   Functional materials and smart materials

It is conventional to view materials as either *structural* or *functional*. RCC mentioned at the start of this chapter is a structural material. It bears load and maintains the integrity of the structure. Changes in the environment, unless they are far too drastic, do not affect it significantly. In a general sense, any *linear-response* application of a material may be described as a structural application (Wang and Kang 1998).

A functional material, by contrast, is significantly sensitive to, say, temperature, external field(s), chemical environment, or any other parameter for which it has been designed to be sensitive in a certain desired way (Wang and Kang 1998). Very often, even the designing is not necessary. A blob of mercury in a thermometer is being a functional material: it changes its volume with temperature significantly enough to function as an indicator of temperature.

A succinct definition of functional materials is as follows (Shenoy, Lookman and Saxena 2005): Functional materials are those with technologically useful properties that can be sensitively controlled by external parameters.

Functional materials may be *passively* functional, or *actively* functional (Cao 1999). The former exhibit strong anomalies in their properties; for example because there is a phase transition nearby. The latter engender transduction of energy; for example from electrical to mechanical through the piezoelectric effect.

Smart materials can be defined as that subset of functional materials which satisfy Culshaw's (1996) information-reduction criterion, described above. They exhibit nonlinear response, and therefore have field-tuneable properties. Such a definition excludes the blob of mercury mentioned above, and deservedly so.

By the same token, a bimetallic strip is *not* a smart material. When strips of two different metals are bonded together, the combination bends with change of temperature because the thermal expansivities of the two metals are different. This can be exploited for making or breaking a circuit. The bimetallic strip is being a sensor and an actuator, but there is nothing smart about it. There is no nonlinearity here for achieving any field-tuneability of response, and the Culshaw criterion is not satisfied. Thus, contrary to what some authors have been writing, a sensor feature plus an actuator feature does not always make a material smart.

The term 'smart materials' is in danger of becoming useless because of excessive and indiscriminate use, particularly in the popular press. We give a couple of examples.

' "Smart" Materials to Fight Smog' was a recent newspaper headline. When ultra-violet rays hit very small particles of titanium dioxide (TiO_2), they trigger a catalytic reaction (*photocatalysis*) that can destroy molecules of nitrogen oxides and some other pollutants. The catalytic reaction also prevents bacteria and dirt from sticking to a surface, so the surface can be cleaned easily (e.g. by rain). So we end up having 'self-cleaning' surfaces (e.g. windows, bathroom tiles, walls of buildings) which also help fight pollution. Who or what is being smart? The material (sub-micron-sized titania particles painted on surfaces), or the human beings who have found a smart use for the material? There is nothing smart about the material in this case. It is just being itself; i.e. just a functional material, with fixed properties.

Here is another example. Properties of materials becomes different at nanometre sizes, and there is nothing smart about that *per se*. Nanosized silver particles have antibacterial properties, and are therefore finding uses in wound dressings, refrigerators, and mobile phones. Stain-proof shirts use nanowhiskers which can repel liquids, just like tiny hairs found on lotus leaves. Smart ideas, but not smart materials.

There is also some tendency in the literature to use the terms 'smart materials' and 'functional materials' in an interchangeable manner, particularly when it comes to *multi*functional materials. Admittedly, a somewhat liberal definition of the term 'smart materials' is quite prevalent, and is as follows: 'Any functional or multifunctional material which is a component of a smart structure or system is a smart material'. The functional material enables the smart structure or system to achieve the 'smart' purpose for which it has been designed. This may happen because of the nonlinear response function(s) of the functional material, or because of its energy transduction capability.

1.7   Classification of structures

Wada, Fanson and Crawley (1990) have given a comprehensive classification of *functional structures*. Smart/intelligent structures constitute a subset of functional structures.

In this scheme, suppose A and B are the two basic types of structures: *Sensory structures* (A) have sensors (for example optical fibres) that provide data about the environment and about the state of the system in which they may be embedded. *Adaptive structures* (B) possess actuators that can alter the state of the system in a pre-designed manner.

Controlled structures (C) are those which combine features of A and B, i.e. those which have both sensors and actuators, with information given by the sensors determining the response of the actuators through a controller. These assemblies may frequently be controlled *systems*, rather than structures, because they may have the controller separate or distinct from the rest of the structure.

Active structures (D) are a subset of controlled structures (C), with a high degree of *integration* of the sensor, actuation, and control functions. Thus, structural functionality and control functionality are integrated into a single whole in the case of D.

In this scheme, *intelligent structures* (E) are a subset of D, and are defined as highly integrated active structures, with sensor, actuator, control, microprocessor, and learning components embedded in a distributed or hierarchic architecture.

An attempt has been made (Rogers 1990) to gain acceptance for the use of the term *intelligent material systems and structures* for what we are simply calling smart structures in this book.

Smartness of structures can be of the *closed-loop* type or *open-loop* type (Srinivasan and McFarland 2001). The latter implies a design that enhances the structural integrity. The former is more in the nature of functional smartness: the structure senses the changes in the environment and takes action to maintain a specified configuration, or to change the configuration in a programmed way. An example of a closed-loop structure (or rather a system) is that of a spacecraft in outer space (in near-zero-gravity conditions), so designed and programmed as to ensure that its antennas point accurately in a specified direction.

1.8   Self-similarity in smart structures

Artificial smart structures mimic biological structures to a small or large extent. Like biological structures, they have a purpose; and adaptability is the main feature to achieve that purpose. Five distinct components can be envisioned for achieving a defined purpose (Spillman 1992): sensing; actuation; structure; communication; and an adaptive control algorithm.

Adaptive ability in the more evolved biological structures exists at several levels of hierarchy, and not only at the highest centralized level of organization, namely the brain. For reasons of efficiency, it is distributed all over the structure, i.e. at each level of hierarchy (although certain decisions must be taken only at the highest-level command centre). At each level of the hierarchy, the basic action plan is the same, namely that shown in Fig. 1.2. In other words, there is *self-similarity* at each level of complexity. This led Spillman (1992) to make the following postulate: *The most efficient smart structure (biological or not) for a given purpose should have a self-similar level of functioning at every hierarchical level of organization; in other words, it should have a fractal character.*

The basket starfish is a particularly vivid and obvious example of fractal design in a living being (Moravec 1999a). The conception of a *bush robot* by Moravec (1999a) is another example of fractal design in a smart structure. The envisioned bush robot is a three-dimensional structure, having repeatedly branching arms that terminate in trillions of microscopic and nanoscopic ultra-dextrous fingers, capable of handling matter at the atomic level.

As discussed in the appendix on chaos, fractal dimensionality is also a characteristic feature of the so-called *strange attractors* associated with chaotic systems. Research has established that the presence of chaos can be beneficial for an efficient and sensitive control of smart structures (Ott and Spano 1995).

1.9   The question of scale

Smart structures are, by and large, *composites*. A composite is made of two or more sub-materials (or constituents, or phases) which are adequately bonded together (cf. the appendix on composites). The fascinating thing about composites is the wide variety of possibilities regarding their composition, symmetry, connectivity, phase transitions, and even connectivity transitions. The question of relative and absolute scales of the constituents is naturally of crucial importance.

RCC is an example of a *structural* composite, with constituent phases occurring on a millimetre to centimetre scale. Wood and particle-board are some other structural composites involving millimetre and sub-millimetre length-scales. Ferrofluids are *functional* composites in which nanometre-sized crystals of a ferromagnetic material are dispersed in a fluid. Thus, several orders of magnitude of length-scales are covered in the wide range of structural and functional composites made by Nature or man.

As one goes down the length-scale, dilemmas regarding some definitions are faced. Crystals, for example, are certainly regarded as single-phase materials, and we speak of transitions in crystals from one phase to another on change of, say, temperature. Yet, we can also view crystals as Nature's own composites (Newnham and Trolier-McKinstry 1990a), with molecules serving as the components from which the crystal composite is constituted.

A composite is made of two or more phases, but the meaning of a phase can become fuzzy at nanometre sizes. There can also be a corresponding blurring of distinction between materials and structures, particularly smart materials and smart structures (and smart systems). This would be all the more so if the idea of self-similarity of smart structures, described above, is used in their design.

1.10   Nanostructured materials

Nanocomposites are becoming increasingly important in the design of smart structures. Therefore it is highly relevant to understand the basics of materials at very small length-scales. Nanostructured materials include atomic clusters, films, filamentary structures, and bulk materials having at least one component with one or more dimensions below 100 nm. The overall progress in the field of smart structures is likely to be influenced very substantially by nanoscience and nanotechnology (Timp 1999; Nalwa 2000; Ying 2001; Koch 2002; Kelsall, Hamley and Geoghegan 2005).

The birth of modern nanotechnology is commonly agreed to have taken place on December 26, 1959, when Richard Feynman gave his famous 'There's Plenty of Room at the Bottom' lecture at the APS Meeting at the California Institute of Technology, Pasadena (Feynman 1960). He argued that it should be possible to assemble molecules and other nanostructures atom-by-atom by self-assembly and other processes. He also envisaged small machines making even smaller machines by manipulating atoms one by one. The scanning tunnelling microscope (STM), which can be used for 'sliding' individual atoms on a surface and taking them to specified locations (Stroscio and Eigler 1991; Birdi 2003; Ng 2004), was invented much later (Binnig and Rhorer 1985).

In 1981 Drexler envisaged great advances in computational devices and in our ability to manipulate biological molecules, based on the projected ability to design protein molecules by positioning reactive groups to atomic precision.

The 1987 Nobel Prize for chemistry was awarded to Donald Cram for his work on the assembly of individual molecules in 'host–guest' chemistry. *Self-assembly* is the buzz word in modern nanotechnology (Wilbur and Whitesides 1999; Kelsall, Hamley and Geoghegan 2005). In Chapter 5 we shall discuss the basics of self-assembly and self-organization in natural phenomena at small length-scales.

The 1997 Nobel Prize for physics was awarded to Chu, Phillips and Tannoudji for their work on laser cooling and trapping of atoms (cf. Metcalf and van der Straten 1999). This has led to the development of *optical tweezers*, a promising micro-manipulation tool. If atoms can be trapped and manipulated, they can also be (often) induced to assemble into molecules and other nanostructures, thus further fulfilling the prophesy of Feynman. Optical tweezers are already finding applications for manipulating biological and other systems (McClelland and Prentiss 1999; McClelland 2000). The living cell is, after all, the smartest and the tiniest of smart structures.

The ultimate use of nanotechnology in the field of smart structures lies in being able to manipulate atoms and molecules individually and placing them exactly where needed, so as to produce a highly integrated structure for a given purpose. In certain ways this is still a distant dream, but there is already implementation of concepts like MEMS, NEMS, and smart dust (MacDonald 1999; Eigler 1999).

MEMS are micro-electromechanical systems comprising of tiny mechanical and other devices like sensors, valves, gears, and actuators, all embedded in semiconductor chips. NEMS go further down in scale from micrometres to nanometres. Large molecules have also been investigated for functioning as nanomachines (Gomez-Lopez and Stoddart 2000). By popular convention, the same acronym MEMS is used for both singular and plural forms of the term; ditto for NEMS.

Advances in digital electronics, laser-driven wireless communication, and MEMS have led to the idea of *motes*. Motes are wireless computing and communicating systems so small (like dust particles) that they can be integrated or introduced into just about anything for creating wireless networks. Through processes like self-assembly of atoms and molecules, as well as by direct micro-manipulation of atoms, sensors the size of dust particles could be mass-produced in very large numbers. They will collect data about the ambient conditions, and transmit information to the network of motes, which, in turn, will initiate corrective or preventive action as programmed.

1.11   The shape of things to come

> *I am convinced that the nations and people who master*
> *the new sciences of complexity will become the*
> *economic, cultural, and political superpowers of the next century.*
> – Heinz Pagels

The subject of smart structures has had modest and down-to-earth beginnings, and this is reflected in our narrative so far, in this introductory chapter of the book. But there are reasons to believe that smart structures of the future may surpass human capabilities, many times over. Although opinion is divided, there are a large number of scientists and engineers who are convinced that smart structures are on a Darwinian-like evolutionary trajectory, and, what is more, their evolution will occur millions of times faster than biological evolution (Moravec 1988, 1999a; Kurzweil 1998).

There are three players in the game of life and evolution: human beings, Nature, and machines. Machines are becoming biological, and the biological is becoming engineered. When the union of the born and the made is complete, our fabrications will learn, adapt, heal themselves, and evolve. The distinction between the born and the made smart structures will become more and more blurred (Kelly 1994; Dyson 1997; Moravec 1988, 1999a; Kurzweil 1998).

Man-made smart structures are on an evolutionary path, in the following sequence:

- Actively smart structures (sensor + feedback + enhanced actuator action).
- Very smart structures (sensor + feedback + enhanced actuator action + one or more other biomimetic features).

- Intelligent structures (actively smart structures with a learning feature, so that the degree of smartness increases with experience).
- *Truly intelligent* structures (with cortical-like intelligence).
- Wise structures capable of taking moral and ethical decisions (Allen, Wallach and Smit 2006).
- Collective intelligence. The internet, a kind of collective intelligence, is already upon us. Moravec and others have envisioned the emergence of a distributed superintelligence, involving robots and humans, or perhaps only robots.
- Man–machine integration. 'Immortality' through repeated repair or replacement of worn out or unserviceable parts (both animate and inanimate) (Lawton 2006). Stem-cell research is already a pointer to what is in store.

In the days to come, we are going to see the coming together of four megatechnologies: advanced materials, nanotechnology, information technology, and biotechnology. Our machines (smart structures included) will evolve, gradually undermining the distinction between technology and Nature. The hardware and the software will produce their own hardware and software, as and when needed and desired (by whom?!).

What will be the role of human beings in such a world? Moravec (1988, 1998, 1999a,b), whose writings have raised some controversy and debate, has given detailed arguments and estimates to make the point that very rapid evolution of smart computers and smart robots (far more advanced and capable than human beings) is inevitable. We shall discuss these arguments at appropriate places in the book.

1.12   Organization of the book

Sensing, decision making, and eventual action or actuation (with appropriate feedback), the basic processes at all levels of hierarchy in any animate or inanimate smart structure, are acts of *information processing*. Biological evolution is a particularly striking example of intelligent processing of information. In fact, evolution and intelligence are thoroughly intertwined. In Chapter 2 we introduce and survey the field of information processing by biological and artificial smart structures, with due emphasis on biological and artificial evolution. The umbrella term *computational intelligence* has come to be used for describing the various approaches to intelligent processing of data. The main sub-fields of the subject of computational science are introduced in Chapter 2, as also the evolutionary approach to computational hardware. An underlying message of this book is the progressively blurring distinction between living and nonliving entities. Salient features of *systems biology* (Alon 2003, 2007; Baker *et al*. 2006), described in Chapter 2, provide an illustration of this situation: The underlying simplicity, modularity, and robustness evolved by Nature in the design of biological circuits has direct parallels in design strategies adopted in electronic engineering and in computational strategies for tackling complex systems.

A typical smart structure comprises of a matrix or supporting structure, in which are embedded sensors, actuators and microprocessors. At least some of the materials constituting a smart structure should possess nonlinear responsivity, so that there is scope for field-tuneability and adaptability of response (Singh 2005). From this point of view, ferroic materials are among the most important materials which can be incorporated into smart structures. We describe them in considerable detail in Chapter 3.

In Chapter 4 we introduce the vocabulary of the subject of soft matter, and discuss briefly the possible role of soft materials in smart structures. Most natural smart structures are, in fact, made of soft matter. Soft matter is inherently nonlinear in response, and its description is continued in Chapter 5, wherein the ubiquitous phenomena of self-assembly and self-organization are considered. After all, the smartest known structures, namely human beings, have evolved through a series of processes of self-assembly and self-organization of molecules. It is becoming increasingly clear that the best approach to develop really smart artificial structures is going to be through a systematic and clever harnessing of the spontaneous processes of self-assembly and self-organization.

As one goes down in size, the surface-to-volume ratio of a material increases. At nanometre sizes, there is not much 'bulk' material left, and a very large fraction is 'surface'. As a result, size-dependent nonlinearity, and even new features not present in the bulk material, can emerge at nanometre sizes (Kelsall, Hamley and Geoghegan 2005). Nanostructured materials therefore merit discussion from the point of view of applications in smart structures. This is done in Chapter 6.

Chapter 7 is devoted mainly to Jeff Hawkins' recently published formalism for human intelligence. He has been particularly emphatic is stating that attempts at artificial intelligence have met with only limited success because they are not modelled on the neocortex of the human brain.

Chapter 8 deals with smart sensor systems. It introduces the basic principles of sensing, and then describes smart sensors, including MEMS. Perceptive networks are also discussed briefly.

In Chapter 9 we consider sensors and actuators for smart structures. There is a sensor–actuator duality when mechanical stress is either the input or the output signal: the same primary material can then be both a sensor and an actuator. For example, the coupling between magnetic and mechanical properties in a magnetostrictive material provides transduction for both actuation, as well as for sensing of a magnetic field (Flatau *et al.* 2000). Similarly, a piezoelectric material can be used as an electrically controlled actuator, and as a sensor of mechanical vibrations (Heffelfinger and Boismier 2000). The basic physics of sensing and actuation by ferroics, soft matter, and nanostructures is described.

Chapter 10 deals with machine intelligence, including distributed intelligence. By its very nature, a good portion of this discussion is somewhat speculative, even controversial. It has never been easy to forecast the future of technology. Brain science is no exception. But, if the optimists can be believed, it is time to get mentally prepared for the future shock that is in store for us when truly intelligent artificial structures make their presence felt.

In Chapter 11 we make some short-term and long-term projections about smart/intelligent structures, after listing briefly their current applications. The next few decades ('short-term projections') will see a continued use of both bulk and nanostructured ferroic materials in smart-structure applications. In the long run, the science and technology of soft matter, as also of nanostructured materials (including ferroic materials), will mature to such an extent that there will be a large overlap with the science and technology of biological systems. Advances in brain science and computation science will have a direct bearing on the development of truly intelligent machines. The ultimate smart structures of the future will be hardly distinguishable from living beings.

Each chapter ends with a very brief 'highlights' section. The highlights singled out are not necessarily of a comprehensive nature; they refer primarily to what is particularly relevant for smart structures.

The appendices provide introduction or review of a number of background topics.

1.13 Chapter highlights

- A smart structure is designed, not only to achieve a purpose, but to do so in an adaptive manner, so as to have features like efficiency, lowered energy costs, speed, survival capability, self-repair, etc.
- Smartness requires that the response of a material or structure be changeable, rather than fixed. Then only can there be an adaptability to changing environmental conditions, with the possibility of pre-designed useful or smart behaviour.
- Nonlinear response is the crux of field-tuneability of properties, so essential for achieving adaptability.
- Three classes of materials offer nonlinear response functions (even for small stimuli): ferroic materials, soft materials, and nanostructured materials.
- The ultimate smart structures of the future are going to be akin to living beings. After all, Nature has been evolving smartness for aeons. Mimicking Nature for building artificial smart structures makes robust sense.
- The biomimetic approach to designing smart structures learns from biological systems and introduces distributed and online sensors, actuators, communication systems, and microprocessors. The idea is to design a structure or system which senses the changing environment at all times, and takes corrective or preventive action globally or locally, before it is too late.
- If smart structures are to graduate to *intelligent* structures, there has to be a provision for learning and decision-making. This requires major inputs from computational science for putting together at least the crude equivalent of the animal brain.
- There is a deep connection between life processes and information processing.
- The toughest challenge for developing truly smart structures is that of mimicking the functioning of the human brain. The burgeoning field of computational intelligence offers hope that this is possible, though not easy.
- There are ample reasons to believe that smart structures of the future will surpass human capabilities. The rate at which low-cost computing power has been

increasing gives hope that truly intelligent structures may be realized, if not by 2050, at least within the present century.
- 'Artificial' smart structures are on a Darwinian-like evolutionary trajectory, and, assisted by us, their evolution will occur millions of times faster than biological evolution.
- The future belongs to 'artificial' smart structures.

References

Abarbanel, H. D. I. (1994). 'Nonlinear systems', in G. L. Trigg (ed.), *Encyclopedia of Applied Physics*, Vol. 11. New York: VCH Publishers.

Allen, C., W. Wallach and J. Smit (July/August 2006). 'Why machine ethics?'. *IEE Intelligent Systems*, 21(4): 12.

Aleksander, I. (19 July 2003). 'I, computer'. *New Scientist*, 179: 40.

Alon, U. (26 September 2003). 'Biological networks: The tinkerer as an engineer'. *Science*, 301: 1866.

Alon, U. (2007). *An Introduction to Systems Biology: Design Principles of Biological Circuits*. London: Chapman and Hall/CRC.

Amato, I. (17 March 1992). 'Animating the material world'. *Science*, 255: 284.

Amit, D. J. (1989). *Modeling Brain Function: The World of Attractor Neural Networks*. Cambridge: Cambridge University Press.

Astrom, K. J. and B. Wittenmark (1995). *Adaptive Control*, 2nd edn. Singapore: Pearson Education.

Baker, D., G. Church, J. Collins, D. Endy, J. Jacobson, J. Keasling, P. Modrich, C. Smolke and R. Weiss (June 2006). 'Engineering life: Building a FAB for biology'. *Scientific American*, 294(6): 44.

Banks, H. T., R. C. Smith and Y. Wang (1996). *Smart Material Structures: Modeling, Estimation, and Control*. New York: Wiley.

Binnig, G. and H. Rhorer (August 1985). 'The scanning tunnelling microscope'. *Scientific American*, 253(2): 40.

Birdi, K. S. (2003). *Scanning Probe Microscopes: Applications in Science and Technology*. London: CRC Press.

Cao, W. (1999). 'Multifunctional materials: the basis for adaptronics'. In H. Janocha (ed.), *Adaptronics and Smart Structures: Basics, Materials, Design, and Applications*. Berlin: Springer-Verlag.

Culshaw, B. (1996). *Smart Structures and Materials*. Boston: Artech House.

Culshaw, B. (1999). 'Future perspectives: Opportunities, risks and requirements in adaptronics', in H. Janocha (ed.), *Adaptronics and Smart Structures: Basics, Materials, Design, and Applications*. Berlin: Springer.

Davidson, R. (April 1992). 'Smart composites: where are they going?' *Materials and Design*, 13: 87.

Di Ventra, M., S. Evoy and J. R. Heflin (eds.) (2004). *Introduction to Nanoscale Science and Technology*. Dordrecht: Kluwer.

Drexler, K. E. (1992). *Nanosystems: Molecular Machinery, Manufacturing and Computation*. New York: Wiley.

Duerig, T. W. (1995). 'Present and future applications of shape memory and superelastic materials', in E. P. George, S. Takahashi, S. Trolier-McKinstry, K. Uchino and M. Wun-Fogle (eds.), *Materials for Smart Systems*. MRS Symposium Proceedings, Vol. 360. Pittsburgh, Pennsylvania: Materials Research Society.

Dyson, G. B. (1997). *Darwin Among the Machines: The Evolution of Global Intelligence*. Cambridge: Perseus Books.

Eigler, D. (1999). 'From the bottom up: Building things with atoms', in G. Timp (ed.), *Nanotechnology*. New York: Springer-Verlag.

Feynman, R. P. (1960). 'There is plenty of room at the bottom'. *Caltech Engineering and Science Journal*, 23: 22. The transcript of this famous lecture given by Feynman in 1959 in also published in Vol. 1 of *J. MEMS* (1992). It is also available at the website http://www.zyvex.com/nanotech/feynmann.html.

Flatau, A. B., M. J. Dapino and F. T. Calkins (2000). 'On magnetostrictive transducer applications', in M. Wun-Fogle, K. Uchino, Y. Ito and R. Gotthardt (eds.), *Materials for Smart Systems III*. MRS Symposium Proceedings, Vol. 604. Warrendale, Pennsylvania: Materials Research Society.

George, E. P., S. Takahashi, S. Trolier-McKinstry, K. Uchino and M. Wun-Fogle (eds.) (1995). *Materials for Smart Systems*. MRS Symposium Proceedings, Vol. 360. Pittsburgh, Pennsylvania: Materials Research Society.

George, E. P., R. Gotthardt, K. Otsuka, S. Trolier-McKinstry and M. Wun-Fogle (eds.) (1997). *Materials for Smart Systems II*. MRS Symposium Proceedings, Vol. 459. Pittsburgh, Pennsylvania: Materials Research Society.

Gomez-Lopez, M. and J. F. Stoddart (2000). 'Molecular and supramolecular nanomachines', in H. S. Nalwa, (ed.), *Handbook of Nanostructured Materials and Nanotechnology*, Vol. 5, page 225. New York: Academic Press.

Grossman, B., T. Alavie, F. Ham, J. Franke and M. Thursby (1989). 'Fibre-optic sensor and smart structures research at Florida Institute of Technology', in E. Udd (ed.), *Proc. SPIE (Fibre Optic Smart Structures and Skins II)*, 1170: 123.

Hawkins, J. and S. Blakeslee (2004). *On Intelligence*. New York: Times Books (Henry Holt).

Heffelfinger, J. R. and D. A. Boismier (2000). 'Application of piezoelectric materials for use as actuators and sensors in hard disc drive suspension assemblies', in M. Wun-Fogle, K. Uchino, Y. Ito and R. Gotthardt (eds.), *Materials for Smart Systems III*. MRS Symposium Proceedings, Vol. 604. Warrendale, Pennsylvania: Materials Research Society.

Horgan, J. (July 1994). 'Can science explain consciousness?' *Scientific American*,: p. 72.

Kelly, K. (1994). *Out of Control: The New Biology of Machines, Social Systems, and the Economic World*. Cambridge: Perseus Books.

Kelsall, R., I. Hamley and Geoghegan (eds.) (2005). *Nanoscale Science and Technology*. Chichester: Wiley.

Kleman, M. and O. D. Lavrentovich (2003). *Soft Matter Physics: An Introduction*. New York: Springer-Verlag.

Koch, C. C. (2002). *Nanostructured Materials: Processing, Properties, and Applications*. New York: Noyes Publications.

Konar, A. (2005). *Computational Intelligence: Principles, Techniques and Applications*. Berlin: Springer-Verlag.

Kurzweil, R. (1998). *The Age of Spiritual Machines: When Computers Exceed Human Intelligence*. New York: Viking Penguin.

Lawton, G. (13 May 2006). 'The incredibles'. *New Scientist*, 190: 32.

MacDonald, N. C. (1999). 'Nanostructures in motion: Micro-instruments for moving nanometer-scale objects', in G. Timp, (ed.), *Nanotechnology*. New York: Springer-Verlag.

McClelland, J. J. (2000). 'Nanofabrication via atom optics', in H. S. Nalwa (ed.), *Handbook of Nanostructured Materials and Nanotechnology*, Vol. 1, page 335. New York: Academic Press.

McClelland, J. J. and M. Prentiss (1999). 'Atom optics: Using light to position atoms', in G. Timp (ed.), *Nanotechnology*. New York: Springer-Verlag, 1999.

McCulloch, W. and W. Pitts (1943). 'A logical calculus of the ideas immanent in nervous activity'. *Bull. Math. Biophys.* 5: 115.

Melton, K. N. (1998). 'General applications of SMA's and smart materials'. In Otsuka, K. and C. M. Wayman (eds.), *Shape Memory Materials*. Cambridge: Cambridge University Press.

Metcalf, H. J. and P. van der Straten (1999). *Laser Cooling and Trapping*. New York: Springer-Verlag.

Moravec, H. (1998). 'When will computer hardware match the human brain?' *J. Evolution and Technology*, 1: 1.

Moravec, H. (1999a). *Robot: Mere Machine to Transcendent Mind*. Oxford: Oxford University Press.

Moravec, H. (December 1999b). 'Rise of the robots'. *Scientific American*,, p. 124.

Nalwa, H. S. (ed.) (2000). *Handbook of Nanostructured Materials and Nanotechnology*. New York: Academic Press.

Newnham, R. E. (1991). 'Tunable transducers: Nonlinear phenomena in electroceramics', in *Chemistry of Electronic Ceramic Materials*, Special Publication 804, National Institute of Standards and Technology. (Proceedings of the International Conference held at Jackson, WY, August 17–22, 1990. Issued January 1991).

Newnham, R. E. (1998). 'Phase transformations in smart materials' *Acta Cryst.*, A54: 729.

Newnham, R. E. and S. E. Trolier-McKinstry (1991a). 'Crystals and composites' *J. Appl. Cryst.*, 23: 447.

Ng, K.-W. (2004). 'Scanning probe microscopes', in M. S. Di Ventra, S. Evoy and J. R. Heflin (eds.), *Introduction to Nanoscale Science and Technology*. Dordrecht: Kluwer.

Nunn, C. (24 June 2006). 'What makes you think you're so special?' *New Scientist*, 190: 58.

Ott, E. and M. Spano (1995). 'Controlling chaos.' *Physics Today*, 48(5): 34.

Phillips, R. and S. R. Quake (May 2006). 'The biological frontier of physics', *Physics Today*, 59: 38.

Rogers, C. A. (1990). 'Intelligent material systems and structures', in I. Ahmad, A. Crowson, C. A. Rogers and M. Aizawa (eds.), *U.S.–Japan Workshop on Smart/Intelligent Materials and Systems*. Lancaster: Technomic Publishing.

Rogers, C. A. (1999). 'Introduction', in H. Janocha (ed.), *Adaptronics and Smart Structures: Basics, Materials, Design, and Applications*. Berlin: Springer.

Rogers, C. A. and V. Giurgiutiu (1999). 'Concepts of adaptronic structures', in H. Janocha (ed.), *Adaptronics and Smart Structures: Basics, Materials, Design, and Applications*. Berlin: Springer.

Shenoy, S. R., T. Lookman and A. Saxena (2005). 'Spin, charge, and lattice coupling in multiferroic materials', in A. Planes, L. Manosa and A. Saxena (eds.), *Magnetism and Structure in Functional Materials*. Berlin: Springer.

Singh, J. (2005). *Smart Electronic Materials: Fundamentals and Applications*. Cambridge: Cambridge University Press.

Spillman, W. B., Jr. (1992). 'The evolution of smart structures', in B. Culshaw, P. T. Gardiner and A. McDonach (eds.), *Proceedings of the First European Conference on Smart Structures and Materials*. Bristol: Institute of Physics Publishing. Also published as SPIE Volume 1777.

Srinivasan, A. V. and D. M. McFarland (2001). *Smart Structures: Analysis and Design*. Cambridge: Cambridge University Press.

Stroscio, J. A. and D. M. Eigler (1991). 'Atomic and molecular manipulation with scanning tunnelling microscope.' *Science*, 254: 1319.

Takahashi, K. (1992). 'Intelligent materials for future electronics', in G. J. Knowles (ed.), *Active Materials and Adaptive Structures*. Bristol: Institute of Physics Publishing.

Thompson, B. S., M. V. Gandhi and S. Kasiviswanathan (1992). 'An introduction to smart materials and structures'. *Materials & Design*, 13(1): 3.

Timp, G. (ed.) (1999). *Nanotechnology*. New York: Springer-Verlag.

Varadan, V. K., X. Jiang and V. V. Varadan (2001). *Microstreolithogrphy and Other Fabrication Techniques for 3D MEMS*. New York: Wiley.

Velmans, M. (25 March 2006). 'In here, out there, somewhere'. *New Scientist*, 189: 50.

Wada, B. K., J. L. Fanson and E. F. Crawley (1990). 'Adaptive structures'. *J. Intelligent Material Systems and Structures*, 1(2): 157.

Wadhawan, V. K. (2000). *Introduction to Ferroic Materials*. Amsterdam: Gordon and Breach.

Wang, Z. L. and Z. C. Kang (1998). *Functional and Smart Materials: Structural Evolution and Structure Analysis*. New York: Plenum Press.

Wilbur, J. L. and G. M. Whitesides (1999). 'Self-assembly and self-assembled monolayers in micro- and nanofabrication', in G. Timp (ed.), *Nanotechnology*. New York: Springer-Verlag.

Wun-Fogle, M., K. Uchino, Y. Ito and R. Gotthardt (eds.) (2000). *Materials for Smart Systems III*. MRS Symposium Proceedings, Vol. 604. Warrendale, Pennsylvania: Materials Research Society.

Ying, J. Y. (ed.) (2001). *Nanostructured Materials*. New York: Academic Press.

Zadeh, L. A. (1975). 'The concept of a linguistic variable and its applications to approximate reasoning. Parts I, II, III'. *Information Sciences*, 8–9: 199–249, 301–357, 43–80.

2

INFORMATION PROCESSING BY BIOLOGICAL AND ARTIFICIAL SMART STRUCTURES

What lies at the heart of every living thing is not a fire,
not warm breath, not a 'spark of life'.
It is information, words, instructions.
– Richard Dawkins, *The Blind Watchmaker*

Sensing, actuation, computation, and control, all involve information processing. This is true for both animate and inanimate systems. Intelligent systems process information intelligently.

There is a deep connection between intelligence and evolution. In biological systems, information about internal and external conditions is processed, and the system evolves towards better and better fitness for survival and efficiency. Therefore, in this chapter, among other things, we shall take a look at the evolutionary aspect of smart structures and computational systems (both living and nonliving).

Some excellent books on information-processing and evolutionary phenomena in machines and self-organizing systems have appeared in recent times: Moravec (1988, 1999a); Kelly (1994); Dyson (1997); Sipper (2002); Konar (2005); Zomaya (2006); Alon (2007). The material in this chapter is only a brief summary of what is covered in more detail in such books.

Human beings have wondered, from times immemorial, about the logical structure of their thinking process. The idea that the brain is a computer has been analysed from several angles. There have been attempts to formalize the correspondence between mathematical systems of symbols and our mental system of thoughts and ideas. This correspondence exists, not only at an abstract level between mathematical logic and the human thinking process, but also at a physical level between machines and the human body and brain. Both have evolved over time. Understanding one helps understand the other. For example, an understanding of the simplicity, modularity, robustness, and hierarchy observed in the design principles evolved by Nature for information processing through 'biological circuits' (Alon 2007) is bound to have an enabling effect on the development of sophisticated artificial smart structures that can 'think'.

2.1   A historical perspective

It is instructive to have at least a nodding acquaintance with the history of logical thought, and with the basics of the interplay of natural processes, that led to the evolution of computational intelligence, both biological and artificial.

- The seeds of a theory of natural evolution were sown as early as the middle of the seventeenth century by Thomas Hobbes (1651) in his book *Leviathan; or, The Matter, Forme, and Power of a Commonwealth Ecclesiasticall and Civill.* Hobbes's irreverent ideas generated an enormous amount of controversy and resistance. Hobbes argued that human society as a whole has evolved as a self-organizing system, possessed of a life and *diffuse* intelligence of its own. He wanted to construct a purely materialistic natural philosophy of the mind. He believed life and mind to be the natural consequences of matter, when suitably arranged. He established that logic and digital computation share common foundations, suggesting a basis common with the mind.
- The coming of the age of digital computers was anticipated by Leibniz (1675, 1714). Although he is better known for his invention of calculus (independently of Newton), he also demonstrated the operation of mechanical calculating machines, based on binary logic. He anticipated that a machine and its parts, no matter how complicated, could be described to the human mind in words (which, in turn, are nothing but an arrangement of the letters of the alphabet). This meant that the mind could know everything about a machine and its parts in terms of just 0s and 1s.
- Jean-Baptiste Lamarck (1744–1829) made lasting contributions to biological science. However, he is best remembered for his mistaken belief in the inheritance of *acquired* characteristics in biological evolution (cf. Francis Darwin 1903). As we shall see later in this book, Lamarckism is certainly relevant today, and is likely to be revived in *artificial* evolution, i.e. for the evolution of *machines.*
- Charles Darwin's (1859) book, *On the Origin of Species*, caused one of the greatest intellectual revolutions of all time (cf. Wilson 2005). He dislodged human beings from their perceived lofty pedestal of being unique, by showing that they share their ancestry with other living creatures. The basic idea was rather simple: In a population there is bound to be some variation among the individuals, making some of them fitter than others for coping with the environment in which they

live. Since they must compete for the limited resources, the fittest are more likely to survive and procreate. Over time, this cumulative process of natural selection must lead to the evolution of the species towards a state of maximum compatibility with the surroundings. If the surroundings change, new species can also evolve. This means that all living beings have a common ancestor, or a few ancestors.

- Samuel Butler (1835–1902) was a contemporary and bitter critic of Charles Darwin. He was offended by the failure of Charles Darwin to give due credit to the earlier evolutionists, particularly Lamarck (1744–1829) and his own grandfather Erasmus Darwin (1731–1802). Butler (1863) had this to say about the evolution of machines: ' ... *It appears to us that we are ourselves creating our own successors ... giving them greater power and supplying by all sorts of ingenious contrivance that self-regulating, self-acting power which will be to them what intellect has been to the human race'*. One cannot help noticing the eerie prophetic quality of these words as one leafs through the pages of Moravec's (1999a) recent book *Robot: Mere Machine to Transcendent Mind*.

- Babbage (1864) fulfilled the prophesy of Leibniz when he introduced, and described in great detail, his *Analytical Engines*. He also introduced what he called his *Mechanical Notation* for laying down the specifications. His work was greatly influenced by the ideas of Leibniz about computation. He anticipated the modern formal languages and timing diagrams in computational science, and was the first to use punched-card peripheral equipment. He was also the first to use reusable coding; the cards, once punched for a given computation, could be used for the same job any time in the future as well. His work was the forerunner of the notion, demonstrated a hundred years later by Alan Turing, that, given an unlimited time and supply of cards or paper tape, even a very simple analytical engine can compute any computable function.

- The vision of Leibniz for the formalization of mental processes and logic was brought closer to accomplishment by the work of George Boole (1854). He developed a precise system of symbolic logic, namely the now well-known *Boolean algebra*. Boolean algebra has provided the mathematical foundations of logic, set theory, lattice theory, and topology. It is an exact, all-or-nothing system of binary logic, completely intolerant of error or ambiguity. Boole's (1854) book also dealt with fuzzy, i.e. probabilistic and statistical, logic, although not in much detail. He showed how individually indeterminate phenomena can be processed digitally to produce logically deterministic results. This was of relevance to how the brain functions. Boole recognized (as did von Neumann later) that the seemingly exact logic of the human brain must be the result of statistical averaging of actions of the very large number of 'imperfect' neurons.

- Smee (1849) carried forward the investigations into the nature of brain and mind. He defined consciousness as follows: 'When an image is produced by an action upon the external senses, the actions on organs of action concur with the actions in the brain; and the image is then a *Reality*. When an image occurs to the mind without a corresponding simultaneous action of the body, it is called a *Thought*. The power to distinguish between a thought and a reality is called *Consciousness*'.

His work had rudiments of the theory of neural networks (NNs). He visualized correctly the role of what we now call 'excitatory synapses' in brain science, but could not think of the equally important inhibitory synapses. Smee also introduced concepts of what we now know as pixelization, bit mapping, and image compression.

- Leibniz (as also Hilbert) had visualized a world in which *all* truths can be calculated and proved by a 'universal coding'. Godel (1931) proved that this is not possible. He proved the theorem (cf. Dyson 1997; Chaitin 2006) that no formal system encompassing elementary arithmetic can be at the same time both consistent and complete (syntactically or semantically). Within any sufficiently powerful and noncontradictory system of language, logic, or arithmetic, it is possible to construct true statements that cannot be proved within the boundaries of the system itself. The famous *Godel sentence* is loosely equivalent to 'This statement is unprovable'. Godel proved another incompleteness theorem, according to which no formal system can prove its own consistency.

- Hilbert had identified the so-called *decision problem* for defining a closed mathematical universe: Does there exist a decision procedure that, given any statement expressed in the given language, will always produce either a finite proof of that statement, or else a definite construction that refutes it, but never both? In other words, can a precisely mechanical procedure distinguish between provable and disprovable statements within a given system? The answer to this question required a way to mathematically define 'mechanical procedure'. Alan Turing (1936) provided an answer by proving that the notions of 'mechanical procedure', 'effectively calculable', and 'recursive functions' were actually one and the same thing. Recursive functions are those that can be defined by the accumulation of elementary component parts; they can be deconstructed into a finite number of elemental steps. Turing constructed an imaginary device, now called *the Turing machine*. It consisted of a black box (it could be a typewriter or a human being) able to read and write a finite alphabet of symbols to and from an unbounded length of paper tape, and capable of changing its own 'm-configuration' or 'state of mind'. In the context of computational science, Turing introduced the all-important notions of *discrete time* and *discrete state of mind*. This made logic a sequence of cause-and-effect steps, and mathematical proof a sequence of discrete logical steps. The Turing machine embodies a relationship between a sequence of symbols in space and a sequence of events in time. Each step in the relationship between the tape and the Turing machine is governed by what is now called a program or algorithm. The program provides instructions to the machine for every conceivable situation. Complicated-looking behaviour does not require a complicated state of mind. Turing's work also established that *all digital computers are equivalent*, this idea being embodied in the concept of the *universal computing machine*.

  In other words, Turing demonstrated that all symbols, all information, all meaning and all intelligence that can be described in words or numbers can be encoded (and transmitted) as binary sequences of finite length (cf. Dyson 1997).

Turing's work marked the advent of modern computing, as well as of the field of artificial intelligence (AI). In 1950 he introduced his famous *Turing test* for deciding whether a computer is exhibiting intelligent behaviour. According to this test, we should consider a machine to be intelligent if its responses are indistinguishable from those of a human being.

As argued recently by Hawkins and Blakeslee (2004), this equating of intelligence with human behaviour was an unfortunate presumption, as it limited our vision of what is possible in machine intelligence. As we shall discuss in Chapters 7 and 10, if we can first understand what intelligence really is, we can develop intelligent machines which can be far more useful and powerful than merely replicating human behaviour.

- In 1951 John von Neumann published his theory of *self-reproducing cellular automata*, which endowed Turing's universal machine with the power of unlimited self-reproduction (cf. Neumann's *Collected Works*, 1963). Cellular automata are discrete dynamical systems, the evolution of which is dictated by *local* rules. They are usually realized on a lattice of cells (in a computer program), with a finite number of discrete states associated with each cell, and with a local rule specifying how the state of each cell is to be updated in discrete time steps. Neumann's idea was to devise a machine that can build any machine that can be described. Once this can be done, it follows logically that, among the machines built, there would be those which are exact replicas of the original. Each machine was envisaged to carry a 'tail' bearing instructions for building the body of the machine, and also how to reprint the instructions. Such considerations led to the conclusion that perhaps it is necessary for a self-replicating machine to carry its own template or negative. This self-description could be used in two ways: as a recipe for self-reproduction, and as data to be copied and attached to the offspring, so that the latter too has the ability to replicate itself. For an update on this subject, see Penrose (1959); Rebek (1994); Sipper and Reggia (2001).

  Von Neumann was also responsible for the development of ENIAC, the first electronic digital computer to be made publicly operational (in 1946). It is not widely realized that Neumann also contributed extensively to parallel computing, evolutionary computing, and neural networks. He also understood correctly the extreme sensitivity of chaotic systems, and how one can exploit this sensitivity to advantage. As quoted by F. Dyson (1988), he stated (around 1950) that one could obtain the desired large-scale changes in weather by introducing small, carefully chosen atmospheric disturbances in a pre-planned manner. Control of chaos is a hot topic of research today (cf. the appendix on chaos).

- The 1940s also saw the publication of McCulloch and Pitts' (1943) work on the description of artificial neural networks (ANNs) that can learn. The simple model of the neuron envisaged by these neurophysiologists was actually an exercise in mathematical biology.

- Barricelli, a mathematical biologist, came to the Institute for Advanced Study (IAS), Princeton, in 1953 to work on a *symbiogenetic* model of the origin of life (the IAS is also where Neumann had done the best of his work). Barricelli's work marked the beginning of *bio-inspired computing*, a field of research that continues to expand ever since (cf. Fogel 2006), particularly after Holland (1975) introduced the notion of *genetic algorithms* (to be discussed later in this chapter).

Investigations on symbiogenesis were actually begun in 1909 by Merezhkovsky in Russia (cf. Khakhina 1992). He ascribed the complexity of living organisms to a series of symbiotic alliances between simpler living forms. Barricelli's work at the IAS carried over these ideas into evolution inside a computer program. The idea that extreme complexity can arise only as a result of serial associations among simpler forms made sense: It is easier to construct sentences by combining words, rather than by combining letters. By the same token, the search space for the emergence of a book from chapters, paragraphs, and sentences is far more restricted (and therefore more probable to be a reality) than that from just an availability of letters of the alphabet (cf. Alon 2007, for a recent discussion of this idea from the vantage point of systems biology). *Efficient search for a solution is, in fact, the essence of intelligence.*

Barricelli (1962) expanded the theory of biological symbiogenesis to a theory of *symbioorganisms*, which are any self-reproducing structures constructed by the symbiotic association of several self-reproducing entities of any kind. This meant that evolution inside a computer is not really very different from biological evolution.

Barricelli pioneered some computer experiments in this direction. He observed, for example, a cooperative self-repair of damage in the self-reproducing structures defined inside a computer. He also found from these computer experiments that sex (or 'crossover') is much more important for achieving evolutionary progress than chance mutations. As Dyson (1997) has remarked, just as Turing had blurred the distinction between intelligence and nonintelligence by introducing the universal machine, *Barricelli's numerical symbioorganisms blurred the distinction between the living and the nonliving* (emphasis added).

• John Koza published his book *Genetic Programming: On the Programming of Computers by Means of Natural Selection* in 1992. This book marked the real beginning of the field of *genetic programming* (GP). In this approach, computer programs *evolve* to their final desired form, rather than having to be written by a programmer. All the basic ingredients of evolution are incorporated into it: One begins with a population of competing computer programs; variations (including mutations) are introduced; the variations are made heritable; crossover is introduced; it is ensured that there are more offspring than the environment can support, so that the individual programs have to compete for survival; and fitness criteria are defined and applied, so that fitter individuals have a higher chance of surviving and reproducing.

In this chapter we shall elaborate on several topics mentioned in this very brief historical survey. We begin by recapitulating the basics of evolutionary processes in Nature.

2.2   Biological evolution

> *Darwin's theory of evolution by cumulative natural selection*
> *is the only theory we know of that is in principle capable of explaining*
> *the evolution of organized complexity from primeval simplicity.*
>   – Richard Dawkins, *The Blind Watchmaker*

Animals and plants are constantly exchanging matter and energy with the environment. There is a dynamic equilibrium between a living organism and its surroundings. The organism cannot survive if this equilibrium is disturbed too much, or for too long. The fact that an organism survives implies that it, in its present form, has been able to *adapt* itself to the environment.

If the environment changes slowly enough, living entities *evolve* (over a long enough time period) a new set of capabilities or features which enable them to survive under the new conditions. Over long periods of evolutionary change, creatures may even develop into new *species*.

It is also found that several similarities exist between, say, lions and tigers, which are therefore said to belong to the same *genus* (namely *Felis*). All lions belong to the same species, *Felis leo*, and all tigers belong to the same species *Felis tigris*, and the two belong to the same *genus*.

In the absence of a theory for explaining the adaptive features of animals and plants, as well as the similarities observed among entire groups of them, it was believed that this was an act of *intelligent design*, which implied the intervention of a *creator* (the *'hand of God'* explanation).

It was against this background that Charles Darwin (1859) proposed his *theory of evolution through cumulative natural selection*. He could demonstrate that adaptation to the environment was a necessary consequence of the exchange processes going on between organisms and their surroundings. According to this theory, all living organisms are the descendents of one or a few simple ancestral forms. In modern genetics, evidence for this statement comes from the fact that almost all life-forms use the same gene for their development pattern. It is called *the Hox gene*; in it, there is a region of conserved nucleotide sequence called *homeobox*. This conserved nucleotide sequence points to the common ancestry of life forms.

Darwin's theory of natural selection started with the observation that, given enough time, food, space, and safety from predators and disease, etc., the size of the population of any species can increase in each generation. Since this indefinite increase does not actually occur, there must be limiting factors in operation. If, for example, food is limited, only a fraction of the population can survive and propagate itself. Since not all individuals in a species are exactly alike, those which are better suited to cope with a given set of conditions will stand a better chance of survival (*survival of the fittest*).

The fittest individuals not only have a better chance of survival, they are also more likely to procreate, or to procreate more frequently. Thus, attributes conducive to survival get naturally selected at the expense of less conducive attributes. This is the process of *natural selection*.

It is also observed that children tend to resemble their parents. The progeny of better-adapted individuals in each generation, which survive and leave behind more offspring, acquire more and more of those features which are suitable for good adaptation to the existing or changing environment. Thus a species perfects itself, or adjusts itself, for the environment in which it must survive, through the processes of cumulative natural selection and *inheritance*.

It is now known that the inherited characteristics of the progeny are caused by *genes*. In sexually reproducing organisms, each parent provides one complete set of genes to the offspring (Ray 1999). Genes are portions of molecules of DNA (deoxyribonucleic acid), and their specificity is governed by the sequences in which their four bases – adenine (A), thymine (T), guanine (G), and cytosine (C) – are arranged. The double helix structure of DNA, together with the restriction on the pairing of bases comprising the DNA molecule to only A-T and G-C, provides a mechanism for the exact replication of DNA molecules. The DNA sequence on a gene determines the sequence of amino acids in the specific proteins created by the live organism (cf. the appendix on cell biology).

Genes programme embryos to develop into adults with certain characteristics, and these characteristics are not entirely identical among the individuals in the population. Genes of individuals with characteristics that enable them to reproduce successfully tend to survive in the *gene pool*, at the expense of genes that fail. This feature of natural selection at the gene level has consequences which become manifest at the organism level. Cumulative natural selection is *not* a random process.

If like begets like (through inheritance of characteristics), by what mechanism do slight differences arise in successive generations so that the species can evolve towards evolutionary novelty? One mechanism is that of *mutations*. Mutations, brought about by radiation or by chemicals in the environment, or by any other agents causing replication errors, change the sequence of bases in the DNA molecules comprising the genes.

In organisms in which sexual reproduction has been adopted as the means for procreation, since the genetic material has two sources instead of one (namely, the two parents), the occurrence of variations in the offspring is higher. The genes from the parents are reshuffled in each new generation. This increases the *evolutionary plasticity* of the species.

However, not all differences in individuals in a population are due to the genetic makeup. Factors such as nutrition also play a role. The observable characteristics of an organism, i.e. its *phenotype*, are co-determined by the genetic potentiality and the environment.

If all living beings have the same or only a few ancestors, how have the various species arisen? The answer lies in *isolation* and *branching*, aided by evolution. *Migrations* of populations also have an important role to play in the evolutionary development of species (Sugden and Pennisi 2006). If there are barriers to interbreeding, geographical or otherwise, single populations can branch and evolve into distinct species over long enough periods of time. Each branching event is called a *speciation*: A population accidentally separates into two, and they evolve independently. When separate evolution has reached a stage that no interbreeding is possible even when there is no longer any geographical or other barrier, a new species is said to have originated.

Can the environmental conditions in which the parents live affect the genetic characteristics of the offspring? 'No' according to the Darwinian theory of evolution, and 'Yes' according to the theory of Lamarck. The Lamarckian viewpoint of *inheritance of acquired characteristics* runs counter to the central dogma of modern molecular genetics, according to which information can flow from nucleic acids to proteins (or from genotype to phenotype), but not from proteins to nucleic acids (or from phenotype to genotype).

The issue of Lamarckism in natural evolution continues to be debated. However, in *artificial evolution* (say, inside a computer), Lamarckism offers viable and interesting scenarios. Even in natural evolution, Lamarckism just refuses to go away completely (cf. Hooper 2006). It appears that changes other than those in the DNA can indeed be inherited. Gene expression or interpretation can be influenced by molecules hitchhiking on genes. This is heritable nongenetic hitchhiking, and is called *epigenetic inheritance*.

Freeman Dyson (1985), better known for his work on quantum electrodynamics, emphasized the need to make a distinction between replication and reproduction. He pointed out that natural selection does not *require* replication, at least for simple creatures. Biological cells can reproduce, but only molecules can replicate. In life as we see it today, reproduction of cells and replication of molecules occur together (cf. the appendix on cell biology). But there is no reason to presume that this was always the case. According to Dyson, it is more likely that *life originated twice*, with two separate kinds of organisms, one capable of metabolism without exact replication, and the other capable of replication without metabolism. At some stage the two features came together. When replication and metabolism occurred in the same creature, natural selection as an agent for novelty became more vigorous.

This *dual-origin hypothesis*, if duly validated, would mean that, for a very extended portion of evolutionary history, only mutations were responsible for evolution, and inheritance played no part. This can bring Lamarckism back into the reckoning: when there are no genes and no replication, acquired characteristics can indeed be transmitted to the offspring to a significant extent. This conclusion is of particular significance for carrying out artificial evolution inside a computer.

Species evolve towards a state of stable adaptation to the existing environment. Then why many of them are extinct, as the fossil records show? The answer lies in the fact that there is never a state of permanent static equilibrium in Nature. There is a never-ending input of energy, climatic changes, terrestrial upheavals, as also mutations and a coexistence with other competing or cooperating species. Rather than evolving towards a state of permanent equilibrium and adaptation, the entirety of species (in fact, the earth as a whole) evolves towards a state of 'self-organized criticality' (SOC) (Bak 1996). This state of *complexity* is poised at the edge of order and chaos. The never-ending inputs just mentioned result in minor or major 'avalanches' (catastrophic events of various magnitudes), which sometimes lead to the extinction of species, or the emergence of new ones. (We shall discuss complexity in Section 2.4 below, and SOC in Section 5.5.2 in the context of self-assembly and self-organization of systems.)

Coevolution of species is also a very important ingredient of the succession of turmoil and relative stability (*stasis*), so that what really unfolds is *punctuated equilibrium* (Holland 1995, 1998; Kauffman 1993, 1995, 2000).

In summary, there are four basic features of evolution:

1. *Variability and variety* in members of a population in the matter of coping with a given environment. Apart from mutations or replication errors, it is sex or crossover that facilitates variety.
2. *Heritability* of this variation.
3. *Limited resources*; i.e. a greater reproductive capability of a population than the environment can sustain, leading to competition for survival.
4. Higher *fitness* leads to higher chances of survival and reproduction.

Variations on the original Darwinism are suggested from time to time. Lamarckism also keeps asserting itself in interesting ways. For example, it is notable that, on evolutionary time-scales, there has been an exceptionally rapid expansion of brain capacity in the course of evolution of one of the ape forms (chimpanzees?) to *Homo sapiens*, i.e. ourselves. The genome of humans is incredibly close to that of chimpanzees. The evolution of language, speech, and culture are believed to be the causative factors for this rapid evolution of the human brain (Dawkins 1998). Similar to the gene, which is the basic unit of biological inheritance, Dawkins (1989, 1998) introduced the notion of the *meme*, which is the unit of cultural inheritance, in his neo-Darwinistic theory. A meme may be some good idea, or a soul-stirring tune, or a logical piece of reasoning, or a great philosophical concept (cf. Kennedy 2006).

In Dawkins' scheme of things, two different evolutionary processes must have operated in tandem, one of them being the classical Darwinian evolution described here so far. The other centred around language, intelligence, and culture. Apes, which have been left far behind by humans in the evolutionary race, lack substantial speech centres in their brains. Emergence of language and speech in one of their species provided a *qualitative* evolutionary advantage to that species, with far-reaching consequences. How did this come about? The answer has to do, not only with gene pools, but also *meme pools*:

The totality of genes of a population comprising a species constitutes its gene pool. The genes that exist in many copies in the population are those that are good at surviving and replicating. Through a reinforcement effect, genes in the population that are good at *cooperating* with one another stand a better chance of surviving. Similarly, memes in a population may jump from one brain to another, and the fittest set of memes has a better chance of surviving to form the meme pool of the population. They replicate themselves by imitation or copying. Cultural evolution and progress occurs through a selective propagation of memes which are good at cooperating with one another (Kennedy 2006).

Memes can evolve like genes. In fact, entities that can replicate, and that have a variation both in their specific features and in their reproductive success, are candidates for Darwinian selection. The coevolution of gene pools and meme pools resulted in a rapid enlargement of the brain size of *Homo sapiens*: Mastery of a particular skill (say language) required a slight increase in brain size. Having developed this larger size, the brain could have got triggered to launch entirely new activities, some of which had evolutionary advantages. This resulted in a relatively rapid increase in brain capacity, because even Lamarckian evolution could contribute to the meme-related part of the coevolution of the gene pool and the meme pool.

In view of the current high level of activity on artificial evolution, which draws parallels from natural evolution but is not identical to it, we list here the three dichotomies one encounters in practice:

- *Hand of God or not*. Natural evolution is blind and has no end-goal. It is open-ended. Instead of planning a structure in advance (as engineers do), Nature tinkers with various possibilities and the one that is better for surviving in the existing conditions gets a preference for possible selection. By contrast, artificial evolution is goal-oriented, and is therefore *not* open-ended; it is *guided* evolution, and there is indeed present the equivalent of the 'hand of God'.
- *Genotype vs. phenotype*. The genotype or genome of an organism is its genetic blueprint. It is present in every cell of the body of the organism. The phenotype, on the other hand, is the end-product (the organism) which emerges through execution of the instructions carried by the genotype. It is the phenotype that is subjected to the battle for survival, but it is the genotype which carries the accumulated evolutionary benefits to succeeding generations. The phenotypes compete, and the fittest among them have a higher chance of exchanging genes among themselves.
- *Darwinism vs. Lamarckism*. Biological evolution is generally believed to be all Darwinian. Variety in a population means that some individuals have a slight evolutionary advantage with respect to a particular characteristic, which is therefore more likely to be passed on to the next generation. Over time, this characteristic gets strengthened and easily noticeable in the phenotype. Lamarckism, on the other hand, is based on two premises: the principle of use and disuse; and the principle of inheritance of *acquired* characteristics (without involving the genotype). Although this is not how biological evolution normally operates, nothing prevents us from using it in artificial evolution and exploiting the much higher speed it may offer for reaching the end-goal.

An effect that is reminiscent of Lamarckism, but is actually quite different from it, is the *Baldwin effect*. It refers to the influence that individual learning in a population can have on the course of its evolution. It does not change the genotype directly, but a few members of a population with better learning capabilities may have a higher probability of survival (say from predators), resulting in their higher chances of survival and procreation. As the proportion of such members grows, the population can support a more diverse gene pool, leading to quicker evolution, suited to meet that challenge. Like Lamarckism, the Baldwin effect is also of special interest to people trying to achieve artificial evolution of smart structures (or evolution of artificial smart structures).

We are now heading for the *post-human* or *post-Darwinian* era (Moravec 1999a): Human beings are on the threshold of using their mastery over genetics and technology to have, and to evolve, better human bodies and brains (the websites *betterhumans.com* and *transhumanism.org* provide some details). *Cyborg* is a term coined by NASA denoting 'cybernetic organism' (cf. Wiener 1965), which is part human, part machine. For example, a person kept alive by a pacemaker is not 100% human; a fraction of him/her is machine. With more and more progress in technology and biology, a sizeable fraction of a human may comprise of either machine or implanted organs. There can, for example, be an implanted chip sending data from the nervous system directly to a computer. We can also look forward to designer babies, with larger brains, higher grades of consciousness, stronger bodies, etc.

The domestication of the new biotechnology by humans in the post-Darwinian era presages an era of *horizontal gene transfer*. It will be possible to move genes from microbes to plants and animals, thus blurring the distinction between species. Like the exchange of software, there will also be an exchange of genes, resulting in *a communal evolution of life* (cf. Dyson 2006).

2.3   Self-organizing systems

> *The world of the made will soon be like the world of the born:*
> *autonomous, adaptable, and creative but, consequently,*
> *out of control.*
> — Kevin Kelly, *Out of Control*

Biological evolution, outlined above, is not only an example of information processing by interacting individuals in a population, it is also an example of self-organization. It appears that the only hope we humans have of developing really sophisticated and smart artificial structures for our benefit is to induce them to self-organize. Therefore it is important that we gain an understanding of the mechanics of self-organization in Nature. In this section, we begin by considering the archetypal example of the beehive as a self-organized superorganism.

2.3.1   *How the bees do it*

What is common among the following: a beehive, an ant colony, a shoal of fish, an evolving population, a democracy, a stock market, and the world economy?

They are all spontaneously self-organizing, interacting, distributed, large systems. They are *adaptive* (i.e. whatever happens to them, they try to turn it to their collective advantage). There is no central command. The individuals (or *agents*) are autonomous, but what they do is influenced strongly by what they 'see' others doing.

Each of these examples is not merely the analogue of an organism; *it is indeed an organism*. A *superorganism*, if you wish (Wheeler 1928), but an organism all the same. Many more instances of such 'out of control', *complex adaptive systems* (CASs) can be found (cf. Waldrop 1992; Kelly 1994; Holland 1995; Schuster 2001; Fewell 2003; Surowiecki 2004).

Here, 'autonomous' means that each member reacts individually according to internal rules and the state of its local environment.

The autonomous or semi-autonomous members are highly connected to one another through communication. There is a *network*, but not necessarily to a central hub.

The population is large, and no one is in control. This means, among other things, that even if a section of the population were to be decimated, the system will adjust and recover quickly, and carry on regardless. Kelly (1994) calls such networked or 'webby' systems *vivisystems*.

To get a feel for the nature of vivisystems, let us see how a beehive functions as a single organism, even when there is no central command. How does it, say, select a new site for setting up a hive?

It has been known for long that, at the time of the year when the sources of honey are aplenty, a large colony of honeybees splits into two by a process called *swarming*: A daughter queen and about half of the population in a beehive stays behind, and the rest (including the queen bee) leave the old hive to start a new one *at a carefully selected site*.

The sequence of events is roughly as follows (Seeley, Visscher and Passino 2006):

1. Typically, out of a total of ~10,000 bees, a few hundred worker bees go scouting for possible sites. The rest stay bivouacked on a nearby tree branch, conserving energy, till the best new site has been decided upon.
2. The 'best' site for nesting is typically a cavity in a tree with a volume greater than 20 litres, and an entrance hole smaller than $30\,cm^2$; moreover, the hole should be several metres above the ground, facing south, and located at the bottom of the cavity.
3. The scout bees come back and report to the swarm about possible nesting sites by dancing a *waggle dance* in particular ways. Typically there are about a dozen sites competing for attention.
4. During the report, the more vigorously a scout dances, the better must be the site being championed.
5. Deputy bees then go to check out the competing sites according to the intensity of the dances.
6. They concur with the scouts whose sites are good by joining the dances of those scouts.
7. That induces more followers to check out the lead prospects. They return and join the show by leaping into the performance of *their* choice.
8. By compounding emphasis (*positive feedback*), the favourite site gets more visitors, thus increasing further the number of visitors.
9. Finally, the swarm takes the daughter queen, and flies in the direction indicated by mob vote.

As Kelly (1994) puts it: 'It's an election hall of idiots, for idiots, and by idiots, and it works marvellously'.

2.3.2   Emergent behaviour

The remarkable thing about distributed complex systems like beehives is emergent behaviour. For them, 2 plus 2 is not necessarily 4; it can be, say, apples. *More is different* (Anderson 1972; also see Williams 1997). In a network of beehives or ant colonies, each member is programmed in a very simple way. Sheer large numbers and effective communication and interaction result in intelligence (*swarm intelligence*), of which no single member is capable alone (Kennedy 2006).

Four distinct features of distributed existence make vivisystems what they are (Kelly 1994): absence of imposed centralized control; autonomous nature of sub-units; high connectivity between the subunits; and webby nonlinear causality of peers influencing peers.

Emergent behaviour in vivisystems is very common in Nature, and has far-reaching consequences in every case (Bhalla and Iyengar 1999; Davies 2005). Human intelligence has also been interpreted as emergent behaviour arising from the interaction and connectivity of the vast number of neurons (or 'agents'), even though each agent, taken individually, is as unintelligent as can be (Minsky 1986; Laughlin and Sejnowski 2003). Thoughts, feelings, and purpose result from the interactions among these basic components.

2.3.3   Positive feedback and pattern formation

Why is it that so many systems head towards order and structure and spontaneous self-organization, in apparent violation of the second law of thermodynamics, according to which a system must minimize its free energy, which generally translates into maximizing its entropy or disorder?

Prigogine addressed this question from the perspective of nonequilibrium thermodynamics, and was awarded the Nobel Prize in 1977 for his work (cf. Prigogine 1998). The central fact is that self-organizing systems are not isolated systems or closed systems. They are *open systems*. There is always an exchange of energy and matter with the surroundings. If this exchange is large enough (large enough to take the system substantially away from equilibrium), entropy *can* decrease locally, and order and pattern formation can emerge. This question has been discussed recently by Avery (2003) from the point of view of information theory.

A central theme of the work of Prigogine is that self-organization or 'spontaneous' pattern formation requires self-reinforcement or positive feedback. In such systems, there is a tendency for small chance events to become magnified because of positive feedback. The increasing returns can lead to a multiplicity of qualitatively different outcomes. The actual outcome depends on the chance initial conditions. In these highly nonlinear systems, the sum total can be very different from the linear superposition of the parts.

Can one formulate an extended (or new) version of the second law of thermodynamics to describe such systems precisely and quantitatively? Efforts have not been entirely successful so far because of the excessive complexity of the problem (see Prigogine 1998). The pioneering work of Tsallis (1988, 1995a, b, 1997) on systems which obey *nonextensive thermostatistics* offers some interesting possibilities. We discuss such systems in an appendix.

2.3.4 *Emergence and evolution*

Emergent behaviour leads to evolution, both natural and synthetic (Holland 1998). The networked swarm comprising a vivisystem is adaptable, resilient, and nurtures small failures so that large failures do not happen frequently. All these features help survival and propagation. Even more importantly, *novelty* is built-in in the scheme of things. Kelly (1994) has listed three reasons for this:

1. Different initial conditions can lead to dramatically different end results (reminiscent of nonlinear systems).
2. The extremely large number of combinations and permutations possible among the interacting individuals has the potential for new possibilities.
3. Individual variation and imperfection is not snuffed out by a central authority; therefore better (more suitable) characteristics can evolve over time.

If heritability is brought in, individual variation and experimentation leads to *perpetual novelty*, the stuff evolution is made of. Before we discuss perpetual novelty (in the next section), let us go back to the site-selection behaviour of honeybees, discussed above in Section 2.3.1.

Seeley *et al.* (2006) have drawn several important conclusions from their exhaustive and sophisticated study of honey bees, using modern techniques of observation, modelling, and analysis. For example, is the decision-making process of the swarm essentially one of consensus building, wherein the process of decision-making continues till all the scout bees agree about one particular site, superior to all others? They found that the answer is 'no'. What happens instead is that, in the final stages, even when there are more than one competing sites, the swarm decides to fly to a particular site if the number of scout bees advocating it exceeds a certain threshold number. In other words, the method adopted is that of *quorum*, rather than consensus.

This threshold number for quorum was found to be ~150 bees in the controlled experiments conducted by Seeley *et al.* (2006). This number is an optimum trade-off between speed and accuracy. A higher number for quorum would mean loss of time and higher energy costs, and a lower number would increase the chances of making excessive errors in choosing a good site. Through a process of Darwinian natural selection and evolution, the species has fine-tuned itself for what is best for its survival and propagation.

Emergent behaviour and evolution go hand in hand (Seredynski 2006).

2.3.5 *Perpetual novelty*

Let us discuss CASs in some more detail, following the work of Holland (1975, 1995). Such systems, though widespread in Nature, have some common characteristics.

(a) The first is the set of features already summarized above (from the book by Kelly 1994). There is a network of large number of individuals or agents, acting in parallel. Each agent constantly reacts to what the others are doing. Therefore, from the vantage point of any particular member of the set, the environment is changing all the time. The control, if any, is highly dispersed. The emergent behaviour is the result of competition as well as cooperation among the agents themselves.

(b) A CAS has many levels of organization. The agents at one level serve as the building blocks for agents at the next higher level of hierarchy (Alon 2007). In the human brain, for example, one block of neurons forms the functional regions for speech, another for vision, and still another for motor action. These functional areas link up at the next higher level of cognition and generalization (cf. Section 7.1).

(c) As a CAS gains experience through feedback, it continuously revises and rearranges the building blocks. The processes of learning, evolution, and adaptation are basically the same. One of their fundamental mechanisms is this revision and recombination of the building blocks.

(d) The CASs are constantly making *predictions*, thus anticipating the future. The predictions are based on various *internal models of the world*, and the models are constantly revised on the basis of new inputs; they are not static blueprints.

(e) The CASs have many niches, each of which can be exploited by an agent which has adapted itself to fill that niche.

(f) Filling up of a niche opens up new niches. The system is just too large to be ever in equilibrium. There is perpetual novelty.

Based on the results of a number of investigations, Kelly (1994) states a universal law for vivisystems: *Higher-level complexities cannot be inferred from lower-level existences*. Actually running or observing a vivisystem is the quickest, the shortest, and the only sure method to discern emergent structures latent in it (Moravec 1988).

2.3.6 *Robust and dependable algorithms based on distributed systems*

We want our computers to solve complex problems we do not know how to solve, but merely know how to state. One such problem, for example, is to create multimillion-line programs to fly airplanes. It is very difficult to make such huge programs bug-free. And, unless an adequate degree of redundancy is built in, just one bug may be enough to cause a plane crash! We cannot afford to allow that.

It has been observed that almost all serious problems encountered in software malfunction can be traced to faults in *conceptual design* that occurred even before the programming started. Design-checking tools such as *Alloy* have therefore been developed to counter this vexing issue (Jackson 2006). Alloy models the design of complex software, and then checks for conceptual and structural flaws in an automated manner, treating designs as massive puzzles to be solved. All possible states that the software can take are simulated, to ensure that none of them results in failure.

Unfortunately, the possible number of states blows up very rapidly with increasing complexity, even for moderately complex problems. Although tools such as 'SAT' (from the word *sat*isfiability) have been developed to overcome this problem, there is always the nagging possibility that not enough checking has been carried out. A radically different approach is therefore needed.

A step in the right direction, of course, is the use of *modular software*, in a distributed-systems configuration, with little or no centralized command, described above as vivisystems.

An active area of research in software design emphasizes the use of object-oriented software. *Object oriented programs* (OOPs) are relatively decentralized and modular. The pieces of an OOP retain integrity as stand-alone units. They can be combined with other OOP pieces into a decomposable hierarchy of instructions (cf. Kelly 1994).

An 'object' in an OOP helps limit the damage a bug can cause. Rather than having to tackle the crashing of the whole program, use of OOPs effectively segregates the functions into manageable units. Huge, very complex, programs can be built up quickly by networking a large number of OOPs.

Use of OOPs results in the creation of a kind of *distributed intelligence* in software, rather like the swarm intelligence of a beehive or an ant colony. Like other distributed systems, it is resilient to errors; it heals faster (just replace the faulty object); and it grows incrementally by assembling *working* and *tested* subunits.

This bio-inspired approach amounts to assembling software from working parts, while continuously testing and correcting the assembly as it grows. One still has the problem of unexpected 'emergent behaviour' (bugs) arising from the aggregation of bugless parts. But there is hope that, as long as one only needs to test at the new emergent level, there is a very good chance of success.

It is becoming increasingly clear that there is only one way to obtain robust and reliable software for extremely complex problems: *through artificial evolution* (we shall describe 'genetic algorithms' in the next section). Kelly (1994) has listed several reasons for this by citing the example of living systems (which are what they are because of their evolutionary history):

- Living organisms are *great survivors*.
- Living organisms are *consummate problem solvers*.
- Anywhere you have many conflicting, interlinked variables and a broadly defined goal, where the solutions are many, evolution is the answer.
- The presence of parasites (the equivalent of viruses in a computer program) actually speeds up the process of evolution. They lead to the formation of more variants (Moravec 1988).
- Predators and death are also necessary for rapid evolution.

- Evolution is a parallel processor. Parallelism is one of the ways around the inherent stupidity and blindness of random mutations.
- Implicit parallelism is the magic by which evolutionary processes guarantee that a system climbs, not just any peak, but the highest peak (or *the global minimum*, in terms of free energy).

One message is that we must head for more and more parallel computing for this. But, for really complex problems, this is easier said than done. As Tom Ray has stated: '*The complexity of programming a massively parallel machine is probably beyond us. I don't think we will ever be able to write software that fully uses the capacity of parallelism.*'

Evolution of computer programs, rather than the actual writing of computer programs, is the way out of this difficulty (Krishnamurthy and Krishnamurthy 2006). We again quote from Kelly (1994) for illustrating one of the bio-inspired approaches being used for evolving complex, massively parallel software, namely by following the example of ant colonies: 'Ants are the history of social organization and the future of computers.'

An army of ants too dumb to measure and too blind to see far can rapidly find the shortest route (for finding food) across a very rugged landscape. An ant colony perfectly mirrors our search for computational evolution: dumb, blind, simultaneous agents trying to optimize a path on a computationally rugged landscape. Real ants communicate with one another through a chemical system called *pheromones* (Holldobler and Wilson 1990; Gordon 1996, 1999; Fewell 2003). These aromatic smells dissipate over time. The odours can also be relayed by a chain of ants picking up the scent and remanufacturing it to pass on to others.

The first ant colony optimization (ACO) procedure was proposed in the early 90s. Recently, Dorigo, Birattari and Stutzle (2006) have given a good survey of this field of research. Dorigo and Gambardella (1997) constructed computational formulas based on ant logic. Their virtual ants (*vants*) were dumb processors in a giant community operating in parallel. Each vant had a meagre memory, and could communicate only locally (pheromone-like). Yet the rewards of doing well were shared by others in a kind of distributed computation. The ant machine was tested on a standard benchmark: the combinatorial (Moravec 1988; Krattenthaler 2005) *travelling-salesman problem* (TSP).

Suppose a salesman has to visit five cities, and then come back home. Each city must be visited only once (except the home city, from which he starts, and to which he returns). Which itinerary will involve the least distance of travelling (and therefore the lowest cost)? It turns out that the answer has to be found from among 120 possibilities: There are five ways of choosing the first destination city. For each of these, there are four different ways of choosing the second city. Having chosen any of these, there are only three ways of picking up the next city (because the salesman cannot touch any city twice). And so on. Thus the total number of possible itineraries is $5 \times 4 \times 3 \times 2 = 120$, or 5! (factorial 5).

For each of these 120 options, one computes the total distance to be covered. The option with the least distance is the best solution.

Next, suppose the number of cities to be visited is 50, rather than 5. The number 50 is not such an outlandish number of cities to visit, after all. But the *search space* now comprises of 50! possibilities; i.e. $\sim 10^{64}$. This is an absurdly large number, and methods of reducing the size of the search space must be found.

In the computer experiments carried out by the Italian group, each vant would set out rambling from 'city' to 'city', leaving a trail of the computer equivalent of 'pheromones' (a suitable time-decaying mathematical function). The shorter the path between two cities, the less the mathematical pheromone function 'evaporated', i.e. decayed with time. The stronger the pheromone signal, the more the other vants followed that route (self-reinforcement of paths). Around 5000 runs enabled the vant group-mind to evolve a fairly optimal global route. Quite an achievement.

Self-evolving solutions and codes are the only answer for massive parallel programming. Their massiveness and distributedness ensures that, even if there is a bug or a local crash, the system carries on after some minor readjustment. For this to happen, absence of a central command is a must. One goes for incremental expansion and testing. Parasites/viruses/adversaries are deliberately self-introduced at the testing stage, so that there is a built-in ability/experience of such eventualities. The end result is robustness and reliability. We do not want that airplane to crash.

2.4 Complexity

I think the next century will be the century of complexity.
– Stephen Hawking

A complex system is a system with large variability (Politi 2006). Such a system usually consists of a large number of simple elements or agents, which interact with one another and the environment. Hurricanes are an example of complexity; so are ant colonies, shoals of fish, and self-organized supramolecular assemblies.

A characteristic feature of complex systems is the emergence of *unexpected* properties or behaviour. It is still debatable whether the new properties can all be described entirely in terms of the known laws of physics (applied to the individual members of the system, each carrying information in the form of internal structure, bonding/interacting capabilities, chemical computational power, etc.). It is conceivable that something totally new can emerge from the interactions among the constituents. Reasons for the latter possibility arise as follows.

Imagine a superintelligent creature who knows at one instant of time the position and momentum of every particle in the universe, as also the forces acting among them. Assuming the availability of a good enough supercomputer, is it possible to predict the future in every detail? The answer would be 'yes', according to the reductionists. But this can be possible only if *unlimited computational power* is available. In reality, there are limits on the speeds of computation, as well as on the extent of computation one can do.

These limits are set by the laws of physics and by the resources available in the universe (cf. Williams 1997; Lloyd and Ng 2004; Davies 2005; Chaitin 2006; Lloyd 2006). The bit (short for 'binary digit') is the basic piece of information, and the bit-flip is the basic operation of information processing. The Heisenberg uncertainty principle puts a lower limit on the time needed for processing a given amount of energy. And the finite speed of light puts an upper limit on the speed with which information can be exchanged among the constituents of a processor. A third limit is imposed by entropy, which is the reverse of information: One cannot store more bits of information in a system than permitted by its entropy.

These limits are imposed by the laws of physics. A different kind of limit arises from cosmology. The universe is believed to have begun in a big bang, some 13.7 billion years ago. Therefore, light cannot have traversed distances greater than 13.7 billion light years. Regions of space separated by larger distances than that cannot have a causal relationship. Therefore the superintelligent creature cannot have a supercomputer larger than the size of the universe (Lloyd 2006).

Thus there is a limit on available computational power. Predictions based on the known laws of physics, but requiring larger computational power than this limit, are not possible. Predictions cannot always have unlimited precision, not even for otherwise deterministic situations.

The implication of this conclusion for complex systems is that, beyond a certain level of computational complexity, new, unexpected, organizing principles can arise (cf. Duke 2006). If the physical laws cannot completely determine the future states of a complex system, then *higher-level laws of emergence* can come into operation.

An example of this type of 'strong' emergence is the origin and evolution of life. From the deterministic point of view, it is a computationally intractable problem. Therefore new laws, different from the bottom-level laws of physics and chemistry, might have played a role in giving to the genes and the proteins the functionality they possess at present.

Complexity, beyond a certain threshold, leads to the emergence of new principles (Prigogine 1998). It is a one-way traffic: The new principles and features are deducible from, but not reducible to, those operating at lower levels of complexity.

2.5   Computational intelligence

In the metaphor of Richard Dawkins, we are the handiwork of a blind watchmaker. But we have now acquired partial sight and can, if we choose, use our vision to guide the watchmaker's hand.

– Hans Moravec (1988), *Mind Children*

Practically all computers to date have been built on the von Neumann model which, in turn, was inspired by the Turing machine. In general, the underlying model of such computing structures works through a *precise* algorithm that works on *accurate* data. But there are innumerable complex systems that cannot be adequately tackled through such a computational approach. One should be able to work with partially accurate or insufficient or time-varying data, requiring the use of suitably variable or tuneable software. One way out is to draw inspiration from Nature, emulating the way Nature handles such computational problems (Zomaya 2006). We would like to work with computational systems which are fault-tolerant, and *computationally intelligent*, making adjustments in the software intelligently and handling imperfect or 'fuzzy' data the way we humans do. The subject of computational intelligence (CI) was mentioned in Section 1.12. Konar (2005) defines CI as *the computational models and tools of intelligence, capable of inputting raw numerical sensory data directly, processing them by exploiting the representational parallelism and pipelining of the problem, generating reliable and timely responses, and withstanding high fault tolerance.*

Bezdek (1994) gave the following definition of CI: '*A system is computationally intelligent when it: deals with only numerical (low level) data, has pattern recognition components, does not use knowledge in the AI sense; and additionally when it begins to exhibit i) computational adaptivity, ii) computational fault tolerance, iii) speed approaching human-like turnaround, and iv) error rates that approximate human performance*'.

CI took birth in the early 1990s. Zadeh (1975) had introduced the notion of *linguistic variables* for making reasoning and computing more human-like. By computing with words, rather than numbers, one could deal with *approximate* reasoning. With increasing emphasis on the use of biomimetics in computational science, Zadeh's fuzzy-logic (FL) approach was joined by *genetic algorithms* (GAs), *evolutionary or genetic programming* (EP or GP), and *artificial life* (AL). ANNs, FL, GAs, GP, and AL constitute the five hard-core components of the subject of CI, although there are also a number of other peripheral disciplines (Fig. 2.1) (Konar 2005; Krishnamurthy and Krishnamurthy 2006). We describe each of these five briefly.

2.5.1 *Artificial neural networks*

As introduced in Section 1.3, ANNs are an attempt at mimicking the human brain (in a crude sort of way) on a computing system. A very brief outline of this vast field of research is given here, based on the book by Khanna (1990). A paper by Lippmann (1987), as also a recent article by Taheri and Zomaya (2006a), provide lucid accounts of the subject.

Computational systems can be made to *evolve* and *learn*. Human beings and some other animals are computing systems that can learn. The neural networks in their brains are designed for the task of learning and remembering. The human brain has almost a trillion nerve cells or *neurons*. Most of the neurons have a pyramidal shaped central body (the *nucleus*); in addition it has an *axon*, and a number of branching structures called *dendrites*. The axon is the signal emitter, and the dendrites are signal receivers. A connection (called a *synapse*) between two neurons is said to be established when a strand of the axon of one neuron 'touches' a dendrite of the other neuron. The axon of a typical neuron makes several thousand synapses.

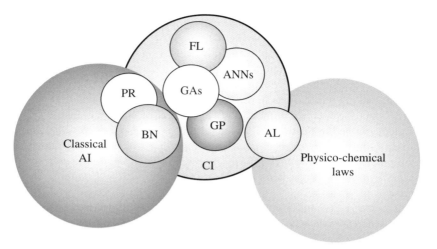

Fig. 2.1   The five main subjects coming under the umbrella term 'computational intelligence' (CI) are: ANNs (artificial neural networks), FL (fuzzy logic), GAs (genetic algorithms), GP (genetic programming), and AL (artificial life). Traditional or classical AI (artificial intelligence) is of a substantially different nature than the modern field of CI, although there is some overlap between the two. PR (probabilistic reasoning) and BN (belief networks) fall in this overlap region, but are not described in this book. The subject of AL (artificial life) is not all about computations only, and draws its sustenance from real physical and chemical laws also. [After Konar (2005). Drawn by permission of the publisher (Springer-Verlag).]

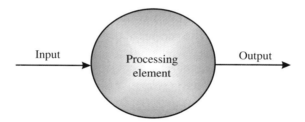

Fig. 2.2   The basic input–output unit in a neural network. A processing element (PE) is that which can transform a synaptic input signal into an output signal. Thus a PE acts as a computational or information-carrying element.

A conventional computer has a central processing unit (CPU), which performs a rigid set of operations on the inputs received by it. ANNs, by contrast, have many inter-connected small *processing elements* (PEs) or neurons. ANNs are computer-based models, designed for simulating pattern recognition and other cognitive (*perceptive*) actions of the mammalian brain. The term *perceptron* is used for the basic input–output device (Fig. 2.2) used in appropriate combinations for achieving this objective (Minsky and Papert 1969).

ANNs have to be *trained*. The training amounts to strengthening or weakening the connections (*synapses*) among the neurons (PEs), depending on the extent to which the output is the desired one. In the so-called *supervised learning* approach, a series of typical inputs are given, along with the desired outputs. After the first cycle of input processing, the ANN calculates the root-mean-square (RMS) error between the desired output and the actual output. The connection weights among the PEs are then adjusted to minimize this error, and the output is recalculated using the same input data. The RMS error is then calculated afresh. This procedure is repeated cyclically, until the connection weights have settled to values which give the least global error. This is an example of a *back propagation network*; in it the information about the error flows opposite to the flow of processing information.

At the end of the training protocol the ANN is *tested*. The difference between training and testing is that, during testing the connection weights are not adjusted after any cycle of processing of input signals.

Parallel distributed processing

Most ANNs are based on the *connectionist* philosophy, according to which the complex connectivity of neurons required for problem-solving can be encoded *locally* in the synapses and synaptic weights, and it is generally not necessary to move around the information over long paths. Memory and processing coexist locally, and this feature is distributed all over the system. Another term for this approach is *parallel distributed processing* (PDP).

PDP models use a large number of PEs (or *units*), which interact with one another via excitatory or inhibitory connections or synapses. The large and distributed nature of the configuration, and the fact that there is mainly a localized connectivity, gives the system a substantial degree of *fault tolerance*. It also has *adaptability*: the system responds to new stimuli by readily changing the nature and degree of interconnectedness.

Contrast this with what happens in a conventional computer. There is a sequential central processor, and a distinct memory unit. It is mandatory that the input information be specified precisely and exactly. The data to be processed must be directed through the sequential processor. The kind of massively parallel processing that goes on in the human brain is impossible to achieve by employing serial processing, involving the retrieval and depositing of data from a centralized memory, with no room for fault tolerance.

In a PDP system like an artificial computing and communication system modelled on the human neural network, there is no central processing unit (CPU). There is no centralized memory either. Both these features are distributed all over the network. The computation is performed, not by a single neuron, but collectively by the entire network. Each neuron receives signals from several neurons through its dendrites, processes the signal, and then may send the output signal through its axon to several neurons. Each neuron computes a weighted sum of signals received from various neurons. If this sum exceeds a certain threshold, an output signal is sent (the neuron 'fires'); otherwise not.

What is continually altered in this process is the strength of the synaptic interaction between any two neurons (the *synaptic weight*). It is this ever-changing pattern of synaptic weights which serves as a distributed memory. It is also the crux of the learning feature. Learning and memory go together: If you do not remember anything, you have learnt nothing!

An important feature of these PDP systems is that, for performing a particular task, the computer is not fed a recipe or a computer program. The system progressively *learns* to perform the task better and better (Resnick 1997).

Some broad types of learning are (Sipper 2002; Konar 2005): supervised learning; reinforcement learning; and unsupervised learning.

In *supervised learning*, the ANN begins with a totally random distribution of synaptic weights, and performs the computation for a given query. The results are compared with what is wanted. A variety of *learning algorithms* have been developed for determining how the gap between what has been computed and what is wanted can be closed by altering the synaptic weights in a certain way. After resetting the weights, the learning cycle is repeated. Several iterations are required, and still the performance may not be perfect if the problem to be solved is too complex.

Reinforcement learning relies on using *environmental feedback*, and works on the principle of reward (for good performance) and punishment (for bad performance).

Unsupervised learning involves feeding a large number of examples (e.g. pictures of animals), so that the system learns to differentiate among them by a process of *generalization*.

A variety of PDP models have been proposed and analysed. They involve formulation of simple, local mechanisms for the modification of connection strengths, such that the system continually adjusts to changes in the environment. The degree of distributed processing varies from one PDP model to another. There are simple models in which the activity of a single PE represents the degree of participation of a known conceptual entity. In other PDP models, the same degree of participation is spread over a large number of PEs. An example of the latter kind is the Hopfield model (Hopfield 1982, 1984), which will be discussed in Chapter 7 in the context of the modelling of human intelligence.

The old debate whether the brain and the mind are one and the same thing or not just refuses to go away. In Chapter 7 we shall discuss Hawkins' framework of human intelligence, from which it follows that the mind and the brain are one and the same. This last statement was actually made by Francis Crick as what he diplomatically called the 'astonishing hypothesis'. There is nothing astonishing about it. 'Consciousness is simply what it feels like to have a cortex' (Hawkins and Blakeslee 2004).

This last statement has got to be correct if we are going to build smart structures which would, at some stage in the future, incorporate cortex-like truly intelligent memories, and can then be called *intelligent structures*.

While this debate goes on, we mention here the work of Smolensky (1986), who argued that both mind and brain *can* be investigated by PDP models: A PDP model can either have a *neural* interpretation, or a *conceptual* interpretation, or both. It is important that the neural and the conceptual interpretations of any mathematical model are entirely independent of each other.

In Section 2.7.1 below, we shall introduce the basics of graph theory or network theory. A neural network is nothing but a *graph*, i.e. a set of *nodes* (namely the PEs) and *edges*. The latter are lines connecting specific nodes. The edges are *directed* (the direction being *from* one node *to* another), and have a weight (which is a measure of the strength of the connection between the nodes).

The following are the major aspects of PDP models (Khanna 1990):

1. Set of PEs, or nodes.
2. State of activation.
3. Output function for each PE.
4. Pattern of connectivity, defined by the edges.
5. Rule for propagating patterns of activities.
6. Activation rule for combining the inputs affecting a PE with its present state to produce an output.
7. Learning rule for modifying interconnections (i.e. the edges) on the basis of experience.
8. Environment within which the learning system has to operate.

Figure 2.3 shows a typical McCulloch–Pitts PE or neuron. The summation node computes the weighted sum of the inputs (with coefficients w_1, w_2, etc., which can be positive or negative). If this sum exceeds a threshold μ, an output of $+1$ is transmitted by the *transfer function*. Otherwise the output is -1.

Learning rules

A learning rule is a recipe for altering the synaptical weights (w_i) in a neural net whenever the net learns something in response to new inputs. The rules can be visualized as a sequence of time steps, a typical general step being

$$w_i(t + 1) = w_i(t) + cr_i(t) \tag{2.1}$$

Here $r_i(t)$ is the reinforcement signal to synapse i at time t, and c is the learning rate.

The best-known learning rule is that enunciated by Hebb (1949), according to which the reinforcement signal in the above general equation is given (approximately) by

$$r_i(t) = x_i(t)y(t) \tag{2.2}$$

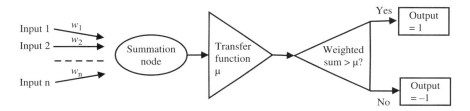

Fig. 2.3   The basic McCulloch–Pitts unit.

Here $x_i(t)$ is one of the inputs to the node or PE, and $y(t)$ is its output.

What the *Hebbian rule* says is: When a neural cell A repeatedly and persistently participates in firing cell B, then the efficiency of A for firing B is increased.

Another well-known learning rule is the *Widrow–Hoff rule*, defined by the reinforcement signal

$$r_i(t) = [z(t) - y(t)]x_i(t) \tag{2.3}$$

where

$$y(t) = \sum_{i=1}^{n} w_i(t)x_i(t) \tag{2.4}$$

It causes the weights to converge so that the response for each stimulus is a particular desired number. Therefore it is also known as the *delta rule*.

Associative memory

In a neural network trained by the Hebbian rule, presentation of a stimulation pattern would cause the net to generate another pattern that it has learnt to associate with the inputted pattern. After this, even when a *portion* of a pattern is presented, since the trained network has associative memory, it would generate the whole pattern. This gives a *content addressability* feature to the net. The address is there in the content itself (or even in a portion of the content).

The key feature of associative memory is the distributed representation of information. Representations of concepts and other types of information are stored as collective states of the network. Distinct items are represented as *patterns* of neural interconnections, reminiscent of *holographic* storage of information.

2.5.2   *Fuzzy-logic systems*

In contrast to the traditional, precise modes of computation, FL computational strategies are able to deal with vague, imprecise and partial knowledge. Such imprecision may result from our limited capability to resolve detail, and may also comprise of partial, noisy, vague and incomplete information about the real world. FL is logic involving a certain amount of inference and intuition (Taheri and Zomaya 2006b). Such logic is used by intelligent systems like human beings, but not by the older, classical AI systems. In it, propositions are not required to be either true or false, but may be true or false to different degrees (unlike classical binary logic). Here is an example: If A is (HEAVY 0.8) and B is (HEAVY 0.6), then A is 'MORE HEAVY' than B. By contrast, in classical binary logic, A and B may be members of the same class (both in HEAVY class or both in NOT-HEAVY class) or different classes. FL is a fundamental tool for reasoning with approximate data and knowledge.

Zadeh (1965, 1975) introduced the notion of *linguistic variables* for making reasoning and computing more human-like. By computing with words, rather than numbers, one can deal with approximate reasoning. For dealing with problems beset with uncertainty and vagueness, one works with *fuzzy-rule-based systems* (FRBSs) (Kosko and Isaka 1993; Ross 1997; Cordon *et al.* 2001). In an FRBS, one deals with IF–THEN rules, the antecedents and consequents of which are composed of fuzzy-logic statements, rather than classical logic statements. Fuzzy logic is suitable for an environment of uncertainty and imprecision; in it the underlying modes of reasoning are approximate rather than exact.

In an FRBS, representation of knowledge is enhanced with the use of linguistic variables and their linguistic values, which are defined by context-dependent fuzzy sets, the meanings of which are specified by gradual membership functions.

Zadeh has also introduced what he calls *f.g-generalization* (cf. Pedrycz and Gomide 1998). f-generalization means fuzzifying of any theory, technique, method or problem by replacing the concept of a crisp set by a fuzzy set. And g-generalization means *granulating* it by partitioning its variables, functions and relations into *granules* or information clusters. One can combine f-generalization with g-generalization, and call it f.g-generalization: It amounts to ungrouping an information system into components by some set of rules, and regrouping them into clusters or granules by another set of rules. This can give rise to new types of information subsystems.

2.5.3   *Genetic algorithms*

For finding good solutions to complex problems, often the best way out is to consider all possible solutions and choose the best. But the number of possible solutions can become embarrassingly large even for moderately complex problems. One speaks of a *search space* or a *phase space* in this context. It comprises all possible solutions: good, bad, indifferent, absurd. In Section 2.3.6 we got a feel for how fast the search space can grow with increasing complexity by considering the example of the archetypal combinatorial problem, namely the TSP (Penna 1995). This is where the usefulness of genetic algorithms (GAs) or evolutionary computation (EC) comes in (Seredynski 2006). It turns out that Nature has been there before us. Evolution of a species best suited for survival and flourishing in a given set of conditions is indeed a problem of search space reduction. GAs draw on ideas from Darwinian and Lamarckian evolution.

Let us be quick to point out a difference first. As discussed above in the section on biological evolution, Darwinian evolution is an open-ended, 'blind', process, whereas GAs have a goal. GAs are meant to solve particular pre-conceived problems. Thus, GAs are closer to the 'hand of God' scenario, the 'God' in this case being the software specialist or the scientist.

GAs were introduced by Holland in the early 1960s (cf. Holland 1975). They have been particularly successful for solving problems involving intelligent search, machine learning, or optimization. One begins by creating (in a computer) a population of bit-strings ('chromosomes'); e.g. (1101010101). Each of them is a representative of a potential solution to the problem.

The next step is the evaluation of each string for *fitness*. This is done by using some suitable functions which represent the constraints of the problem.

The fittest individuals are then *selected* for operation by *genetic operators*. The two main genetic manipulations are: sex or *crossover*, and *mutations*. These operations bring in variety, thus increasing the chances of hitting upon the global minimum in the search space, rather than getting trapped in some shallow minimum. Crossover amounts to recombining the bits (genes) of selected pairs of strings. And in mutation, bits at one or more randomly selected sites on the chromosomes are altered.

The offspring resulting from the actions of the genetic operators are then tested for fitness, and the fittest are programmed to have a higher chance of getting selected for the next cycle of operations. And so on, until the optimum solution is reached.

As an illustration, we consider here the example of efficient *drug design*, based on evolutionary principles. Very often, for a drug to be effective, its molecular structure should be such that it can fit snugly into a relevant cleft in a relevant protein molecule. It can be very expensive to actually synthesize all those trial drugs and test their compatibility with the protein–molecule cleft. Bring in GAs. The computer code generates billions of random molecules, which it tests against the cleft in the protein. One such imaginary molecule may contain a site which matches one of, say, six sites on the cleft. This molecule is selected, and a billion variations of it are created, and tested. So on. 'Breeding drugs is the future of biotechnology' (Kelly 1994).

GAs are also being used for solving difficult problems of optimization and discovery in materials science (Holland 1975; Goldberg 1989; Mitchell 1996). As an illustration, we mention here the work of Giro, Cyrillo and Galvao (2002), who have designed conducting polymers by the GA approach. They combined the use of GAs with the so-called negative-factor-counting (NFC) technique. The latter can determine the eigenvalues of large matrices without the conventional lengthy diagonalization procedure. Copolymerization is one way of developing new polymers. Because of the huge number of possible configurations, and because of the ever-present possibility of structural disorder, the task becomes daunting when one wants to investigate theoretically the various ways of combining two or more monomers to design a copolymer with desired properties. Giro, Cyrillo and Galvao (2002) have described the successful implementation of the task of designing binary and ternary disordered copolymers of conducting polyaniline by employing GAs.

2.5.4 *Genetic programming*

GP aims at evolving the computer programs themselves, rather than evolving just solutions to problems based on supplied computer programs.

Smart (evolutionary) searches through a phase space can be made, not only for finding quick solutions using a suitably written computer program, but also for evolving the programs themselves. GP is the art and science of making computers evolve their own algorithms (Koza 1992, 1994; Banzhaf *et al.* 1998; Konar 2005; Seredynski 2006). Evolution involves crossover and mutations. Each of the competing algorithms is normally a *sequence* of commands. What is being attempted, in effect, is that during the process of artificial evolution the sequence of commands is shuffled and the commands are tempered with in other ways. This process is continued till the right combination of everything clicks.

But conventional algorithms are *fragile*. Even a minor crossover or *mutation* of any command in an algorithm can really be a *mutilation*, resulting in a totally nonsensical or illegal statement, making the computer program to crash. In principle, one could get over this problem by assigning zero fitness to all illegal algorithms, but then most of the members of the population would be declared unfit, and the evolution process would come to a halt because it would not have the all-important benefit of continual *variety*. A way must be found for making computer programs less fragile. The answer lies in giving them a *tree structure*, rather than a sequential structure. We refer the reader to the books by Sipper (2002) and Konar (2005) for some details of this, as also to the original works of Koza (1992, 1994) and Banzhaf *et al.* (1998). The tree structure makes the algorithm robust enough to withstand the depredations of crossover, etc. with a certain degree of resilience.

Bio-inspired computing has also been used for evolving sophisticated analogue electrical circuits (Koza *et al.* 1999).

2.5.5 *Artificial life*

AL is '... an inclusive paradigm that attempts to realize lifelike behaviour by imitating the processes that occur in the development or mechanics of life' (Langton 1989). This subject overlaps, not only with CI, but also with real-life phenomena governed by the laws of physics and chemistry.

In the field of artificial *intelligence* (AI) (cf. the appendix on AI), one uses computers to model thought processes. Likewise, in the field of artificial *life* (AL), one uses computers to model the basic biological mechanisms of evolution and life (Heudin 1999). This term was coined around 1971 by Chris Langton, who did the early, groundbreaking work, although some other persons did similar work at about the same time. The brief description we give here is summarized from the book on complexity by Waldrop (1992), who quotes extensively from the work of Langton (1989) and other scientists at the Santa Fe Institute who worked in this area.

The credo of AL is that life is not a property of matter per se, but the *organization* of that matter. Its operating principle is that the laws of life must be laws of dynamical form, independent of the details of a particular carbon-based chemistry that happened to arise here on earth. It attempts to explore other possible biologies in new media, namely computers and robots. The idea is to view *life-as-we-know-it* in the context of *life-as-it-could-be*.

Alan Turing's notion of a universal computing machine is essentially an expression of the fact that what is responsible for the performance of a computer is not the material that has gone into making it, but rather the abstract control structure, namely a program that can be expressed as a set of rules. By the same token, the aliveness of an organism is also in the software, in the *organization* of the molecules, and not merely in the molecules themselves.

Living systems are complex machines, which have evolved by bottom-up processes from large populations of much simpler 'agents'. For example, a brain consists of neurons (Laughlin and Sejnowski 2003), and an ant colony (Holldobler and Wilson 1990; Fewell 2003) or a beehive (Kelly 1994; Seeley, Visscher and Passino 2006) has very simple agents, all in large numbers. Complex behaviour need not have complex roots.

The way to achieve lifelike behaviour is to simulate large populations of simple units, with local, and not global, decision-making. At present this is being done mainly inside computers, although evolutionary or adaptive robotics is taking the action to the real world (Moravec 1988, 1999a; Krichmar *et al.* 2005; Fox 2005).

What makes life and mind possible is a certain kind of balance between the forces of order and the forces of disorder. In other words, there should be *an edge-of-chaos existence.* Only such systems are both stable enough to store information, and yet evanescent enough to transmit it.

Life is not just *like* a computation, in the sense of being a property of the organization rather than the molecules: *Life literally is computation.*

One of the most striking characteristics of a living organism is the distinction between its genotype and phenotype. The genotype can be thought of as a collection of little computer programs, running in parallel, one program per gene. When activated, each of these programs enters into the logical fray by competing and/or cooperating with the other active programs. And collectively these interacting programs carry out an overall computation that is the phenotype. As we have seen above, *artificial* GAs are already finding applications in a number of problem-solving strategies (see, e.g. Mitchell (1996) for more examples). The system *evolves* towards the best solution of a posed problem.

The term GTYPE is introduced in the field of AL to refer to any collection of low-level rules. Similarly, PTYPE means the structure and/or behaviour that results (*emerges*) when these rules are activated in a specific environment. For example, in Langton's self-reproducing cellular automata, the GTYPE is the set of rules specifying how each cell interacts with its neighbours, and PTYPE is the overall pattern that results.

Once a link has been made between life and computation, an immense amount of theory can be brought in. For example, the question 'Why is life full of surprises?' is answered in terms of *the undecidability theorem of computer science*, according to which, unless a computer program is utterly trivial, the fastest way to find out what it would do (does it have bugs or not) is to run it and see. This explains why, although a biochemical machine is completely under the control of a program (the GTYPE), it still has surprising, spontaneous behaviour in the PTYPE. It never reaches equilibrium. (See Heudin (1999) for a partial update on the implementation of these ideas.)

Darwin's theory of evolution of new species by natural selection continues to be challenged occasionally because of the seeming unlikelihood of evolution of complex organisms from very simple beginnings. Among other things, the theory has received very convincing support from a variety of AL simulations. As stated above, these are designed, not just to simulate evolution, but to actually *be* evolutionary processes, albeit inside a computer. One of the latest in this class, and still continuing, is a series of computer experiments carried out with the computer program called **Avida** (Lenski *et al.* 1999, 2003, Zimmer 2005). In it, digital organisms (strings of commands, akin to computer viruses) are created and then allowed to evolve. Each individual has a ('circular') sequence of instructions that are executed sequentially (except when the execution of one instruction causes the instruction pointer to jump to another position); it also has a virtual CPU with two stacks and three registers that hold 32-bit strings. Each item in a sequence is one of 26 possible instructions.

Experiments were begun in Avida with an ancestor that could replicate (by producing tens of thousands of copies of itself in a matter of minutes, thus providing a highly speeded up and fully recorded version of natural evolution). The replicated digital bits can undergo mutations (imperfect replication) the way DNA does. They not only replicate and mutate, they also compete for resources, thus meeting all the three requirements for Darwinian evolution. The resources in this case are a supply of numbers. Most often the individual strings of self-replicating code are not able to do anything (perform any logic function) to a presented number. But, once in a while, a mutation (change of command line) in a replicated individual may give it the ability to, say, read the number and then produce an identical output. Or, in a more advanced (more *evolved*) case, the individual may be able to add two numbers correctly. There is a reward system for such evolved behaviour; e.g. such mutants are allowed to replicate faster. The more complex the evolved behaviour, the higher is the reward.

Emergent behaviour is observed. Avida is not only able to confirm the soundness of Darwin's theory, it provides answers to a variety of other evolutionary puzzles as well. For example: Why does a forest have more than one kind of plant? Why sexual reproduction is an important aid to the evolution of advanced species (Ray 1999)? Why altruism (i.e. regard for other members of the species, as a principle of action) should exist at all (Zimmer 2004)?

More details about this computer-based approach to evolution can be found on the websites http://dllab.caltech.edu/avida, or http://myxo.css.msu.edu/papers/nature2003.

The current interest in computer-mediated evolution, working on self-replicating filaments of code, brings to the fore Freeman Dyson's (1985) assertion that metabolism and replication are logically separable. Avida works on replication (software), without involving metabolism (hardware).

The smart structures of the future will be an embodiment of *real* artificial life, i.e. artificial life which is not just inside a computer.

The following are some of the websites related to artificial life and virtual worlds:

- http://www.biota.org/nervegarden
- http://www.digitalspace.com/avatars
- http://www.2nd-world.com
- http://www.fl.aec.at/~watson
- http://www.digitalworks.org/rd/p5
- http://www.fraclr.org/howareyou.htm

2.5.6 *Combinations of soft-computing tools*

In soft computing one uses basic methodologies like fuzzy logic, neural networks, evolutionary or genetic computing, and 'probabilistic reasoning'. Each has its strong and weak points, and a combined use of two or more of them in a synergistic manner can draw from the strong points of each (Konar 2005; Zomaya 2006). We mention just a couple of possibilities here.

Neuro-GA synergism

ANNs are good for machine learning, whereas GAs are a powerful technique for an intelligent search for global minima in a given large search space. In neuro-GA synergism, one generally aims at using GAs for determining the best-adapted synaptic weights for the ANN. Another use for GAs in this context can be for determining the optimum structure of the ANN.

Neuro-fuzzy synergism

FL is good for approximate reasoning, but not suitable for optimization problems or machine learning. The latter task is best handled by ANNs. Their combination enables one to achieve machine learning from approximate or imprecise inputs. Together, they can also be used for approximate reasoning, using the knowledge acquired by machine learning.

2.5.7 *Data mining and knowledge discovery*

Over the last few decades, the field of data mining and knowledge discovery (DMKD) has been growing steadily (cf. Pal and Jain 2005; Schwartz 2006). Enormous amounts of data are generated by modern high-speed automatic devices. In DMKD, the techniques of CI are employed to make sense of the data, and to discover the knowledge hidden in them.

The DMKD process runs through the following six steps (Pal and Jain 2005):

1. *Understanding the problem domain.* This involves working closely with the domain experts, so as to define the DMKD goals in concrete terms.
2. *Understanding the data,* so as to create a target data set after assessing the relative importance and usefulness of the various attributes of the data.
3. *Preparation of the data.* This step may involve data cleaning and pre-processing; data reduction and projection; matching the DMKD goal to the particular data mining method; and exploratory analysis, modelling, and hypothesis selection.
4. *Data mining,* by using the appropriate CI tools.
5. *Evaluation of the discovered knowledge.* This step may include an iteration of the DMKD process to identify the models more successful at knowledge discovery.
6. *Use of the discovered knowledge* by the owners of the database.

2.6   Machines that can evolve

In a typical device having a microprocessor, e.g. a washing machine or a digital camera, the control part comprises a hardware and a software. The hardware is usually fixed and unchangeable. The software has been written by somebody and is therefore fixed (although new software can be loaded), and the user can only key-in his/her preferences. Computational science has been progressively moving towards a scenario in which the hardware is no longer absolutely hard; and it is even evolvable (Gordon and Bentley 2006; Hornby 2006). Evolutionary ideas have permeated a whole host of scientific and engineering disciplines, and computational science is no exception. We discussed the evolutionary aspects of software in the previous section on CI. We now take up the hardware possibilities.

2.6.1   *Reconfigurable computers*

Conventional general-purpose stored-program computers have a fixed hardware, and they are programmable through software. Their microprocessors can be led through just about any conceivable logical or mathematical operations by writing a suitable set of instructions. Since their hardware configuration is not fine-tuned for any specific task, they tend to be relatively slow.

For major specialized jobs involving a very large amount of number-crunching, it is more efficient to design *application-specific ICs* (ASICs), but then the overall long-term cost goes up, particularly if there is a need for upgradation or alteration from time to time.

An intermediate and very versatile approach is to have reconfigurable computers (RCs) (cf. Villasenor and Mangione-Smith 1997; Gordon and Bentley 2006). A computer has a memory and a processor. The software is loaded into the memory; and the processor normally has a fixed, unalterable, configuration. By 'configuration' we mean the way the various components of the processor are interconnected, and the logical and mathematical operations they perform. If we can make these interconnections and operations (*gate arrays*) alterable, without having to physically change them by hand, we get an alterable or reconfigurable hardware. This is equivalent to having customized hardware, which can be reconfigured at will (within limits), without incurring additional costs.

In RCs one makes use of *field-programmable gate arrays* (FPGAs), the logic structure of which can be altered and customized by the user. The generic architecture of such RCs has four major components connected through a *programmable* interconnect:

1. Multiple FPGAs;
2. Memories;
3. Input/output channels; and
4. Processors.

FPGAs are highly tuned hardware circuits that can be altered at almost any point during use. They comprise of arrays of configurable logic blocks that perform the functions of logical gates. The logic functions performed within the blocks, as well as the connections among the blocks, can be changed by sending control signals.

A single FPGA can perform a variety of tasks in rapid succession, reconfiguring itself as and when instructed to do so. For example, Villasenor and Mangione-Smith (1997) made a single-chip video transmission system that reconfigures itself four times per video frame: It first stores an incoming video signal in the memory; then it applies two different image-processing transformations; and then becomes a modem to send the signal onward.

FPGAs are ideally suited for algorithms requiring rapid adaptation to inputs. Such softening of the hardware raises visions of autonomous adaptability, and therefore *evolution*. It should be possible to make the *soft hardware* evolve to the most desirable (*fittest*) configuration. That would be a remarkable Darwinian evolution of computing machines, a marriage of adaptation and design.

2.6.2 *Self-healing machines*

Biological systems are robust because they are fault-tolerant and because they can heal themselves. Machines have been developed that can also heal themselves to a certain extent.

The new field of *embryonic electronics* (Mange and Tomassini 1998; Tempesti, Mange and Stauffer 1998) has drawn its sustenance from ontogeny and embryogenesis observed in Nature (cf. the appendix on cell biology). One works with a chessboard-like assembly of a large number of reconfigurable computer chips described above. This is like a multicellular organism in Nature. The chips or *cells* are blank-slate cells to start with. We specify a task for the assembly; say, to show the time of the day. We also wish to ensure that the artificial organism is robust enough to heal itself; i.e. it should repair itself if needed. Such a BioWatch has indeed been built (cf. Sipper 2002).

Ontogeny in Nature involves cell division and cell differentiation, with the all-important feature that *each cell carries the entire genome*. The BioWatch borrows these ideas. The genome, of course, comprises of the entire sequence of instructions for building the watch. One starts by implanting the genome (the zygote) in just one cell. Cell division is simulated by making the zygote transfer its genome to the neighbouring cells successively. When a cell receives the genome from one of its neighbours, information about its relative location is also recorded. In other words, each of the cells knows its relative location in the assembly. This information determines how that cell will specialize by extracting instructions from the relevant portion of the genome. Thus each cell or chip, though specialized ('differentiated') for doing only a part of the job, carries information for doing everything, just as a biological cell does.

The BioWatch is now ready to function. Its distinctive feature is that it can *repair* itself. Suppose one of the cells malfunctions, or stops functioning. There are kept some undifferentiated (and therefore unused) cells in the same assembly. Repair or healing action amounts to simply ignoring the dead cell after detecting its relative position, and transferring its job to one of the fresh cells which already has the complete genome, and has to only undergo differentiation for becoming operational.

One can do even better than that by making the system *hierarchical*. Each cell can be given a substructure: Each cell comprises of an identical set of 'molecules'. When one of the molecules malfunctions, its job is transferred to a fresh molecule. Only when too many molecules are nonfunctional does the entire cell become dead, and the services of a fresh cell are requisitioned.

2.6.3   *Hardware you can store in a bottle*

Adamatzky and coworkers have been developing chemical-based processors that are run by ions rather than electrons (Adamatzky, Costello and Asai 2005; also see Graham-Rowe 2005). At the heart of this approach is the so-called *Belousov–Zhabotinsky* or *BZ reaction*.

A BZ reaction is a repeating cycle of three sets of chemical reactions. After the ingredients have been brought together, they only need some perturbation (e.g. a catalyst, or a local fluctuation of concentration) to trigger the first of the three sets of reactions. The products of this reaction initiate the second set of reactions, which then set off the third reaction, which then restart the first reaction. And so on, cyclically. The BZ reaction is self-propagating. Waves of ions form spontaneously and diffuse through the solution, inducing neighbouring regions to start the reactions.

Adamatzky has been developing *liquid logic gates* (for performing operations like 'not' and 'or') based on the BZ reaction. It is expected that an immensely powerful parallel processor (*a liquid robot brain*) can be constructed, which will be a blob of jelly, rather than an assembly of metal and wire. Such a system would be highly reconfigurable and self-healing.

A possible host material for this *blobot* is a jelly-like polymer called PAMPS, an electroactive gel. It expands or contracts when an electric field is applied (cf. Graham-Rowe 2002). BZ waves can travel through it without being slowed down substantially, and they (the waves) can be further manipulated interactively by internal or external electric fields. One day we may end up having an intelligent, shape-changing, crawling blob based on such considerations.

2.7   Design principles of biological networks

The similarity between the creations of evolution and engineering raises a fundamental challenge: understanding the laws of nature that unite evolved and designed systems.

– Uri Alon

The biological cell is a marvel of an integrated information-processing device, perfected by Nature over billions of years of evolutionary optimization, constrained by the conditions in which an organism had to evolve. The chemical processes occurring in the aqueous slurry enclosed in the cell membrane appear to be extremely complex, involving interactions among thousands of molecules. Can we understand them in terms of some general design principles of molecular networking, rather like the way we do network analysis in electronics? The task is difficult because we do not yet have a thorough understanding of how the large number of molecules involved interact (or do not interact) with one another, and what are the strengths of the interactions. The subject of *systems biology* is still in an early stage of development, but there has been substantial progress in the last couple of decades (Alon 2003, 2007; Palsson 2006).

We have described the pioneering work of Kauffman (1993) in this direction in the appendix on cell biology, which should be treated as supplementary reading.

It turns out that one can indeed discern (and model) some general features in the *biochemical circuitry* evolved by Nature for the effective and efficient functioning of cells (Bray 1995). The 'wiring' in these circuits comprises predominantly of diffusion-limited interactions between molecules. Such circuitry handles at least three types of interactions: protein–protein; protein–DNA; and protein–metabolite. Certain protein molecules are involved in each of these three types. In fact, an important information-processing task inside a cell is the determination of the identity and amount of protein needed to be produced for a specific job. This involves the use of what are called *transcription networks* (TNs) (Lee *et al*. 2002). Such networks are a set of transcriptional interactions in the cell. An example is that of the interactions involved whereby the genetic information encoded in a sequence of nucleotides (namely a gene) is copied from the gene for the synthesis of a molecule of mRNA.

The cell monitors the environment all the time, and keeps estimating the amount and the rate of production of each type of protein needed for every job. This information-processing activity is mostly carried out through the mechanism of *information networks*.

Each of the thousands of interacting proteins in a cell can be viewed as a molecular machine, designed to carry out a specific task. The tasks can involve sensing, computation, production, control, actuation, etc. The sensing can be for parameters such as temperature, osmotic pressure, signalling molecules from other cells, nutrient molecules, intruder or parasitic molecules, the assessment of damage caused by such molecules, and so on. Actuation action may involve: transport of nutrient or fuel molecules where they are needed, transport of waste products or debris out of the cell membrane, repair of damaged parts, etc. For example, when sugar is sensed, the cell starts synthesizing protein molecules for transporting and consuming the sugar molecules.

The sensor inputs are used by the cell for producing proteins which can take suitable action on the internal or external environment. Special proteins called *transcription factors* are used for representing the environmental states. The production of proteins is governed by the genetic coding on the DNA in the cell. A transcription factor can bind to the DNA molecule for regulating the rate at which the specific target is read or 'transcribed'. The transcription is into a polynucleotide strand, and thence into the protein to be synthesized. In the transcription networks describing such processes, the nodes of the network are the genes, and the edges represent the transcriptional regulation of one gene by the protein produced by another gene.

Growth factors are an example of transcription factors. These are hormone proteins which stimulate cells to undergo cell division (cells usually do not divide in the absence of this stimulation).

Engineers plan structures in advance, and draw up blueprints and flow charts. By contrast, Nature is more of a tinkerer (operating through blind evolutionary processes), rather than a conscious designer of effectively functioning biological structures (Alon 2003). It is quite a challenge at present to understand how and why Nature has evolved biological networks as if some design engineer were at work. We list here a few such design principles discovered by biologists. We begin with a recapitulation of some basic concepts and definitions in network theory or graph theory.

2.7.1 *Elements of network theory*

Network theory has been put to good use for gaining an understanding of what goes on inside the biological cell. Many of the essential characteristics of the processes can be represented in an abstract way as biological *networks*. A network is a discrete set of points (called nodes), some or all of which may be joined by lines (called edges). In biological networks, the nodes are the relevant molecules, and the edges or arrows connecting specific nodes are the interactions or reaction pathways between those molecules. For example, for transcription networks, each node represents a gene. Consider two nodes X and Y. If there is a transcription interaction between them, it is represented by a directed edge (arrow) between them: $X \rightarrow Y$.

In general, more than one arrows may point at a particular node in some networks, in which case a suitable input function has to be defined (e.g. 'AND' or 'OR'). Similarly, two nodes may be simultaneously involved in more than one types of interaction (e.g. on different time-scales). One can then use different *colours* for the edges joining them. The strength of the interaction between two nodes can be represented by assigning a *weight* to the edge joining them.

Naturally, there are a huge number of possibilities in which a given set of nodes can be interconnected (Barabasi 2002). We mention just three here (cf. Bray 2003), in anticipation of the discussion that will follow in this section. These three networks are classified as: *regular*, *random*, and *scale-free*.

In a regular network or graph, the nearest-neighbour nodes are connected in a regular manner. Such networks have local groups of highly interconnected nodes.

A random network has randomly selected connections or edges. Such a network is easily traversed because the average number of steps in going from a node to any other node (called the 'distance') is relatively small.

In a scale-free network, a few of the nodes have a much larger number of connections than others.

For a given set of nodes and a given fixed number of connections among them, the global characteristics of these three types of networks are very different. For regular and random networks, the number of edges per node has an approximately Gaussian distribution, with a mean value that is a measure of the 'scale' of the network.

In a scale-free network, by contrast, there are a few strongly connected nodes and a large number of weakly connected ones. Typically, there is a power-law decay of the number of nodes, N, having a given number, k, of connections:

$$N(k) \sim k^{-g} \tag{2.5}$$

The exponent g generally lies between 2 and 3. Since the distribution of $N(k)$ does not show a characteristic peak (unlike for the other two networks defined above), it is described as *scale-free*. For it, the average distance between any two nodes is almost as small as for a random network. And yet, its 'clustering coefficient', namely the extent to which neighbours of a particular node are themselves interconnected, is practically as large as for a regular network. This last property is called *cliquishness*: there are regions of the network in which the nodes form cliques. The cliques are the equivalent of friends in a human population. Therefore the term *small-world networks* is also used in sociology for scale-free networks.

Networks may be *isotropic*, or otherwise. Isotropic networks do not have well-defined inputs and outputs. By contrast, neural networks discussed earlier in this chapter have well-defined inputs, which they are supposed to convert into certain desired outputs.

In spite of the fact that evolutionary processes involve a fair amount of random tinkering, Nature has zeroed-in again and again on certain recurring circuit elements, called *network motifs* (NMs). Each such motif in the molecular network performs a specific information-processing job (Milo *et al.* 2004). Examples of such jobs in the case of the transcriptional networks of the cell comprising the bacterium *E. coli* are (cf. Alon 2003): filtering out spurious input fluctuations; generating temporal programs of gene expression; and accelerating the throughput of the network. Similar NMs have also been found in the transcription networks of yeast (Milo *et al.* 2002). Operational amplifiers and memory registers in human-made electronic circuits are some of the equivalents of NMs.

NMs are formally defined as follows (Milo *et al.* 2002). Given a network with N nodes and E edges, one can construct a *randomized network* from it which has the same single-node characteristics as the actual network (i.e. the numbers N and E are unchanged), but where the connections between nodes are made at random. NMs are those patterns that occur in the actual network significantly more frequently than in the randomized network.

2.7.2 *Network motifs in sensory transcription networks*

Sensory transcription networks have been so far found to contain mostly four families of NMs:

1. Negative autoregulation NMs;
2. Feedforward loop (FFL) NMs;

3. Single-input modules (SIMs); and
4. Dense overlapping regulons (DORs).

Negative autoregulation NMs, as the name implies, are like negative-feedback electronic circuits. Why did Nature evolve such networks? There must be a good reason, because such NMs are found in large numbers in biological circuits. In fact, the reason is the same as that for which we humans design, say, a temperature controller for controlling a set temperature by negative-feedback circuitry (Becskei and Serrano 2000) (Fig. 2.4a). The difference between the sensed and the set temperature constitutes the error signal (which has a positive or negative sign). The negative of this error signal determines the correction applied to the power being fed to the system for sustaining the set temperature. This arrangement enables the circuit to apply large heating power in the beginning, which is then progressively decreased, so that a temperature close to the set temperature is reached *quickly*. In the absence of such a negative-feedback loop, it would take much longer to reach the set temperature because power would be increased only in small steps all the way. Similarly, the emergence of negative autoregulation NMs provided an evolutionary advantage to organisms because they could adapt to the environment more quickly (Alon 2007).

There are 13 possible ways of connecting three nodes with directed edges. Only one of them (called the FFL) (Fig. 2.4b) is found to be an NM (Milo *et al.* 2002). The FFL has eight possible types, only two of which are very common. The most common one is the so-called *coherent type-1 FFL*. It is a sign-sensitive delay motif responsible for protection against unwanted responses to fluctuating inputs (Alon 2007).

The other commonly occurring FFL type, namely the *incoherent type-1 FFL*, can function as a pulse generator and a response accelerator (Mangan and Alon 2003).

The FFL is also an NM in several *nonsensory* types of biochemical networks.

The SIM is an NM in which one regulator controls a group of genes. It can generate gene-expression programs which are turned on in a specific temporal order. An important (economical) design principle involved here is that of producing proteins only when needed, and not beforehand.

Lastly, the DOR family is a dense array of regulators that control output genes in a combinatorial manner. The inputs to the various genes determine the decision-making computations carried out by DORs.

Fig. 2.4   Three-node feedback loop (a), and feedforward loop (b).

2.7.3 *Network motifs in developmental transcription networks*

A fertilized egg cell develops into a multicellular entity. Cell differentiation (i.e. change from one cell type to another) is an important feature of this developmental process. It involves the expression of specific sets of proteins which distinguish, say, a nerve cell from a kidney cell. Developmental TNs have been identified which play a crucial role in this process (see, e.g. Davidson *et al.* 2002; Levine and Davidson 2005).

Whereas sensory TNs have to make rapid and reversible decisions on time-scales shorter than a cell generation time, developmental TNs often make irreversible decisions, on slower time-scales spanning one or more cell generations. This important difference leads to the emergence of additional NMs in developmental TNs, which are not found in sensory TNs. These NMs carry out operations which include the following: two-node positive feedback loops for decision making; regulation of feedback; long transcription cascades; and interlocked FFLs. We refer the reader to the book by Alon (2007) for details.

2.7.4 *Signal transduction network motifs*

NMs have been discovered, not only in the TNs operative in a cell, but also in other interaction networks, like signal transduction transcription networks, and metabolic networks.

Signal transduction TNs operate on rapid time-scales. The edges in these networks represent interactions between signalling proteins. Such networks process environmental information, and then regulate the activity of, for example, transcription factors like growth factors. When a particular growth factor is sensed, it triggers a signal transduction pathway that finally activates genes leading to cell division.

Two NMs encountered in signal transduction TNs are: bi-fan NMs, and diamond NMs. The bi-fan is also found in transcription networks, where it generalizes to single-layer patterns called DORs. The diamond, however, is not generally found in transcription networks. It generalizes to form *multi*layer patterns: There is a cascade of DOR-like structures, with each DOR receiving inputs from an upper-level DOR, rather like in multilayer perceptrons (Bray 1995).

2.7.5 *Intercellular networks*

Not only intracellular networks, but also intercellular networks show NMs. The much-modelled neural network (NN) is an example of an intercellular network. The human cortical network, of course, is extremely complex. It has been easier to investigate the NN of the nematode *C. elegans*, a worm made up of just ∼ 1000 cells, 300 of which are neurons (Bargmann 1998).

In an NN, the nodes are the neurons, and a directed edge between two nodes is the synaptic connection from one neuron to the other. The NN of *C. elegans* has several NMs which occur *intra*cellularly as well (in transcription networks and in signal-transduction networks). This has been found to be so in spite of the fact that there is *spatial* confinement in the latter case, and the time-scales are also very different. NNs (as we shall see in Chapter 7) operate at millisecond response times, which are much faster than the time-scales involved in intracellular processes.

The most notable three-node NM in the neural network of *C. elegans* is the FFL. According to Alon (2007), the reason why the FFL occurs both intracellularly and intercellularly is that in both cases information-processing is carried out on noisy signals using noisy components. In both cases, information is conveyed between input sensory components and output motor components. Evolutionary processes converged on the same strategy for both situations for doing jobs like sign-sensitive filtering. However, the FFLs in NNs are connected differently than those in transcription networks.

In the NN of *C. elegans*, patterns with four or higher number of nodes have multilayer perceptrons as the most common NM. These are similar to what is found in signal transduction networks, although in the case of *C. elegans* the multilayer perceptrons have a higher abundance of feedback loops between pairs of nodes in the same layer.

2.7.6   *Hallmarks of biological design*

Modularity

Nature seems to prefer a modular architecture in the design of molecular networks (Ravasz *et al.* 2002), rather like the use of subroutines or OOPs by software engineers, or replaceable plug-in parts by hardware engineers.

In an overall molecular network, a module is a subset having the features of a scale-free network described above. It is a set of nodes that have strong connectivity among themselves, and a common function to perform. It also has defined input and output nodes that control the interactions with the rest of the network. There are internal nodes which do not interact significantly with nodes outside the module; these are the highly connected nodes of the scale-free network defined above. Molecular modules also have features like low output impedance, so that introduction of additional modules downstream does not cause a major upheaval in the integrity of the overall network, and the new modules are able to merge seamlessly with the existing network.

Experience with ANNs and other networks tested on a computer indicates that modular connectivity does not yield the most optimum results, as it amounts to forbidding the occurrence of possibly more optimal but nonmodular connections among the nodes. Why does it occur in biological circuits then? The answer lies in the fact that, if design specifications *change* from time to time (as happens in the ever-changing evolutionary scenario), modular design is more advantageous. In such situations it is more efficient and economical to regroup and rewire the existing time-tested modules, rather than having to start all over again in search of the most optimal design. In the biological evolution of molecular circuits, if environmental conditions change too frequently and rapidly (on evolutionary time-scales), the chances of the survival and flourishing of that species are higher which takes the route of modularity of design. Of course, if there is also a long enough period of stability of environmental conditions, the gene pool can further optimize itself by experimenting with nonmodular design of molecular networking to improve its fitness.

Robustness

A born or a made smart structure must be robust against moderate fluctuations in internal and external conditions. An organism must, for example, be able to survive in a temperature *band*, rather than only at a particular temperature. Similarly, no part of the molecular network in the cell should depend critically on having, say, exactly 100 copies of a particular protein, and not 103 or 95. This is indeed found to be the case in biological circuits. In fact, there is often a statistical variation of tens of percent from cell to cell in the expression level of a protein in genetically identical cells. There is a *robustness principle* in operation here (Alon 2007): *Biological circuits have robust designs such that their essential function is nearly independent of biochemical parameters that tend to vary from cell to cell.*

There is, however, a price to pay for this robustness with respect to component tolerances. Only a small fraction of possible networks for a given task can perform it robustly. Nonrobust circuit designs get eliminated by evolutionary filtering.

The robustness involved is property-specific and parameter-specific. Of necessity, some properties have to be *fine-tuned*, rather than robust. Such properties must change significantly with internal or external conditions, and they do.

Kinetic proofreading

Evolutionary processes have resulted in robustness of some properties, not only with respect to fluctuations in the biochemical parameters, but also against errors of *molecular recognition* (Elowitz *et al.* 2002). Suppose a biochemical molecule X is to be recognized and interacted with by a molecule Y. There are bound to be present molecules which are chemically similar to X, so that the molecular recognition process can be confounded by one or more of them. Let Z be one such molecule, and ΔG_{YZ} the difference in the free energies of two possible reaction pathways, one in which Y recognizes X (the correct process), and the other in which Z recognizes X (an erroneous process). Under equilibrium conditions, the probability of error is greater than or equal to $e^{-\Delta G_{YZ}/RT}$, where R is the gas constant in the ideal-gas law $pV = nRT$. This probability of error turns out to be far higher than what is actually observed in real-life situations. For example, error rates of only 10^{-4} are observed in protein synthesis, both in the case of codon–anticodon binding and in amino-acid recognition (Hopfield 1974). Similarly, error rates are as low as 10^{-9} for DNA replication.

Hopfield (1974) suggested the mechanism of kinetic proofreading or editing for explaining this. Nature has indeed evolved a mechanism in which, in addition to the initial molecular recognition process (error-prone by the factor just defined), there is a further recognition test for the product of the initial recognition step. If the second testing process occurs with the same precision as the first one, the probability of error would drop to $[e^{-(\Delta G_{YZ}/RT)}]^2$.

There is a certain degree of reversibility in chemical binding and unbinding reactions in equilibrium. Therefore, a strong degree of irreversibility has to be present in the proofreading mechanism just described. This becomes possible if the biosynthetic pathway is kinetically complex, is far away from equilibrium, and is strongly driven. Nature uses such reaction pathways, somewhere in the translation process, to achieve a high degree of transcription fidelity and recognition processes in the presence of thermal noise. Thus several additional chemical steps, which may otherwise appear to be superfluous and wasteful, become integral parts of the overall evolutionary success of the living organism. This is achieved at the expense of energy and delays.

Hierarchical design

In addition to the widespread occurrence of network motifs and modularity, hierarchical design is another hallmark of design principles in biological systems (Ravasz *et al.* 2002). The most striking example of this is the neocortical network of the mammalian brain (cf. Chapter 7). This is again like in engineered systems, particularly in telecommunications technology.

2.8   Chapter highlights

- Sensing, actuation, computation, and control, all involve information processing. This is true for both animate and inanimate systems. Intelligent systems process information intelligently.
- There is a deep connection between intelligence and evolution. In biological systems, information about internal and external conditions is processed, and the system evolves towards better and better fitness for survival and efficiency. Efficient search for a solution is the essence of intelligence.
- Humans have rapidly developed into something unique on the evolutionary landscape. The meme pool, and not just the gene pool, has played a major role in this. Evolution of sophisticated language skills gave us tremendous evolutionary advantages. Innovations made by individuals in the population got transmitted by the written word over large distances and times, and this has been a self-accelerating phenomenon so far as innovations go.
- The born and the made are progressively acquiring the characteristics of one another. Both have an unmistakable evolutionary character. Emergent behaviour and evolution go hand in hand.
- Life is not just *like* a computation: Life literally *is* computation. Therefore, advances in computational science in general, and in the field of computational intelligence in particular, have a direct bearing on the progress we achieve in developing artificial smart structures.
- Sheer large numbers and effective communication and interaction result in intelligence (swarm intelligence), of which no single member is capable alone.
- Four distinct features of distributed existence make such swarms, or vivisystems, or complex adaptive systems what they are: absence of imposed centralized

control; autonomous nature of subunits; high connectivity between the subunits or 'agents'; and webby nonlinear causality of peers influencing peers.

- Higher-level complexities cannot be inferred from lower-level existences. Actually running or observing a vivisystem is the quickest, the shortest, and the only sure method to discern emergent structures latent in it.

- Self-organizing systems are open systems. There is always an exchange of energy and matter with the surroundings. If this exchange is large enough to take the system substantially away from equilibrium, order and pattern-formation can emerge. Self-organization or spontaneous pattern-formation requires self-reinforcement or positive feedback.

- Evolution, both 'natural' and 'artificial', has an inherent parallel-processing underpinning.

- Artificial parallel processing on a massive scale is impossible to achieve, except through an evolutionary approach. The bio-inspired approach to complex computing relies on assembling software from working parts, while continuously testing and correcting the assembly as it grows. There is still the problem of unexpected emergent behaviour (bugs) arising from the aggregation of bugless parts. But there is hope that, as long as one only needs to test at the new emergent level, there is a very good chance of success. There is only one way to obtain robust and reliable software for extremely complex problems: through artificial evolution.

- There is a lot one can still learn from beehives, ant colonies, and other social insect networks.

- Complexity, beyond a certain threshold, leads to the emergence of new principles. The new principles and features are deducible from, but not reducible to, those operating at the lower levels of complexity.

- The burgeoning field of computational intelligence (CI) aims at working with computational systems which are fault-tolerant, making adjustments in the software intelligently and handling imperfect or 'fuzzy' data the way we humans do. The main branches of CI are: artificial neural networks; fuzzy-logic systems; genetic algorithms; genetic programming; and artificial life.

- The evolutionary approach for developing software has been extended to hardware also. Reconfigurable computers and self-healing machines are some examples.

- Liquid logic gates (for performing operations like 'not' and 'or') based on the Belousov–Zhabotinsky (BZ) reaction have been developed. It is expected that a powerful parallel processor (a liquid robot brain) can be constructed, which will be a blob of jelly, rather than an assembly of metal and wire. Such hardware would be highly reconfigurable and self-healing. A possible host material for this blobot is the polymer PAMPS. We can look forward to having an intelligent, shape-changing, crawling blob based on such an approach.

- Investigations on information-processing carried out by biological structures, and by artificial smart structures and computational systems, nourish one another. Understanding one helps understand the other.

- Network theory is now routinely employed for gaining an understanding of what goes on inside the biological cell. Many of the essential characteristics of the processes can be represented in an abstract way as biological networks, in which the nodes of the network are the relevant molecules, and the edges or arrows connecting specific nodes are the interactions or reaction pathways between those molecules.

- In spite of the fact that evolutionary processes involve a fair amount of random tinkering, Nature has zeroed-in again and again on certain recurring circuit elements, called network motifs. Each such motif in the molecular network performs a specific information-processing job. Operational amplifiers and memory registers in human-made electronic circuits are some of the equivalents of network motifs in the biological cell.
- Some of the hallmarks of biological design have been found to be: modularity; robustness against reasonable variations; kinetic proofreading for minimizing the chances of error during production or replication; and a hierarchical structure.

References

Adamatzky, A., Ben de Lacy Costello and T. Asai (2005). *Reaction–Diffusion Computers*. London: Elsevier.

Alon, U. (26 September 2003). 'Biological networks: The tinkerer as an engineer'. *Science*, 301: 1866.

Alon, U. (2007). *An Introduction to Systems Biology: Design Principles of Biological Circuits*. London: Chapman and Hall/CRC.

Anderson, P. W. (4 August 1972). 'More is different'. *Science*, 177: 393.

Avery, J. (2003). *Information Theory and Evolution*. Singapore: World Scientific.

Babbage, C. (1864). *Passages from the Life of a Philosopher*. London: Longman and Green. Facsimile reprint: New York: A. M. Kelly (1969).

Bak, P. (1996). *How Nature Works: The Science of Self-Organized Criticality*. New York: Springer.

Banzhaf, W., P. Nordin, R. E. Keller and F. D. Francome (1998). *Genetic Programming, An Introduction: On the Automatic Evolution of Computer Programs and Its applications.* San Francisco, California: Morgan Kaufmann.

Barabasi, A,-L. (2002). *Linked: The New Science of Networks*. Cambridge, MA: Perseus Books.

Bargmann, C. I. (11 December 1998). 'Neurobiology of the *Caenorhabditis elegans* genome'. *Science*, 282: 2028.

Barricelli, N. A. (1962). 'Numerical testing of evolution theories. Part 1'. *Acta Biotheoretica*, 16: 94.

Becskei, A. and L. Serrano (2000). 'Engineering stability in gene networks by autoregulation', *Nature*, 405: 590.

Bezdek, J. C. (1994). 'What is computational intelligence?' In J. M. Zurada, R. J. Marks and C. J. Robinson (eds.), *Computational Intelligence Imitating Life*. New York: IEEE Press.

Bhalla, U. S. and R. Iyengar (15 January 1999). 'Emergent properties of networks of biological signalling pathways', *Science*, 283: 381.

Boole, G. (1854). *An Investigation of the Laws of Thought, on which are Founded the Mathematical Theories of Logic and Probabilities*. London: Macmillan.

Bray, D. (27 July 1995). 'Protein molecules as computational elements in living cells', *Nature*, 376: 307.

Bray, D. (26 September 2003). 'Molecular networks: the top down view'. *Science*, 301: 1864.

Butler, S. (1863). *Darwin Among the Machines*. Canterbury Press. Reprinted in H. F. Jones (ed.) (1923), *Canterbury Settlement and other Early Essays*, Vol. 1, of *The Shrewsbury Edition of the Works of Samuel Butler*. London: Jonathan Cape.

Chaitin, G. (March 2006). 'The limits of reason'. *Scientific American,* 294: 74.

Cordon, O., F. Herrera, F. Hoffmann and L. Magdalena (2001). *Genetic Fuzzy Systems: Evolutionary Tuning and Learning of Fuzzy Knowledge Bases*. Singapore: World Scientific.

Darwin, C. (1859). *On the Origin of Species*. Reprint of the first edition, 1950, London: Watts and Co.

Darwin, F. (1903) (ed.). *More Letters of Charles Darwin*, Vol. 1, p. 125. London: John Murray.

Davidson, E. H., J. P. Rast, P. Oliveri, A. Ransick, C. Calestani, C.-H. Yuh, T. Minokawa, G. Amore, V. Hinman, C. Arenas-Mena, O. Otim, C. T. Brown, C. B. Livi, P. Y. Lee, R. Revilla, A. G. Rust, Z. jun Pan, M. J. Schilstra, P. J. C. Clarke, M. I. Arnone, L. Rowen, R. A. Cameron, D. R. McClay, L. Hood and H. Bolouri (1 March 2002). 'A genomic regulatory network for development'. *Science*, 295: 1669.

Davies, P. (5 March 2005). 'The sum of the parts'. *New Scientist*, 185: 34.

Dawkins, R. (1989). *The Selfish Gene*. Oxford: Oxford University Press.

Dawkins, R. (1998). *Unweaving the Rainbow*. London: Allen Lane, Penguin.

Dorigo, M., M. Birattari and T. Stutzle (November 2006). 'Ant colony optimization: Artificial ants as a computational intelligence technique'. *IEEE Computational Intelligence Magazine*, 1(4): 28.

Dorigo, M. and L. M. Gambardella (1997). 'Ant colonies for the travelling salesman problem'. *Biosystems*, 43(2): 73.

Duke, C. (July 2006). 'Prosperity, complexity and science'. *Nature Physics*, 2: 426.

Dyson, F. J. (1985). *Origins of Life*. Cambridge: Cambridge University Press.

Dyson, F. J. (1988). *Infinite in All Directions*. New York: Harper and Row.

Dyson, F. J. (11 February 2006). 'Make me a hipporoo'. *New Scientist*: p. 36.

Dyson, G. B. (1997). *Darwin Among the Machines: The Evolution of Global Intelligence*. Cambridge: Perseus Books.

Elowitz, M. B., A. J. Levine, E. D. Siggia and P. S. Swain (16 August 2002). 'Stochastic gene expression in a single cell'. *Science*, 297: 1183.

Fewell, J. H. (26 September 2003). 'Social insect networks'. *Science*, 301: 1867.

Fogel, D. B. (2006). 'Nils Barricelli – Artificial life, coevolution, self-adaptation'. *IEEE Computational Intelligence Magazine*, 1: 41.

Fox, D. (5 November 2005). 'Brain Box'. *New Scientist*, p. 28.

Giro, R., M. Cyrillo and D. S. Galvao (2002). 'Designing conducting polymers with genetic algorithms'. In Takeuchi, I., J. M. Newsam, L. T. Willie, H. Koinuma and E. J. Amis (eds.), *Combinatorial and Artificial Intelligence Methods in Materials Science*. MRS Symposium Proceedings, Vol. 700. Warrendale, Pennsylvania: Materials Research Society.

Godel, K. (1931). Translated from the original German by E. Mendelson as 'On formally undecidable propositions of *Principia Mathematica* and related systems. I' in Davis, M. (ed.) (1965), *The Undecidable*. Hewlett, New York: Raven Press.

Goldberg, D. E. (1989). *Genetic Algorithms in Search, Optimization and Machine Learning*. New York: Wesley.

Gordon, D. (1996). 'The organization of work in social insect colonies'. *Nature*, 380: 121.

Gordon, D. (1999). *Ants at Work*. New York: The Free Press.

Gordon, T. G. W. and P. J. Bentley (2006). 'Evolving hardware'. In Zomaya, A. Y. (ed.), *Handbook of Nature-Inspired and Innovative Computing: Integrating Classical Models with Emerging Technologies*. New York: Springer, p. 387.

Graham-Rowe, D. (6 July 2002). 'Minibeast muscles in on robot limbs'. *New Scientist*, 175: 33.

Graham-Rowe, D. (26 March 2005). 'Glooper computer'. *New Scientist*, p. 33.

Hawkins, J. and S. Blakeslee (2004). *On Intelligence*. New York: Times Books.

Hebb, D. O. (1949). *The Organization of Behaviour: A Neuropsychological Theory*. New York: Wiley.

Heudin, J.-C. (1999) (ed.). *Virtual Worlds: Synthetic Universes, Digital Life, and Complexity*. Reading, Massachusetts: Perseus Books.

Hobbes, T. (1651). *Leviathan; or, The Matter, Forme, and Power of a Commonwealth Ecclesiasticall and Civill*. London: Andrew Crooke.

Holland, J. H. (1975). *Adaptation in Natural and Artificial Systems*. Ann Arbor: University of Michigan Press.

Holland, J. H. (1995). *Hidden Order: How Adaptation Builds Complexity*. Reading, Massachusetts: Addison Wesley.

Holland, J. H. (1998). *Emergence: From Chaos to Order*. Cambridge, Massachusetts: Perseus Books.

Holldobler, B. and E. O. Wilson (1990). *The Ants*. Berlin: Springer-Verlag.

Hooper, R. (4 March 2006). 'Inheriting a heresy'. *New Scientist*, p. 53.

Hopfield, J. J. (1974). 'Kinetic proofreading: A new mechanism for reducing errors in biosynthetic processes requiring high specificity'. *Proc. Nat. Acad. Sci. (USA)*, 71: 4135.

Hopfield, J. J. (1982). 'Neural networks and physical systems with emergent collective computational abilities'. *Proc. Nat. Acad. Sci. (USA)*, 79(8): 2554.

Hopfield, J. J. (1984). 'Neurons with graded response have collective computational properties like those of two-state neurons'. *Proc. Nat. Acad. Sci. (USA)*, 81: 3088.

Hornby, G. S. (2006). 'Evolvable hardware: Using evolutionary computation to design and optimize hardware systems'. *IEEE Computational Intelligence Magazine*, 1: 19.

Jackson, D. (June 2006). 'Dependable software by design'. *Scientific American*, 294: 68.

Kauffman, S. A. (1993). *The Origins of Order*. Oxford: Oxford University Press.

Kauffman, S. A. (1995). *At Home in the Universe*. Oxford: Oxford University Press.

Kauffman, S. A. (2000). *Investigations*. Oxford: Oxford University Press.

Kelly, K. (1994). *Out of Control: The New Biology of Machines, Social Systems, and the Economic World.* Cambridge: Perseus Books.

Kennedy, J. (2006). 'Swarm intelligence'. In Zomaya, A. Y. (ed.), *Handbook of Nature-Inspired and Innovative Computing: Integrating Classical Models with Emerging Technologies*. New York: Springer, p. 187.

Khakhina, L. N. (1992) (ed.). *Concepts in Symbiogenesis: A Historical and Critical Study of the Research of Russian Botanists*. Translated into English by S. Merkel, and edited by L. Margulis and M. McMenamin, Yale Univ. Press, New Haven, Connecticut.

Khanna, T. (1990). *Foundations of Neural Networks*. Reading: Addison-Wesley.

Konar, A. (2005). *Computational Intelligence: Principles, Techniques and Applications*. Berlin: Springer-Verlag.

Kosko, B. and S. Isaka (July 1993). 'Fuzzy logic'. *Scientific American,* 269: 62.

Koza, J. R. (1992). *Genetic Programming: On the Programming of Computers by Means of Natural Selection*. Cambridge, MA: The MIT Press.

Koza, J. R. (1994). *Genetic Programming II: Automatic Discovery of Reusable Programs*. Cambridge, MA: MIT Press.

Koza, J. R., F. H. Bennett III, D. Andre and M. A. Keane (1999). *Genetic Programming III: Darwinian Invention and Problem Solving.* San Francisco, CA: Morgan Kauffmann.

Krattenthaler, C. (2005). 'Combinatorics: Overview', in J.-P. Francoise, G. L. Naber and T. S. Tsun (eds.), *Encyclopaedia of Mathematical Physics*, Vol. 1, p. 553. Amsterdam: Elsevier.

Krichmar, J. L., D. A. Nitz, J. A. Gally and G. M. Edelman (8 February 2005), 'Characterizing functional hippocampal pathways in a brain-based device as it solves a spatial memory task'. *PNAS, USA*, 102: 2111.

Krishnamurthy, E. V. and V. Krishnamurthy (2006). 'Multiset rule-based programming paradigm for soft computing in complex systems'. In Zomaya, A. Y. (ed.), *Handbook of Nature-Inspired and Innovative Computing: Integrating Classical Models with Emerging Technologies*. New York: Springer, p. 77.

Langton, C. G. (ed.) (1989). *Artificial Life*. Santa Fe Institute Studies in the Sciences of Complexity, Proceedings Vol. 6. Redwood City, CA: Addison Wesley.

Laughlin, S. B. and T. J. Sejnowski (26 September 2003). 'Communication in neural networks'. *Science*, 301: 1870.

Lee, T. I., N. J. Rinaldi, F. Robert, D. T. Odom, Z. Bar-Joseph, G. K. Gerber, N. M. Hannett, C. T. Harbison, C. M. Thompson, I. Simon, J. Zeitlinger, E. G. Jennings, H. L. Murray, D. B. Gordon, B. Ren, J. J. Wyrick, J.-B. Tagne, T. L. Volkert, E. Fraenkel, D. K. Gifford and R. A. Young (25 October 2002). 'Transcriptional regulatory networks in *Saccharomyces cerevisiae*'. *Science*, 298: 799.

Leibniz, G. W. von (1675), to H. Oldenburg, 18 December 1675, in Turnbull, H. W. (1956) (ed.), *The Correspondence of Issac Newton*, Vol. 1, p. 401. Cambridge: Cambridge Univ. Press.

Leibniz, G. W. von (1714), to N. Remond, 10 January 1714, in Loemker, L. E. (1956) (editor and translator), *Philosophical Papers and Letters*, Vol. 2, p. 1063. Chicago: University of Chicago Press.

Lenski, R. E., C. Ofria, T. C. Collier and C. Adami (12 Aug. 1999). 'Genome complexity, robustness and genetic interactions in digital organisms'. *Nature*, 400: 661.

Lenski, R. E., C. Ofria, R. T. Pennock and C. Adami (2003). 'The evolutionary origin of complex features'. *Nature*, 423: 139.

Levine, M. and E. H. Davidson (5 April 2005). 'Gene regulatory networks for development'. *Proc. Nat. Acad. Sci. USA*, 102: 4936.

Lippmann, R. (April 1987). 'An introduction to computing with neural nets'. *IEEE ASSP Magazine*, p. 4.

Lloyd, S. and Y. J. Ng (Nov. 2004). 'Black hole computers'. *Scientific American,* 291(5): 30.

Lloyd, S. (2006). *Programming the Universe*. Alfred A. Knopf.

Mangan, S. and U. Alon (2003). 'Structure and function of the feed-forward loop network motif'. *PNAS USA*, 100: 11980.

Mange, D. and M. Tomassini (eds.) (1998). *Bio-Inspired Computing Machines: Toward Novel Computational Architectures*. Lausanne: Presses Polytechniques et Universitaires Romandes.

McCulloch, W. and W. Pitts (1943). 'A logical calculus of the ideas immanent in nervous activity'. *Bull. Math. Biophys.* 5: 115.

Milo, R., S. Shen-Orr, S. Itzkovitz, N. Kashtan, D. Chklovskii and U. Alon (2002). 'Network motifs: Simple building blocks of complex networks'. *Science*, 298: 824.

Milo, R., S. Itzkovitz, N. Kashtan, R. Levitt, S. Shen-Orr, I. Ayzenshtat, M. Sheffer and U. Alon (5 March 2004). 'Superfamilies of evolved and designed networks'. *Science*, 303: 1538.

Minsky, M. (1986). *The Society of Mind*. New York: Simon and Schuster.

Minsky, M. and S. Papert (1969). *Perceptrons*. Cambridge, MA: MIT Press.

Mitchell, M. (1996). *An Introduction to Genetic Algorithms*. Cambridge, Massachusetts: MIT Press.

Moravec, H. (1988). *Mind Children: The Future of Robot and Human Intelligence*. Cambridge: Harvard University Press.

Moravec, H. (1999a). *Robot: Mere Machine to Transcendent Mind*. Oxford: Oxford University Press.

Neumann, J. von (1963). *Collected Works*. Oxford: Pergamon Press.

Pal, N. R. and L. Jain (2005) (eds.). *Advanced Techniques in Data Mining and Knowledge Discovery*. London: Springer-Verlag.

Palsson, B. O. (2006). *Systems Biology: Properties of Reconstructed Networks*. Cambridge: Cambridge University Press.

Pedrycz, W. and F. Gomide (1998). *An Introduction to Fuzzy Sets: Analysis and Design*. Cambridge Massachusetts: MIT Press.

Penna, T. J. P. (1995). 'Travelling salesman problem and Tsallis statistics'. *Phys. Rev. E*, 51: R1.

Penrose, L. S. (June 1959). 'Self-reproducing machines'. *Scientific American,*, p. 105.

Politi, A. (2006). 'Complex systems'. In G. Fraser (ed.), *The New Physics for the Twenty-First Century*. Cambridge, U. K.: Cambridge University Press, p. 334.

Prigogine, I. (1998). *The End of Certainty: Time, Chaos, and the New Laws of Nature*. New York: Free Press.

Ravasz, E., A. L. Somera, D. A. Mongru, Z. N. Oltvai and A.-L. Barabasi (30 August 2002). 'Hierarchical organization of modularity in metabolic networks'. *Science*, 297: 1551.

Ray, T. S. (1999). 'An evolutionary approach to synthetic biology: Zen and the art of creating life'. In Heudin, J.-C. (1999) (ed.), *Virtual Worlds: Synthetic Universes, Digital Life, and Complexity*. Reading, Massachusetts: Perseus Books.

Rebek, J. (July 1994). 'Synthetic self-replicating molecules'. *Scientific American,*, p. 34.

Resnick, M. (1997). *Turtles, Termites, and Traffic Jams: Explorations in Massively Parallel Microworlds*. Cambridge Massachusetts: MIT Press.

Ross, T. J. (1997). *Fuzzy Logic with Engineering Applications*. New York: McGraw-Hill.

Schuster, H. G. (2003). *Complex Adaptive Systems*. Scator Publisher. See http://www.theo-physik.uni-kiel.de

Schwartz, D. G. (ed.) (2006). *Encyclopedia of Knowledge Management*. Hershey, PA, USA: Idea Group Reference.

Seeley, T. D., P. K. Visscher and K. M. Passino (May–June 2006). 'Group decision making in honey bee swarms'. *American Scientist*, 94(3): 220.

Seredynski, F. (2006). 'Evolutionary paradigms'. In Zomaya, A. Y. (ed.), *Handbook of Nature-Inspired and Innovative Computing: Integrating Classical Models with Emerging Technologies*. New York: Springer, p. 111.

Sipper, M. (2002). *Machine Nature*. New Delhi: Tata McGraw-Hill.

Sipper, M. and J. A. Reggia (Aug. 2001). 'Go forth and replicate'. *Scientific American,* 285: 35.

Smee, A. (1849). *Principles of the Human Mind Deduced from Physical Laws*. London: Longman, Brown, Green and Longmans. Reprinted in Elizabeth Mary (Smee) Odling, *Memoir of the late Alfred Smee, F.R.S., by his daughter; with a selection from his miscellaneous writings*. London: George Bell and Sons (1978).

Smolensky, P. (1986). 'Neural and conceptual interpretation of PDP models'. In Rumelhart, D., J. McClelland and the PDP Research group (eds.), *Parallel Distributed Processing: Explorations in the Microstructure of Cognition*. Vol. 2. Cambridge, MA: MIT Press.

Sugden, A. and E. Pennisi (1 August 2006). 'When to go, where to stop'. *Science*, 313: 775.

Surowiecki, J. (2004). *The Wisdom of Crowds*. New York: Doubleday.

Taheri, J. and A. Y. Zomaya (2006a). Artificial neural networks'. In Zomaya, A. Y. (ed.), *Handbook of Nature-Inspired and Innovative Computing: Integrating Classical Models with Emerging Technologies*. New York: Springer, p. 147.

Taheri, J. and A. Y. Zomaya (2006b). 'Fuzzy logic'. In Zomaya, A. Y. (ed.), *Handbook of Nature-Inspired and Innovative Computing: Integrating Classical Models with Emerging Technologies*. New York: Springer, p. 221.

Tempesti, G., D. Mange and A. Stauffer (1998). 'Self-replicating and self-repairing multicellular automata'. *Artificial Life*, 4(3): 259.

Tsallis, C. (1988). 'Possible generalizations of Boltzmann–Gibbs statistics'. *J. Stat. Phys.* 52: 479.

Tsallis, C. (1995a). 'Some comments on Boltzmann–Gibbs statistical mechanics'. *Chaos, Solitons and Fractals*, 6: 539.

Tsallis, C. (1995b). 'Non-extensive thermostatistics: brief review and comments'. *Physica A*, 221: 277.

Tsallis, C. (July 1997). 'Levy distributions'. *Physics World*: p. 42.

Turing, A. (1936). 'On computable numbers, with an application to the Entscheidungsproblem'. *Proceedings of the London Mathematical Society,* 2nd ser. 42 (1936–1937); reprinted with corrections in Martin Davis (ed.), *The Undecidable*. Hewlett, New York: Raven Press.

Villasenor, J. and W. H. Mangione-Smith (June 1997). 'Configurable computing'. *Scientific American,* 276: 54.

Waldrop, M. M. (1992). *Complexity: The Emerging Science at the Edge of Order and Chaos*. New York: Simon and Schuster.

Wheeler, W. M. (1928). *Emergent Evolution and the Development of Societies*. London: W. W. Norton and Co.

Wiener, N. (1965), 2nd edition. *Cybernetics: Or Control and Communication in Animal and the Machine*. Cambridge MA: MIT Press.

Williams, N. (25 July 1997). 'Biologists cut reductionist approach down to size'. *Science*, 277: 476.

Wilson, E. O. (5 November 2005). 'Can biology do better than faith?' *New Scientist*, p. 48.

Zadeh, L. A. (1965). *Fuzzy Sets, Information, and Control*, 8: 338.

Zadeh, L. A. (1975). 'The concept of a linguistic variable and its applications to approximate reasoning. Parts I, II, III'. *Information Sciences*, 8–9: 199–249, 301–357, 43–80.

Zimmer, C. (April 2004). 'Whose life would you save?'. *Discover*, 25, 60.

Zimmer, C. (February 2005). 'Testing Darwin'. *Discover*, 26: 28.

Zomaya, A. Y. (ed.) (2006). *Handbook of Nature-Inspired and Innovative Computing: Integrating Classical Models with Emerging Technologies*. New York: Springer.

3

FERROIC MATERIALS

A large number of ferroic materials are already in use in smart-structure applications, as sensors and actuators. Their importance arises from their field-tuneability, which is a consequence of their inherently nonlinear response. This comes either because of the vicinity of a ferroic phase transition (FPT), or because of their domain structure, or because of the occurrence of a field-induced phase transition. To understand these properties, we begin with a quick recapitulation of the basics of phase transitions and critical phenomena in condensed matter.

3.1   Phase transitions and critical phenomena

Water becomes ice on cooling to a certain temperature (the freezing point). Similarly, it becomes steam on heating to a certain characteristic temperature (the boiling point). Steam, liquid water, and ice are different *phases* of water. A phase is characterized by a definite atomic structure and composition.

The phase a material exists in depends on parameters like temperature (T), environmental pressure (p), and composition (x). Later in this chapter, we shall also bring in the effect of electric field ($\mathbf{E}$), magnetic field ($\mathbf{H}$), and directed or uniaxial mechanical stress (σ) on the phase of a material.

To start with, let us consider the effect of temperature alone. It is impossible to have a totally isolated system. There is always at least the effect of contact with the surroundings (a *thermal bath*). This contact introduces a *thermal noise* (or *thermal fluctuations*) in the dynamical equations describing the system.

The dynamical equations determine the evolution of a system with time (t) as a trajectory in phase space. The effect of the random thermal fluctuations is that the exact dynamical trajectories get mixed up randomly.

Let us consider a system characterized by N microscopic variables $\{s_i\}$, ($i = 1, 2, \ldots N$). Any 'observable' A is a function of these variables. What is generally of interest is the *average value*, or the *expectation value*, of this observable:

$$\langle A \rangle = \lim_{t \to \infty} (1/t) \int_0^t dt' A[\mathbf{s}(t')] \tag{3.1}$$

Here $\mathbf{s} = \{s_1, s_2, \ldots s_N\}$. It is normally assumed that enough time is given to the system in equilibrium to 'visit' all its microscopic states many times. The reason for making this *ergodicity hypothesis* is that, given long enough observation time, the thermal fluctuations may cause a thorough mixing of the dynamical trajectories, so that all microscopic states may become accessible. Of course, the relative frequency of the visits to the various states is determined by the *probability distribution function* of the states, i.e. some function $P(s_1, s_2, \ldots s_N)$.

Under the ergodicity hypothesis, the above time average can be replaced by the following *ensemble average*:

$$\langle A \rangle = \int ds_1 ds_2 \cdots ds_N A[\mathbf{s}] P(s_1, s_2, \ldots s_N) \tag{3.2}$$

subject to the constraint:

$$\int ds_1 ds_2 \cdots ds_N P(s_1, s_2, \ldots s_N) = 1 \tag{3.3}$$

If we bring in the first law and the second law of thermodynamics, namely the conservation of energy E and the minimization of free energy F ($F = E - TS$, where S is the entropy), it can be shown that P, for any microscopic state labelled by, say, α, must be given by the well-known Boltzmann distribution function (see, e.g. Dotsenko 2001):

$$P_\alpha = (1/Z) \exp(-\beta H_\alpha) \tag{3.4}$$

Here Z is the so-called *partition function*, β is inverse temperature ($\beta = 1/T$), and H_α is the Hamiltonian for the state α:

$$Z = \sum_\alpha \exp(-\beta H_\alpha) \tag{3.5}$$

$$E \equiv \langle H \rangle = \sum_\alpha P_\alpha H_\alpha \tag{3.6}$$

If we vary the temperature of the system, its free energy will vary. It can happen that at and below a certain temperature T_c, another competing phase of the system has a lower free energy than the phase the system is in. It will therefore make a *phase transition* and adopt the configuration of the new phase. To make our discussion concrete, let us assume that we are dealing with a (large) crystal, and above T_c the crystal is in a paramagnetic phase, characterized by a zero average magnetic moment in the absence of an applied magnetic field. Let us assume that below T_c the disordering thermal fluctuations are weak enough to be overpowered by the ordering tendency of the crystal, so that there can exist a nonzero average magnetization, even when no external magnetic field is present. So, a *spontaneous* magnetization emerges below T_c, and we speak of a *ferromagnetic phase transition* when this happens.

It is necessary to make a distinction between *structural* phase transitions and non-structural ones. A structural phase transition in a crystal involves a change of the crystal structure: the atoms in the crystal move to new locations. By contrast, certain ferromagnetic phase transitions may involve only an ordering of spins on some or all atoms, with no significant change in the *positions* of the various atoms in the crystal; these are nonstructural phase transitions.

There are various models for understanding the nature of ferromagnetic phase transitions. The simplest and the best-known is the *Ising model*. In it, the relevant microscopic variables are the magnetic spins $\mathbf{S}$, which can take only two values: 'up' and 'down'. The long-range ferromagnetic ordering, resulting in the emergence of a macroscopic spontaneous magnetic moment, is mediated by the *magnetic exchange interaction* between the spins. For spins $\mathbf{S}_i$ and $\mathbf{S}_j$ on sites i and j of the crystal lattice, the spin Hamiltonian describing the interaction is written as

$$H_{ij} = -J_{ij}\mathbf{S}_i\mathbf{S}_j \qquad (3.7)$$

In this equation for the so-called *Ising Hamiltonian*, summation over repeated indices is implied for all the lattice sites; and only nearest-neighbour interactions are taken as nonzero.

The Ising model was originally proposed as a simple description of magnetic systems. But this model and its variants have now become the central paradigms in the statistical physics of phase transitions in a variety of systems.

Direct exchange interaction between valence electrons on two neighbouring lattice sites consists of the Coulomb repulsion between them and the overlap of their wavefunctions. Unlike the former part, the latter can be effective only over short distances (it falls off exponentially with distance). Further, a configuration in which the two electron spins are close-by and parallel (positive J) is forbidden by the Pauli exclusion principle, whereas antiparallel spin interaction (negative J) is permitted by the principle. *Exchange energy* is the difference between the parallel and antiparallel configurations. There are several mechanisms other than direct exchange which result in ferromagnetic, rather than antiferromagnetic, ordering below T_c.

Landau (1937) introduced the concept of the *order parameter* for building up a general theory of an important class of phase transitions in crystals, namely the *continuous phase transitions*. The order parameter is a thermodynamic quantity, the emergence of which at the phase transition results in a lowering of the symmetry of the parent phase of the crystal. Let S_0 denote the space-group symmetry of the parent phase, and S that of the daughter phase. The order parameter, $\boldsymbol{\eta}$, has a symmetry of its own. Suppose that it is described by a symmetry group Γ. Since the emergence of the order parameter heralds the emergence of the daughter phase from the parent phase, it follows that S should have only those symmetry elements which are common to both S_0 and Γ (in accordance with the Curie principle of superposition of symmetries; cf. the appendix on the symmetry of composite systems).

The order parameter, like other thermodynamic parameters, is subject to thermal fluctuations. It is envisaged as having a zero mean or expectation value for temperatures above T_c. As the temperature is lowered from values above T_c, the mean value becomes nonzero below T_c. A *continuous* phase transition is defined as one for which the mean value of the order parameter approaches zero continuously as T_c is approached from below, i.e. as we approach the parent phase. (We assume that the phase transition to the ordered phase occurs on cooling, rather than on heating; this is usually the case.)

In the Landau theory, the Gibbs free energy density, g, of the crystal is expressed as a Taylor series in powers of the order parameter, and minimized with respect to it:

$$g = g_0 + \alpha\eta + (a/2)\eta^2 + \beta\eta^3 + (b/4)\eta^4 + (c/6)\eta^6 \cdots \qquad (3.8)$$

Here g_0 is the free energy per unit volume in the parent phase. The pre-factors for the various powers of the order parameter η depend on temperature, pressure and certain other parameters (see, e.g. Wadhawan 2000).

For minimizing g with respect to η, we equate its first derivative to zero:

$$\frac{\partial g}{\partial \eta} = \alpha + a\eta + 3\beta\eta^2 + b\eta^3 + c\eta^5 \cdots = 0 \qquad (3.9)$$

In addition, the second derivative of g with respect to η must be positive (so that the free energy is minimum, rather than maximum):

$$\frac{\partial^2 g}{\partial \eta^2} = a + 6\beta\eta + 3b\eta^2 + \cdots > 0 \qquad (3.10)$$

If eqn 3.10 is not satisfied, the system is unstable. It is therefore a *stability condition*.

Since eqn 3.9 must hold even when $\eta = 0$, we must have $\alpha = 0$. Equation 3.9 can be therefore rewritten as

$$\eta(a + 3\beta\eta + b\eta^2 + \cdots) = 0 \qquad (3.11)$$

Thus there are two solutions. One of them, namely $\eta = 0$, corresponds to the parent (or disordered) phase. The other, with a nonzero mean value of the order parameter, corresponds to the daughter phase or ordered phase.

Suppose X is the field conjugate to the order parameter. It is defined by

$$\frac{\partial g}{\partial \eta} = X \tag{3.12}$$

The stability condition, eqn 3.10, can therefore be written as

$$\frac{\partial^2 g}{\partial \eta^2} = \frac{\partial X}{\partial \eta} \equiv \frac{1}{\chi} > 0 \tag{3.13}$$

χ is a *generalized susceptibility*. Thus, for a phase to be stable, its inverse generalized susceptibility must be positive.

The disordered phase is characterized by $\eta = 0$. Putting this in eqn 3.10, we get the following as the condition for the stability of the disordered phase:

$$\chi^{-1} = a > 0 \tag{3.14}$$

The inverse susceptibility is a function of temperature. Let T_0 be the temperature below which the parent phase stops being stable:

$$\chi^{-1}(T = T_0) = 0 \tag{3.15}$$

In the vicinity of T_0, we can write

$$a = \left(\frac{\partial a}{\partial T}\right)(T - T_0) \equiv a'(T - T_0) \tag{3.16}$$

Substituting this into eqn 3.14,

$$\chi^{-1} = a'(T - T_0) \tag{3.17}$$

or

$$\chi = \frac{C}{T - T_0} \tag{3.18}$$

for $T > T_0$. This is the well-known *Curie–Weiss law*.

Likewise, the stability limit of the ordered phase can also be worked out. It turns out that this stability limit coincides with T_0, provided we are dealing with a *continuous* (or 'second-order') phase transition (cf. Wadhawan 2000). We can then replace T_0 in eqn 3.18 by T_c, the temperature of the phase transition:

$$\chi = \frac{C}{(T - T_c)} \tag{3.19}$$

For *discontinuous* phase transitions, by contrast, the stability limits of the two phases do not coincide. This means that *there is a range of temperatures in which the parent phase and the daughter phase can coexist.*

For continuous phase transitions it is enough to stop at the fourth-degree term in the Landau expansion (eqn 3.8); i.e. we can assume that $c = 0$. However, for discontinuous phase transitions, the sixth-degree term must be included.

Equation 3.19 (or, for that matter, eqn 3.18) embodies a very important result for smart-structure applications of materials. It tells us that *the response function χ corresponding to the order parameter becomes extremely large and nonlinear in the vicinity of a phase transition.*

The order parameter can be visualized as a vector in a certain phase space. It can have two or more allowed orientations, each corresponding to the free-energy minimum. For simplicity, let us assume that there are only two configurations, corresponding to up and down spins (if we are dealing with a ferromagnetic phase transition). We have to skip details, but it can be shown that the two orientations of the order parameter are separated by an energy barrier, which must be overcome by an external force if the crystal is to switch from one allowed configuration to another.

The barrier between the two free-energy minima is proportional to the volume V of the crystal. Therefore, in what is called the *thermodynamic limit*, namely $V \rightarrow \infty$ (corresponding to the extrapolation of *microscopic* results to *macroscopic* systems), the barrier separating the up and down spins becomes infinitely high. This is an instance of *ergodicity breaking* at and below T_c: The phase space of the system is now divided into two equal valleys, with an infinitely high hill separating them, and therefore not all the microscopic states are available for 'visiting' and ensemble-averaging; either spin-up or spin-down states are available for *observable* thermodynamics. In fact this is the reason why, in spite of the fact that the Ising Hamiltonian (eqn 3.7) is symmetric (remains the same) with respect to a global change of the signs of all the spins, we end up with *two* ground states, one for each global sign of spin. This is *spontaneous breaking of symmetry* (Strocchi 2005).

The degree of positivity of the second derivative of free energy changes with temperature, and it approaches the value zero as the stability limit T_c is approached. Thus, at T_c the free-energy versus η curve acquires a 'flat bottom' (i.e. infinite radius of curvature). This means that there can be large, scale-free, fluctuations in η which do not cause changes in free energy. Since T_c is called the *critical point*, these are called *critical-point fluctuations*, or just *critical fluctuations*.

A flaw of the original Landau theory of phase transitions was that it ignored the fluctuations of the order parameter, i.e. it assumed it to be spatially homogeneous. A fairly good correction to this assumption was introduced by Ginzburg (1961) by including in the Landau expansion of the free energy terms which are functions of the *gradient* of the order parameter. This modified theory is known as the *Ginzburg–Landau theory* of phase transitions (cf. Als-Nielsen and Birgeneau 1977; Wadhawan 2000).

Critical fluctuations of the order parameter at and in the immediate vicinity of T_c are just one of the manifestations of a host of complex *critical phenomena*, which occur when the crystal is very close to the critical point (Ball 1999). Another manifestation, which is actually a consequence of the critical fluctuations of the order parameter, is the occurrence of *universality classes*: There can exist a whole class of phase transitions, involving a variety of interatomic interactions, which exhibit a common or universal temperature dependence of the order parameter and several other physical properties.

We discussed complexity in Section 2.4. In general, complex behaviour is shown mainly by *open systems* far from equilibrium. For closed equilibrium systems being considered here, complexity can exist only under very specific conditions. Critical phenomena at continuous phase transitions in crystals are an example of this. At the critical point, the system passes from a disordered state to an ordered state. The system has to be brought very close to the critical point to observe complex behaviour, namely scale-free fluctuations of the order parameter, giving transient ordered domains of all sizes. Such complex criticality is *not robust*; it occurs only at the critical point, and not at other temperatures (cf. Section 5.5.2 on self-organized criticality).

3.2 Ferroic phase transitions

In this book, we are interested in a special class of phase transitions called *ferroic phase transitions* (FPTs). Ferromagnetic phase transitions considered above, characterized by the emergence of spontaneous magnetization (which is a macroscopic tensor property), are an example of such transitions. Another example is ferroelectric phase transitions, in which there emerges a spontaneous polarization when the crystal is cooled to a temperature below T_c.

Let us consider phase transitions involving a change of the space-group symmetry of the crystal. Space-group symmetry has a translational part and a directional part (cf. the appendix on crystallographic symmetry). The translational part is described by the Bravais group, and the directional part by one of the crystallographic point groups.

If there is a change of the point-group symmetry, we have an FPT. This is the definition of an FPT (Wadhawan 2000). If only the translational part changes and the point-group remains the same, we speak of a nonferroic phase transition.

A *ferroic material* is that which can undergo at least one FPT.

Across an FPT, the parent phase has a higher point-group symmetry than the daughter phase or ferroic phase. The orders of the two point groups must differ by at least a factor of 2. This means that the ferroic phase must have *domains* (or *variants*). For example, if there is a ferromagnetic phase transition, one type of domains will have spin up, and the other spin down.

Similarly, in the case of the ferroelectric phase transition which occurs in a crystal of BaTiO$_3$ at $T_c = 130°C$, six ferroelectric domain types or variants can arise. This is because the point-group symmetry above T_c is cubic ($m\bar{3}m$) (with no preferred or 'polar' direction), and the point-group symmetry in the ferroelectric phase below T_c is $4mm$, which is a polar group belonging to the tetragonal crystal system. The four-fold axis is the polar axis, and it can point along any of six possible directions: $+x$, $-x$, $+y$, $-y$, $+z$, $-z$. Each of these possibilities gives rise to a distinct ferroelectric *domain type*.

The concept of *prototype symmetry* is crucial for a proper description of FPTs (Aizu 1970; Wadhawan 2000). A crystal may undergo a sequence of symmetry-lowering phase transitions, say on cooling. Any phase in this sequence has a 'parent' phase, from which it arose on cooling. Prototype symmetry is not just the symmetry of the next higher 'parent' phase. It is the highest symmetry conceivable for that crystal structure, so that all the daughter phases can be considered as derived from it by the loss of one or more symmetry operators. For example, $BaTiO_3$ undergoes a sequence of FPTs on cooling: cubic to tetragonal to orthorhombic to rhombohedral. For each of the lower phases, it is the cubic phase which has the prototypic symmetry. For a rigorous definition of prototype symmetry, see Wadhawan (1998, 2000).

3.3   Ferroic materials

3.3.1   *Classification of ferroic materials*

Tensor properties like magnetization, polarization, strain, and permittivity have translational invariance. Therefore it is not necessary to invoke the full space-group symmetry of the crystal for describing these properties, and it is sufficient and appropriate to work at the point-group level. A transition to a ferroic phase always involves a lowering of the point-group symmetry. This means that in the ferroic phase there must be at least one tensor-property coefficient which was zero because of the higher point-group symmetry of the prototypic phase, and which is permitted to be nonzero by the lower point-symmetry of the ferroic phase. What is more, such tensor coefficients can differ from one domain type to another.

Ferroic phases can be adequately classified in terms of their tensor properties.

Consider a crystal under the influence of an external electric field (E_i), a magnetic field (H_i), and a uniaxial stress (σ_{ij}) (i, $j = 1, 2, 3$). Its generalized Gibbs free energy per unit volume can be written as follows (Wadhawan 2000):

$$g = U - TS - E_i D_i - H_i B_i - \sigma_{ij} e_{ij} \qquad (3.20)$$

Here S is entropy, D_i the electric displacement, B_i the magnetic induction, and e_{ij} the strain.

The reduction of point symmetry on entering the ferroic phase leads to the occurrence of two or more types of domain states or *orientation states*. We have seen above that there are six of them for the phase transition in $BaTiO_3$, corresponding to the six directions along which the spontaneous polarization vector can point. Let us denote orientation states of a ferroic phase as S_1, S_2, S_3, etc.

Let us pick up any pair of these orientation states, say S_1 and S_2. For this domain pair,

$$D_{(s)i}(1) = \varepsilon_0 E_i + P_{(s)i}(1) \qquad (3.21)$$

$$D_{(s)i}(2) = \varepsilon_0 E_i + P_{(s)i}(2) \qquad (3.22)$$

Therefore, the difference between the displacement vectors for states S_1 and S_2 is given by

$$D_{(s)i}(2) - D_{(s)i}(1) = P_{(s)i}(2) - P_{(s)i}(1) = \Delta P_{(s)i} \qquad (3.23)$$

Similarly for the magnetic and elastic counterparts of this.

It can be shown that the difference $\Delta g (= g_2 - g_1)$ between the free-energy densities of the two orientation states is given by (Wadhawan 2000):

$$-\Delta g = \Delta P_{(s)i} E_i + \Delta M_{(s)i} H_i + \Delta e_{(s)ij} \sigma_{ij}$$
$$+ \left(\tfrac{1}{2}\right) \Delta \varepsilon_{ij} E_i E_j + \left(\tfrac{1}{2}\right) \Delta \mu_{ij} H_i H_j + \left(\tfrac{1}{2}\right) \Delta s_{ijkl} \sigma_{ij} \sigma_{kl} \qquad (3.24)$$
$$+ \Delta \alpha_{ij} E_i H_j + \Delta d_{ijk} E_i \sigma_{jk} + \Delta Q_{ijk} H_i \sigma_{jk} + \cdots$$

This equation is of central importance in the subject of ferroic materials, and forms the basis for their tensor classification.

If there is at least one pair of domains for which the first term on the right-hand side of eqn 3.24 is nonzero (for one or more values of i), the crystal is said to be in a *ferroelectric* phase. What this means is that the two domains differ in spontaneous polarization.

Similarly, if the second term is nonzero, the crystal is in a *ferromagnetic* phase.

A nonzero third term implies, by definition, that the crystal is in a *ferroelastic* phase.

Ferroelectricity, ferromagnetism, and ferroelasticity are the three *primary* forms of ferroicity.

If any of the next six terms on the right-hand side of eqn 3.24 is nonzero, the crystal is said to be in a *secondary* ferroic phase. Still higher-order terms define *tertiary* and *quaternary* ferroics, etc. The six types of secondary ferroics are: *ferro-bielectrics* (defined by the fourth term); *ferrobimagnetics* (fifth term); *ferrobielastics* (sixth term); *ferromagnetoelectrics* (seventh term); *ferroelastoelectrics* (eighth term); and *ferromagnetoelastics* (nineth term) (see Fig. 3.1).

A word of caution

Sometimes one comes across statements to the effect that, for example, a ferromagnetoelectric material is that which is simultaneously ferromagnetic and ferroelectric in the same phase (Hill 2002; James 2005). This is a wrong perception, because it need not be the case always (at least from symmetry considerations alone). It is conceivable that, in eqn 3.24, the terms $\Delta P_{(s)i} E_i$ and $\Delta M_{(s)i} H_i$ are both zero, and the term $\Delta \alpha_{ij} E_i H_j$ is nonzero. This will happen if the point-group symmetries of the prototype phase and the ferroic phase demand it. Such a phase is ferromagnetoelectric, but neither ferroelectric nor ferromagnetic.

Similarly, the occurrence of the magnetoelastic effect or the magnetostrictive effect does not necessarily mean that the material is also a ferroelastic and a ferromagnetic. In eqn 3.24:

1. If the term $\Delta Q_{ijk} H_i \sigma_{jk}$ is nonzero, the material is a ferromagnetoelastic.
2. If the term $\Delta M_{(s)i} H_i$ is nonzero, the material is a ferromagnetic.
3. If the term $\Delta e_{(s)ij} \sigma_{ij}$ is nonzero, the material is a ferroelastic.
4. Only when all three terms are nonzero is the material a ferromagnetic, a ferroelastic *and* a ferromagnetoelastic.

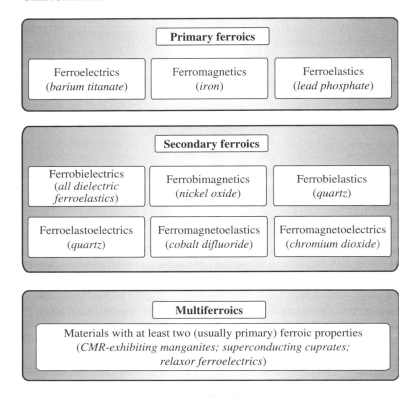

Fig. 3.1   Primary and secondary ferroics, and multiferroics.

In the light of the discussion so far, we can now list the key features of ferroics:

- Existence of ferroic orientation states or domain states, resulting from long-range ordering with respect to at least one macroscopic tensor property.
- Existence of domain boundaries (and also phase boundaries, if the FPT is a discontinuous one). The domain boundaries exist in the ferroic phase and disappear in the parent phase. They can be moved by applying a field, the movement being easier in the close vicinity of the FPT.
- Enhancement of certain properties in the vicinity of the FPT. One example is the temperature dependence of the order parameter. Another is that of the response function corresponding to the order parameter.
- A coupling between the order parameter and other macroscopic properties. Because of this coupling, these other properties may also exhibit anomalous and nonlinear behaviour in the vicinity of the phase transition. A particularly important example of this are the enhanced electro-optical effects observed in a number of ferroic crystals (cf. Fousek 1986).

• Field-tuneability of certain properties. As we saw in Section 1.4, nonlinear response implies field-tuneability. In a ferroic material, the nonlinear response can arise either because the material is in the vicinity of the FPT, or because of its domain structure. Another type of field-tuneability will be discussed below in Section 3.8 when we consider field-induced phase transitions.

3.3.2   Primary ferroics

Ferromagnetics

We have already discussed above the direct exchange interaction. If no other mechanisms are operative, only antiparallel ordering can occur between close-by embedded spins. Therefore, the parallel-spin ordering in ferromagnets must involve other mechanisms as well. There are several of them.

One, of course, is the classical *dipolar interaction*; it is weak but always present. And it is anisotropic: For example, if the spins are oriented *along* the line connecting their positions, they would couple parallel; if they are oriented *perpendicular* to this line, they would couple antiparallel.

In magnetic alloys, the availability of conduction electrons gives rise to a strong and relatively long-range indirect-exchange interaction, called the *RKKY interaction* (Ruderman and Kittel 1954; Kasuya 1956; Yosida 1957). The isotropic form of its Hamiltonian can be expressed symbolically as

$$H = J(r)\mathbf{S}_i \cdot \mathbf{S}_j \tag{3.25}$$

where r is the distance between the two embedded spins. The presence of embedded magnetic atoms in a sea of conduction electrons, with itinerant spins $\mathbf{s}(\mathbf{r})$, causes damped oscillations in the susceptibility of the conduction electrons, and this leads to an indirect coupling interaction between the embedded spins at sites i and j in the crystal lattice.

When conduction electrons are not available as significant via-media for magnetic interaction, as is the case with insulators or semiconductors, and direct exchange is not a strong interaction mechanism, *superexchange* provides the main ferromagnetic or antiferromagnetic interaction, over relatively longer distances than is possible with the exponentially decaying direct exchange. If between two magnetic cations there is situated an anion or a ligand, which can transfer its two electrons, one each to the neighbouring magnetic cations, two covalent-type bonds can form. Since the two electrons donated by the anion have opposite spins, the net coupling will be antiferromagnetic. Ferromagnetic coupling by superexchange is also possible if the exchange polarization of the anion orbitals is substantial.

The long-range cooperative interaction leading to ferromagnetic ordering may be either nearest-neighbour exchange, or through itinerant electrons. Fe, Co and Ni are examples of itinerant-electron ferromagnetic crystals.

A crystal in a ferromagnetic phase has at least one pair of domain states which differ in spontaneous magnetization. Application of a sufficiently strong magnetic field, directed so as to favour the existence of one domain over the other, can lead to a switching of one domain to the other, resulting in a characteristic magnetization *vs.* magnetic field *hysteresis* curve when the applied field is cycled through positive and negative values.

Ferroelectrics

Ferroelectrics exhibit the characteristic hysteresis curve when the net polarization is plotted against an applied ac electric field.

Since ferroelectricity requires the existence of spontaneous polarization, it can occur only in those crystal classes (i.e. for only those point-group symmetries) which are polar. The parent phase may be either centrosymmetric, or noncentrosymmetric–nonpolar, or even polar in certain conceivable situations. An example of the first type is the cubic-to-tetragonal phase transition in $BaTiO_3$, mentioned above. An example of the second type occurs in KDP (KH_2PO_4) crystals, in which the parent phase has tetragonal nonpolar symmetry $\bar{4}2m$, and the ferroelectric phase has the polar symmetry $mm2$.

An example of the third type is conceivable. Suppose the parent phase has the polar symmetry $m_x m_y 2_z$, and the ferroelectric phase has the polar symmetry m_x. In this case, there is a spontaneous dipole moment P_z even in the parent phase. In the ferroelectric phase, there can be a dipole moment pointing along some direction in the plane m_x. It will have components P_y' and P_z' in one domain, and $-P_y'$ and $-P_z'$ in the other domain. Thus there is a reversible part $-P_y'$ and an irreversible part $-P_z'$, the net result of ferroelectric switching being a *reorientation* of the spontaneous polarization, rather than complete reversal.

Ferroelastics

Let us consider again the example of the ferroelectric phase transition in $BaTiO_3$. There are six ferroelectric domain states. Consider any of them, say the one having spontaneous polarization P_z. When this spontaneous polarization arises, there is, naturally, a deformation (elongation) along the z-axis. This elongation of the initially cubic unit cell along one of its axes lowers its symmetry to tetragonal. Thus, not only spontaneous polarization, but also spontaneous *strain* arises at the phase transition. Similarly, the domain type with P_x has elongation of the unit cell along the x-axis, and the one with P_y has elongation along the y-axis. If we take any two of these three types of domains, they differ in spontaneous strain, so the tetragonal phase of $BaTiO_3$ is ferroelastic. It is also ferroelectric, of course.

In this phase transition, the primary instability is ferroelectric, so it is called a *proper ferroelectric* phase transition. Since the ferroelasticity arises only as a coupling of strain to the primary order parameter, namely the polarization, it is called an *improper ferroelastic* phase transition.

Lithium ammonium tartrate monohydrate, by contrast, is an example of a material with a *proper* ferroelastic phase transition (cf. Wadhawan 2000).

There is an important class of phase transitions, called *martensitic phase transitions*, the kinetics of which is dominated by strain energy arising from 'shear-like' displacements of atoms. They are usually diffusionless, first-order, structural phase transitions. Since spontaneous shear strain is one of their essential features, it is natural to compare and contrast them with ferroelastic phase transitions (cf. Wadhawan 2000).

Martensitic phase transitions can be divided into three types: M_1, M_2, M_3, depending on the magnitude of the spontaneous strain involved (cf. Izyumov, Laptev and Syromyatnikov 1994). Type M_1 are the least 'disruptive' (cf. Wadhawan 2000), and are the same as quasi-continuous ferroelastic phase transitions; the spontaneous strain involved is small. The other two types are too drastic or disruptive in nature, entailing a change of coordination numbers of atoms.

The lowering of point-group symmetry of a crystal across a ferroelastic phase transition leads to a characteristic domain structure, governed by the need to minimize the overall strain energy. The emergence and propagation of the 'distorted' or ferroelastic phase in the restrictive parent-phase matrix creates a strain field. After a certain amount of growth of the ferroelastic phase has occurred, the strain energy in the surrounding matrix becomes so large that the system has to reduce it by undergoing a self-induced ferroelastic switching of the daughter phase to a domain of opposite spontaneous strain. The growth of the new ferroelastic domain keeps increasing the strain field in the surrounding matrix. When that too becomes unbearable, the system makes another ferroelastic switching, this time back to the original ferroelastic state. This occurs repeatedly, so that what we have is a *polysynthetic twin* or *polytwin*: a plane-parallel plate comprising of alternating plane-parallel ferroelastic domains (James 1989, 2005).

The same reasoning applies, not only to the occurrence of interfaces (domain boundaries) within the martensitic/ferroelastic phase, but also to phase boundaries. One of the approaches in this context puts forward the notion of *hierarchical domain structures* and *polydomain phases* (Roytburd 1993), or *adaptive phases* (Khachaturyan, Shapiro and Semenovskaya 1991). The basic idea is that, although there is a discontinuous change of lattice parameters at such a phase transition, there must be a continuity of the two lattices at a phase boundary, as also at a domain boundary. It is postulated that the polytwin and/or the polydomain phase, which is the result of self-induced deformation twinning, is subject to the constraint that the overall strain is an *invariant-plane strain*. This assumption, based on a large number of experimental results, means that, not only does the product phase or domain meet the parent phase or domain in a coherent fashion (i.e. across a continuous lattice), the intervening plane is also an undistorted and unrotated plane (Wechsler, Lieberman and Read 1953; Bowles and Mackenzie 1954). It is an *invariant plane*.

The notion of symmetry-adaptive phases has been recently applied explicitly to ferroelectrics also (Viehland 2000; Jin *et al.* 2003).

Another approach to the microstructure of ferroelastics is based on treating explicitly the anisotropic and long-range strain interaction as the main driving factor for the kinetics and other features of the structural phase transition (cf. Khachaturyan,

Semenovskaya and Long-Qing Chen 1994; Bratkovsky *et al.* 1995; Kartha *et al.* 1995). Structural changes occurring in any unit cell of the crystal result in a local stress field, which is felt by even distant unit cells through a 'knock-on' effect, i.e. a transmission of stress through successive cells. The effective elastic interaction over long distances has a strongly anisotropic component. One of the striking manifestations of this elastic interaction is the occurrence of cross-hatched *tweed patterns* in crystals having some degree of disorder (e.g. composition fluctuations in an alloy) (cf. Castan *et al.* 2005). The intrinsic statistical variation of composition leads to the occurrence of pre-transitional tweed-structure patterns even for first-order phase transitions, and these patterns can start appearing even hundreds of degrees above the nominal bulk transition temperature: A dense mass of *embryos* of the ferroelastic phase is present as thermodynamic fluctuations (concentration inhomogeneities) in the parent phase. As the temperature is lowered, the embryos become more and more prominent, and freeze into metastable structures on quenching the crystal through the phase-transition temperature. The metastable tweed pattern sharpens and coarsens with the passage of time.

Kartha *et al.* (1995) have shown that long-range, cooperative, nonlinear processes give rise to the tweed structure, which is really a distinct thermodynamic phase, straddled by the parent phase and the ferroelastic phase. It also has features of a quasi-glassy phase: slow relaxation, a diverging nonlinear susceptibility, and glassy dynamics (cf. Mazumder 2006). There is also a similarity between tweed structure and spin glasses (cf. Section 3.5).

The martensitic phase

Ni-Ti is the best known alloy exhibiting the martensitic phase transition. Its high-temperature (or *austenitic*) phase (also known as B2 phase, or β phase) has cubic symmetry. For device applications, the martensitic transition should occur in the temperature range of application. This temperature can be easily manipulated for Ni-Ti by introducing even small changes of composition. Naturally, the occurrence of various phases as a function of temperature also depends on the composition. For example, $Ni_{50.7}Ti_{49.3}$ exhibits the so-called R-phase (rhombohedral phase), straddled by the cubic austenitic phase on the high-temperature side, and the martensitic monoclinic phase on the low-temperature side (Saburi 1998; Khalil-Allafi, Schmahl and Reinecke 2005). The monoclinic phase is also known as the B19' phase.

Fully annealed, near-equiatomic NiTi alloy transforms directly from B2 to B19' structure martensitically. Thermally cycled, or thermomechanically treated, or slightly Ni-rich NiTi alloys undergo two martensitic transitions: First from B2 to R, and then from R to B19'.

3.3.3   *Secondary ferroics*

We discuss only a few types of secondary ferroics here. For a more comprehensive discussion the book by Wadhawan (2000) should be consulted.

Ferroelastoelectrics

A ferroelastoelectric is a ferroic which has at least one pair of orientation states, or domain states, which differ in their piezoelectric response. That is, referred to a common frame of coordinate axes, there is at least one component of the piezoelectric tensor (d_{ijk}) which is not the same for that pair of domain states.

Piezoelectrics, along with electrostrictive materials, are the most popular actuator materials. This is because electrical input or output signals are easier to handle than mechanical signals or magnetic signals. They also generate a fast and large actuation stress ($\sim$tonnes cm^{-2}).

Large single crystals are generally expensive and/or difficult to grow, so one would like to use the piezoelectrics in polycrystalline form wherever possible and sufficient. It is a happy circumstance that most of the crystalline materials that are piezoelectric are also ferroelastoelectric. There is an important practical consequence of this. An as-prepared piezoelectric ceramic exhibits little or no piezoelectric effect. This is because its grains (crystallites) are randomly oriented, so the strains induced in the various grains by an applied electric field get cancelled out to practically a zero value. Similarly, in an inverse transduction application, application of a directed mechanical stress induces dipole moments in various grains which get cancelled out because the grains are randomly oriented. Therefore, to get a usable output, one must first *pole* the material.

Electrical poling

The poling process gives a preferential direction to the orientation of the grains. To understand how this happens, let us take note of another circumstance, namely the fact that the commonly used piezoelectrics are also ferroelectrics. (Out of the 32 crystal classes, 20 allow piezoelectricity, and out of these 20, there are 10 which are 'polar' classes and therefore allow ferroelectricity.)

So if a material is both ferroelectric and ferroelastoelectric, we can use electric field to make the ferroelectric domains point preferentially along a particular direction (the direction along which the poling field is applied). When this is done, the entire polycrystalline ceramic behaves like a single crystal to some extent, in the sense that there is anisotropy and a preferred direction. Poling would not be possible if we were not dealing with a ferroic material. Only in a ferroic material there can be alternative orientation states, or domain states available, to which a domain can switch under the action of a poling field.

Figure 3.2 illustrates the basic process of poling, brought about an electric field E_z (applied along the z-direction). Typically, the specimen is heated to the temperature of the ferroelectric phase transition, and a poling field E_z of sufficient strength is switched on (Fig. 3.2b). The specimen is then cooled slowly to room temperature under the action of the electric field. The field induces domain switching, as indicated by shifted domain walls in Fig. 3.2(c), giving rise to a net nonzero value for the spontaneous dipole moment of the entire specimen.

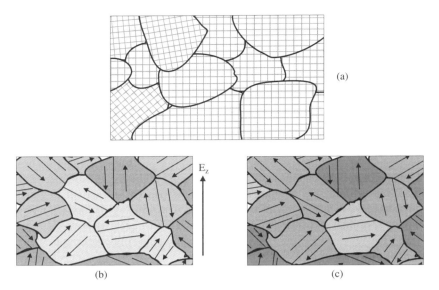

Fig. 3.2 (a) A nonferroic polycrystalline ceramic (e.g. porcelain). Each grain is a tiny crystal, as indicated schematically by a mesh drawn in it. The grains are oriented randomly. (b) A ferroic (to be more specific, a ferroelectric) polycrystalline ceramic. Each grain is crystalline, but it can also have domain walls, which separate domains having differently oriented spontaneous polarization (indicated by arrows). In an as-prepared specimen, the net dipole moment may be zero because the grains are oriented randomly, even though each grain may have a nonzero dipole moment (indicated by the degree of grey shading, which is, crudely, a measure of the net spontaneous polarization along the z-axis (P_z) for the grain). Under the action of a poling field, E_z, the sum total of P_z values for the entire specimen acquires a nonzero value, as indicated by the higher degree of grey shading in (c).

The reorientation of domains by the poling electric field is not entirely a process of domain switching; it may also be facilitated by field-induced phase transitions.

Needless to say, no amount of poling of a piezoceramic can produce a material with a response as large as that of a single crystal of the same size and shape.

For efficient transduction of energy, one would like to use materials which have large d-coefficients. That apart, there are other details of practical importance. For example, how good is the *electromechanical coupling*? The symbol k is used for this property, and the square of the electromechanical coupling factor, i.e. k^2, is defined as that fraction of the input electrical energy that gets stored as mechanical energy.

In a typical work cycle, a fraction of the input electrical energy gets converted to mechanical energy, and the rest gets stored as electrostatic energy, to be used in the next cycle, provided there are no losses. Typically, the dielectric losses are 1–3% in the commonly used material PZT (lead zirconate titanate, a solid solution of lead zirconate and lead titanate), so that the *efficiency*, η, is close to 100%. (η is defined as the ratio of the output mechanical energy and the consumed electrical energy.)

Generally speaking, k^2 varies directly with d^2, and inversely with the dielectric permittivity and the elastic compliance (Uchino 1997).

PZT is the best known piezoelectric transducer material. In its temperature vs. composition phase diagram (*T–x diagram*), there is a nearly vertical phase boundary for the composition $x = 0.52$, i.e. for $Pb(Zr_{0.52}Ti_{0.48})O_3$ (PZT(52/48), for short). Such a vertical phase boundary is called a *morphotropic phase boundary* (MPB), and has been a subject of much investigation. *The MPB implies the presence of two solid phases that coexist in a near-equilibrium state over a very wide temperature range.*

In the vicinity of the MPB a large number of phase boundaries and domain boundaries can exist in the material, which makes the material very responsive to external fields, the reason being that, since there are so many orientation states available, the ferroic switching of the material becomes more easy (Wadhawan 2000). Apart from domain switching, there can also be field-induced phase transitions (to be discussed below), which further add to the already large and nonlinear response function. More recently, the opinion has been gaining ground that the reason for the large electromechanical response in the vicinity of the MPB is the occurrence of a monoclinic phase, which allows the polarization vector to *rotate* easily in its mirror plane of symmetry (cf. Noheda and Cox 2006, for a review).

As discussed by Uchino (1997), not only is the dielectric and piezoelectric response of PZT(52/48) very high, the electromechanical coupling coefficient, k, is also very large because of the vicinity of the MPB.

Ferrobielastics

Quartz is the best known ferrobielastic material. It is also a ferroelastoelectric. Its piezoelectric nature accounts for its extensive applications as a transducer material.

The ferroic nature of quartz stems from the phase transition from the high-temperature phase, called β-*quartz*, to the phase α-*quartz* stable at room temperature. The β-phase has a six-fold axis of symmetry, which reduces to a three-fold axis across the phase transition. The two orientation states in the ferroic α-phase are the two twin states, better known as Dauphine twins.

The two states or twins do not differ in spontaneous strain, so the crystal is not a ferroelastic. But they differ in their elastic response to stress: Referred to a common system of coordinates, some components of the elastic compliance tensor are different. As a result of this, when a common stress is applied to both of them, the strain induced by the applied stress is different for the two twins. This can be expressed as follows (cf. eqn 3.24):

$$-\Delta g = \left(\tfrac{1}{2}\right)\Delta s_{ijkl}\sigma_{ij}\sigma_{kl} = \left(\tfrac{1}{2}\right)\Delta e_{kl\,(\text{induced})}\sigma_{kl} \qquad (3.26)$$

This offers a way for detwinning of Dauphine twins (cf. Wadhawan 2000).

3.4   Designer domain patterns

A transition to a ferroic phase entails a loss of one or more point-group symmetry operations. This results in the occurrence of domains in the ferroic phase. Domains are regions in the crystal which were identical before the FPT occurred, but which are now distinct in terms of mutual orientation and/or mutual placement (although they all have the same crystal structure, conforming to the symmetry group of the ferroic phase).

How many types of domains are possible in a ferroic phase? The answer is determined entirely by the prototype point-group symmetry and the ferroic point-group symmetry. Let G denote the symmetry group of the prototype, and H the symmetry group of the ferroic phase in question. Let $|G|$ be the order of the point group underlying G, and $|H|$ the corresponding number for H. Then the possible number of ferroic domain types is simply the ratio $|G|/|H|$. Let us denote this number by n. For the cubic-to-tetragonal phase transition in $BaTiO_3$, this number is 48/8 or 6: The order of the point group of the cubic (prototypic) phase is 48, and that of the tetragonal phase of point-symmetry 4mm is 8. The six domain types correspond to the six directions $(+x, -x, +y, -y, +z, -z)$ along which the spontaneous-polarization vector can point in the ferroic (ferroelectric) tetragonal phase.

If all the six domain types were present randomly and in equal volume fractions, the net spontaneous polarization of the ferroelectric specimen would cancel out to zero. The cubic phase is centrosymmetric, and the tetragonal phase is not. But, unless we do something about it, the specimen would behave as if it were centrosymmetric (on a crude enough scale) even in the ferroelectric phase. For example, it would not exhibit a net piezoelectric effect. The effects produced by individual domain types would cancel out.

Suppose we apply an electric field along the z-axis as the crystal is being cooled though the phase transition. Such a biasing field will favour the creation and growth of z-domains (domains with the polarization vector pointing along the z-direction), at the cost of occurrence of other domain types, so that the net macroscopic spontaneous polarization of the specimen would be nonzero, and it would also exhibit the piezoelectric effect. This is the basic process of poling, described above.

This is an example of a designer domain pattern, created to achieve a superior practical application or device performance.

A striking example of creating tailor-made domain patterns in a ferroic crystal is that of *periodic domain inversion* (PDI) in a ferroelectric crystal, like KTP (potassium titanyl phosphate) or lithium niobate, for achieving *quasi-phase matching* (QPM) in nonlinear optics (cf. Fejer 1994; Joannopoulos, Meade and Winn 1995; Hu *et al.* 1998). The spontaneous polarization of a single-domain ferroelectric crystal is periodically reversed, to obtain a regularly spaced array of domains with a half-period of domain inversion equal to the coherence length L_c of the laser beam for which one wants to obtain the second harmonic in an efficient manner.

There are three possible approaches by which one can engineer the domain pattern of a ferroic crystal (Fousek and Cross 2001; Fousek, Litvin and Cross 2001).

The first approach is to do *domain-shape engineering* (DSE), and we have just seen an example of this for achieving QPM for a birefringent crystal.

The second approach is that of *domain-average engineering* (DAE) (Fousek, Litvin and Cross 2001; Litvin, Wadhawan and Hatch 2003). For this, one should first ensure that the specimen crystal has a large number of small-sized domains (for a good averaging of properties). Suitable processing techniques can then be used for ensuring that, by and large, only m out of the possible number n of domain types are present ($m < n$). As an example, we have seen above how the presence of a biasing electric field during the cubic-to-tetragonal phase transition in $BaTiO_3$ tends to hamper the creation and growth of domain types for which the spontaneous-polarization vector is not favourably inclined with respect to the direction of the biasing field. The domain-averaged macroscopic tensor properties of such a 'biased' specimen naturally correspond to a lower overall symmetry than for the undoctored specimen.

In this context, it is relevant to mention the notion of *latent symmetry* (Wadhawan 2000; Litvin and Wadhawan 2001, 2002) (cf. the appendix on the symmetry of composite systems). Consider an object (e.g. a ferroic domain) A of symmetry described by the group **H**. Let us make m copies of A and superimpose them all (this is like the DAE of the m domain types considered here). We have in hand a composite system $S = \{A, g_2 A, \ldots g_m A\}$, the components of which are generated from the object A by the set of symmetry operations $\{1, g_2, \ldots g_m\}$. One may tend to expect that the symmetry of the composite object (which in the present context is the superposition of the m types of DAE domains) should be the product of the symmetry group **H** and the set of operations $\{1, g_2, \ldots g_m\}$. Latent symmetry, by definition, is that unexpected symmetry which is higher than that defined by this product. It has been pointed out by Litvin, Wadhawan and Hatch (2003) that, because of the possibility of latent symmetry manifesting itself when equal or identical objects are superimposed, the symmetry of a DAE specimen can indeed be unexpectedly high sometimes.

In the third approach for creating designer domain patterns, one resorts to *domain-wall engineering* (DWE). In DSE and DAE it is usually assumed that the walls separating the ferroic domains are thin, small in number, and insignificant in terms of overall strain energy and potential energy. The fact is that only specific ('permissible') wall orientations can result in strain-free configurations (Sapriel 1975). Any other interface between a pair of ferroic domains may have serious strain incompatibility and/or charge incompatibility (cf. Wadhawan 1988, 2000). DWE becomes an important approach when the wall thickness is large (relative to average domain thickness) and the number of walls per unit volume is large. Another thing to remember is that, in certain situations, domain walls may carry a dipole moment even for centrosymmetric ferroic phases (cf. Wadhawan 2000).

For first-order FPTs, there is a range of temperatures in which there is a coexistence of the two phases. Then we have not only domain boundaries, but also phase boundaries. This can provide further scope for engineering, not only domains and domain boundaries, but also phases and phase boundaries for achieving some desirable macroscopic behaviour.

3.5   Spin glasses

Spin glasses are an example of a very special type of magnetic ordering in crystals in which the spins on embedded magnetic atoms are randomly oriented and located. Mydosh (1993) has given the following definition of a spin glass: *A random,* *mixed-interacting, magnetic system characterized by a random, yet cooperative, freezing of spins at a well-defined temperature T_f below which a highly irreversible, metastable frozen state occurs without the usual long-range spatial magnetic order.*

One class of materials which exhibit spin-glass properties are alloys made from nonmagnetic metals and small amounts of 'good-magnetic-moment' atoms like Mn, Fe, Gd, Eu. Two simple examples are $Cu_{1-x}Mn_x$ and $Au_{1-x}Fe_x$.

Let us consider a pure Cu crystal (at $T = 0$ K), in which we progressively increase the doping of Mn. The dopant Mn atoms substitute for some of the host Cu atoms at their official crystalline sites. When there are only a few Mn atoms, and they are distributed randomly throughout the host Cu lattice, their magnetic spins do not interact with one another significantly, and just point along random directions. If the distribution is truly random, the sum total of all the magnetic moments is zero. This is an example of frozen or *quenched disorder*; the location of randomly distributed Mn atoms is frozen in space.

If more and more Mn atoms are brought in, their degree of isolation would decrease, and there would be regions with Mn atoms on neighbouring lattice sites. Naturally, the question of parallel (ferromagnetic) or antiparallel (antiferromagnetic) alignment of spins will arise because of a variety of possible interactions (direct exchange, RKKY, superexchange, dipolar interaction). There are thus *competing interactions* for deciding whether the spin at a particular site should point up or point down.

The inherent randomness of the configuration means that there can be *frustration* because of these competing interactions. In the example chosen by us, namely $Cu_{1-x}Mn_x$, although the net ordering is predominantly ferromagnetic (and weak), the antiferromagnetic interaction is also present. To see the consequences of these mixed ordering tendencies, imagine a triangle ABC in the lattice, all corners of which are occupied by Mn atoms, and only nearest-neighbour interaction is significant. Let us assume that the couplings AB, BC and CA are all ferromagnetic (Fig. 3.3a). Then there is no frustration because all the three spins can point up or down in a self-consistent manner.

Next, suppose that the interaction is *anti*ferromagnetic, and spin A is pointing up (Fig. 3.3b). Then both spins B and C should point down. Suppose, spin B indeed points down (as shown in Fig. 3.3b). Spin C should also point down, but this is inconsistent with the fact that the interaction between B and C is also antiferromagnetic. So spins B and C are frustrated, not knowing whether to point up or down! Spin A is also frustrated, because we could as well start from spin B and go around the triangle. Thus, there is frustration all around.

A consequence of the frustration is that the system has a large number of competing ground states, all of similar total energy, but with *activation barriers* separating them.

Although the system has competing interactions, cooperative coupling also occurs: Not all the spins may be frustrated. Thus there is *both competition and cooperation*.

In spite of the underlying randomness, there is also *anisotropy*. This comes partly from the very weak dipolar interaction, but more importantly from the mixture of this interaction with superexchange, etc. (Mydosh 1993). This anisotropy provides a preferred direction along which the spins can orient in an average sense.

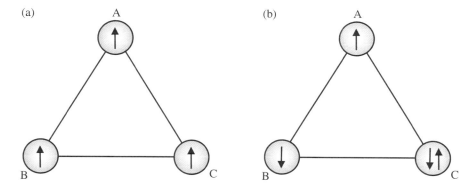

Fig. 3.3 The triangle of frustration (b). In (a) there is no frustration because the interaction between spins sitting at the corners A, B, C is ferromagnetic. By contrast, if the interaction is antiferromagnetic (Fig. 3.2b), there is frustration because the spin at any of the corners does not know whether to point up or down.

Let us examine this in some more detail by starting from a very high temperature $T \gg T_f$. At this temperature, all the spins are oriented totally randomly because the thermal fluctuations flip them around. This is the paramagnetic state.

At sufficiently low temperatures (but still above T_f), some spins will form interacting *clusters*, while others will still be unattached. The clusters can have dynamically changing sizes (partly because of the frustration feature), and can also rotate as a whole. They are the building blocks of the spin-glass state towards which the system is heading as we lower the temperature. They may have giant magnetic moments because all spins may be lined up the same way, and behave like *superparamagnets*. They may also have antiferromagnetic ordering through the RKKY coupling, etc.

We can introduce a spin-glass *correlation length*, ξ_{SG}, which is a measure of the rigid coupling of two spins i and j which are a distance r_{ij} apart. There is a competition between the disordering effect of temperature and the short-range exchange forces $J(r_{ij})$. A given group of spins form a cluster if $J(r_{ij}) > k_B T$.

In addition to the *spatial* magnetic clustering, there is also a *temporal* aspect. There is a wide distribution of temperature-dependant *relaxation times*, τ, which vary with cluster size.

As the temperature decreases, i.e. as $T \rightarrow T_f$, the average size of these clusters grows (ξ_{SG} increases), although clusters of practically all sizes and shapes still exist. Depending on local anisotropy conditions, the system tends to make a *cooperative freezing transition*, at $T = T_f$, into a particular set of spin orientations and exchange interactions. But, because of the underlying randomness and frustration, it gets trapped in some metastable, local free-energy minimum in phase space.

At $T = T_f$, the correlation lengths are so large that a kind of *percolation transition* occurs, which creates (in the thermodynamic limit) an infinitely large cluster of rigidly frozen spins (Mydosh 1993). The random anisotropy creates random freezing directions within this large percolation cluster. There are smaller clusters within this supercluster. But the small clusters are no longer able to react individually to an external magnetic field because they are embedded within the supercluster.

Thus, the freezing transition is like a percolation transition, with the infinite cluster composed of a large number of randomly frozen small clusters of almost all shapes and sizes. *There are features of self-similarity and fractal structure in this scenario, something which is always of interest from the point of view of smart structures.*

Dotsenko (2001) has given an update on this description. According to him, below T_f there is *a continuous sequence of transitions* with decreasing temperature. There is a macroscopically large number of states below T_f into which the system can get frozen. In the thermodynamic limit, the rugged landscape in phase space has infinitely high energy barriers separating valleys of low-energy minima. With decreasing temperature, each of these valleys further splits into a rugged landscape, with new barriers emerging within each valley. Therefore, any temperature below T_f is a critical temperature.

This is not at all like a regular FPT discussed so far in this chapter. Nor is it fully like a glass transition (cf. the appendix on the glass transition). Characteristic features are observed depending on whether the system is cooled under an external magnetic field (*field-cooling*, FC), or when no such field is present (*zero-field-cooling* ZFC). The observed behaviour also depends on the *rate* of cooling, and on the dwell time at a particular temperature, features reminiscent of a conventional or *canonical glass* (like window glass).

Not all clusters are enveloped by the percolation supercluster. Those which are not a part of it go on exhibiting a wide range of sizes and relaxation times even for $T < T_f$.

We end this very brief introduction to spin glasses by mentioning their *memory feature*. A hysteresis curve (*BH* curve) of a field-cooled spin glass is found to be *shifted* along the H-axis, rather than being symmetrically placed about the B-axis. This indicates that the specimen has a memory of the direction in which the field was applied while cooling it. Such behaviour is linked, among other things, to the nonergodic nature of the energy landscape. The memory feature is of great interest for smart-structure applications.

3.6   Dipolar and quadrupolar glasses

An electric dipole moment is the electrical analogue of magnetic moment or spin. If a crystal has such dipole moments randomly distributed (and/or oriented) at a small fraction of lattice sites, we speak of a *dipolar glass*.

A more general concept is that of *orientational glasses*. These are crystals, some sites of which are associated randomly with dipole, quadrupole, or higher-order multipole moments which have orientational degrees of freedom, and which interact with one another sufficiently to undergo, at some freezing temperature T_f, a relaxational freezing into a state devoid of spatial long-range correlations (Hochli, Knorr and Loidl 1990).

An example of a random-*site* dipolar glass is a crystal of $KTaO_3$, in which a fraction of K-sites have been replaced randomly by Li or Na dopants. The different size of the dopant atom results in small displacements of atomic positions, giving rise to local, randomly located and oriented, electric dipole moments. The system behaves as the electric analogue of a spin glass (Hochli, Kofel and Maglione 1985) for a certain range of compositions, a typical one being $K_{0.984}Li_{0.016}TaO_3$.

Another type of dipolar-glass behaviour is exemplified by a solid solution of RbH_2PO_4 and $(NH_4)H_2PO_4$. In this case, both are polar materials, but the first one is a ferroelectric and the second an antiferroelectric. In the mixed crystal, these two competing types of ordering result in frustration and spin-glass-like dynamics (Courtens 1984).

When the elastic quadrupole–quadrupole interaction dominates over the electric dipole–dipole interaction, and the quadrupoles are randomly distributed on crystalline sites, we speak of a *quadrupolar glass*. These are the mechanical or elastic analogues of spin glasses. An example is provided by $(KCN)_x(KBr)_{1-x}$ (Sethna and Chow 1985). The Br ions are spherical. Competing interactions and frustration result when some of them are replaced randomly by the cigar-shaped CN ions, which carry not only a small dipole moment, but also a quadrupole moment (with the concomitant elastic dipole moment). The random stress field created by the elastic dipoles provides the via-media for local interactions.

3.7 Transport properties of ferroics

From the point of view of applications, *magnetoresistance* is the most important transport property of ferroics. A magnetoresistive material has a magnetic-field-dependent resistivity, and if the field dependence is substantial, the material can be used as a sensor of magnetic field.

Generally speaking, all metals have this property. The effect arises because switching on of a magnetic field makes the conduction electrons go into a helical path, and the effect becomes substantial when the field is large enough to significantly bend the electron trajectory within a length shorter than the mean free path. At moderate fields the effect is too small in bulk metallic materials to be of much practical use.

Spintronics is already an important field of research and device applications. One tries to exploit the fact that, in addition to the charge that an electron carries, there is also a magnetic spin associated with it, which can have two possible values ('up' and 'down'). Any device application of this has a provision for spin-polarizing the electrons, a provision for manipulating this polarized population, and a provision for sensing the difference between the polarization of the input and output populations (Dragoman and Dragoman 2006). One would like to do this with magnetic semiconductors because of the already established industry for semiconductor-based electronics. The efforts in this direction have been hampered by the difficulties in injecting spin-polarized electrons into semiconductors. By contrast, spintronic devices using metallic *ferromagnets* have been already in use in the magnetic recording industry.

A variety of magnetoresistive effects are possible (see, e.g. Heinonen 2004; Burgler and Grunberg 2005): AMR (anisotropic magnetoresistance); GMR (giant magnetoresistance); TMR (tunnelling magnetoresistance); BMR (ballistic magnetoresistance); CMR (colossal magnetoresistance); EMR (enormous magnetoresistance).

All read- and write-heads in modern magnetic hard drives are based on the use of the GMR effect in certain composite layers (cf. Chapter 8). The magnetic field created by the stored information generates a large electrical signal in the magnetoresistive read-head sensor. This is very convenient to use, but one would like to have more and more sensitivity, so that the devices can be miniaturized further.

Current efforts focus on developing *semiconductor*-based spintronic devices. The ensuing advantages in terms of ready integration with existing semiconductor electronics will be enormous. New effects are also expected to arise.

3.8   Field-induced phase transitions

A more general form of eqn 3.20 for the thermodynamic potential can be written as follows:

$$g = U - TS + pV + \mu_i N_i - E_i D_i - H_i B_i - \sigma_{ij} e_{ij} \qquad (3.27)$$

Here we have taken note of the presence of an ambient hydrostatic pressure p. N_i is the number of molecules of type i, and μ_i is the corresponding chemical potential.

Two distinct types of control parameters can be identified in eqn 3.27: scalars (T, p, μ), and nonscalars ($\mathbf{E}$, $\mathbf{H}$, $\boldsymbol{\sigma}$). Each of these influences the free-energy density, and can cause a phase transition at a certain critical value. Phase transitions caused by variation of the scalars (temperature, pressure or composition) are more familiar, and better investigated.

Phase transitions caused by any of the three nonscalars listed above are called *field-induced phase transitions*. Some of the most spectacular effects of practical importance are caused by field-induced phase transitions: CMR; shape-memory effect (SME); giant photoelastic effect; etc. We discuss the SME and the CMR later in this chapter.

Another example of a fairly well investigated field-induced phase transition is the antiferroelectric-to-ferroelectric phase transition in $Pb(Zr, Ti, Sn)O_3$ (PZTS). Long-range ferroelectric and antiferroelectric ordering involves a delicate balance between long-range and short-range interactions. Hydrostatic pressure (a high-symmetry isotropic field) and electric field (a lower-symmetry directional field) are found to have opposite effects on the stability-regimes of the antiferroelectric and the ferroelectric phases in the phase diagram of PZTS (Yang and Payne 1996). Electric field expands the stability-regime of the ferroelectric phase, whereas hydrostatic pressure favours the antiferroelectric phase.

3.9 Multiferroics

As discussed in Section 3.3.1, eqn 3.24 forms the thermodynamic basis for defining different types of ferroic materials. A material in which two or more ferroic properties coexist is called a multiple ferroic, or a multiferroic.

In condensed-matter physics there are three types of interactions to consider: electrical, magnetic, and elastic or mechanical. In a given primary ferroic, at least one of them is responsible for the long-range ordering at and below the FPT. Of special interest are multiferroics in which two or all three of these interactions compete in a delicately balanced manner, and even a very minor local factor can tilt the balance in favour of one or the other of these interactions (cf. Bishop, Shenoy and Sridhar 2003; Mathur and Littlewood 2003; Dagotto 2005; Planes, Manosa and Saxena 2005; Moessner and Ramirez 2006). Dominance of any one of these interactions would give a fairly definite ground-state configuration for the ferroic. But if there is close ('hairy edge') competition between two or all three interactions, there are competing ground states.

Ferroelectric and ferroelastic phase transitions are *structural* phase transitions: there is a change of crystal structure caused by the movement of atomic nuclei to their new equilibrium sites (Fujimoto 2005). By contrast, purely ferromagnetic phase transitions are essentially *nonstructural* in nature: Magnetic ordering *can* occur without significant alteration of the equilibrium positions of the atomic nuclei (i.e. no change of the 'chemical' point group, but certainly a change of the magnetic point group). In a multiferroic it can transpire that the tendency for structural, electronic, and magnetic ordering occurs at the same temperature, or over a narrow range of temperatures. This has two important consequences: (i) There is no longer a unique ground state. Instead, there are competing ground states. (ii) It is no longer possible to assume that electrons in the material, in spite of being much lighter than the nuclei, follow the nuclear motions instantaneously. That is, we cannot make the so-called *adiabatic slaving-limit approximation* (Muller 2003). The many-body wavefunction of such a multiferroic therefore has a 'vibronic' character: It has both a nuclear part and an electronic part:

$$\Psi = \psi_n \psi_e \qquad (3.28)$$

The all-important consequence of the competing-ground-states scenario is that, in the same crystal, different portions may order differently, no matter how competent a job the crystal grower has done in growing a 'perfect' single crystal. Even the slightest of local perturbations (defects, inclusions, voids, composition variations, etc.) can tilt the balance in favour of ferroelastic, ferroelectric, and/or ferromagnetic ordering over mesoscopic length-scales. The single crystal is no longer 'single', in the sense that any randomly picked up unit cell may not necessarily be identical to another unit cell elsewhere. The crystal structure of a multiferroic is *intrinsically inhomogeneous*.

After the discovery of quasi-crystals, a modern definition of crystallinity has been adopted: A material is said to be crystalline if its diffraction pattern has sharp peaks. By this criterion, multiferroics are certainly crystalline materials, even though not all their unit cells are identical. It so transpires that, at least for the perovskite-based systems among them (and that covers most of them), the oxygen atoms constitute the periodic edifice, and the inhomogeneities reside mainly in the distribution of spins, electronic charges, and/or mechanical strains.

Transition-metal oxides (TMOs) are particularly prone to exhibiting multiferroic behaviour involving a strong correlation of spin, charge, and elastic degrees of freedom (Stojkovic *et al.* 2000; Dagotto 2005). Some examples of such systems are: CMR-exhibiting manganites (Moreo, Yunoki and Dagotto 1999; Mathur and Littlewood 2003); superconducting cuprates (Shenoy, Subrahmanayam and Bishop 1997; Sharma *et al.* 2000; Neto 2001); systems exhibiting martensitic phase transitions (Shenoy *et al.* 1999; Rasmussen *et al.* 2001; Ren and Otsuka 2002); relaxor ferroelectrics (Li *et al.* 1996, 1997; Bobnar *et al.* 2000); etc. For them, local and global deviations from stoichiometry are the rule rather than the exception. This happens because any local deficiency in the oxygen content is readily accommodated by an appropriate change in the valency of the transition-metal cation. There are thus varying degrees of dynamic charge transfer between the transition metal and the oxygen ions, apart from the fact that the oxygen ions are highly polarizable.

Multiferroics are a subject of active investigation at present (see, for example, Shenoy, Lookman and Saxena (2005), and references therein). They are particularly interesting for possible applications as smart materials. We summarize their characteristic features here:

1. Multiferroics are *strongly correlated electronic systems*; i.e. there is a strong correlation among the structural and electronic degrees of freedom (cf. Bishop 2003).
2. This strong interaction leads to *nonlinear response* and *feedback*, so central to the subject of smart materials.
3. Nonlinearity has several consequences. For one, it leads to *coherence* of structures on *mesoscopic* scales, which control macroscopic response.
4. Nonlinearity also results in *spatio-temporal complexity*, including the occurrence of *landscapes* of metastable or coexisting phases (in the same single crystal), which can be readily fine tuned, or disturbed, by the slightest of internal or external fields.
5. Very often, there is an underlying competition between short-range and long-range ordering interactions, resulting in *a hierarchy of structures* (of different length-scales) in mesoscopic patterns (Stojkovic *et al.* 2000).
6. Multiferroics are *intrinsically multiscale*, not only spatially, but also temporally: There is usually a whole range of relaxation times.

7. The term *glassy behaviour* is used in the context of systems that exhibit noncrystallinity, nonergodicity, hysteresis, long-term memory, history-dependence of behaviour, and multiple relaxation rates. Multiferroics usually display a variety of glassy properties.

8. A likely scenario for most multiferroics is that the same intra-unit-cell distortions are responsible for both *local* anisotropic chemistry and *long-range* anisotropic elastic behaviour. Any local atomic ordering occurring in a unit cell of the crystal has a push–pull effect on neighbouring unit cells; i.e. it creates a local displacement field, which then influences further neighbouring unit cells, thus having a knock-on effect, resulting in an elastic propagation of the disturbance or ordering to distant parts. This coexistence of short- and long-range interactions is responsible for the self-organization of patterns of hierarchies (Bratkovsky *et al.* 1995; Shenoy *et al.* 1999, 2003; Saxena *et al.* 2003; Lookman *et al.* 2003; Ahluwalia *et al.* 2004; Zeng *et al.* 2004).

9. The notion of proper and improper ferroelastics was introduced earlier in this chapter. An improper-ferroelastic phase arises as a result of an FPT for which the primary interaction driving the transition is something other than the elastic interaction. An example is that of $BaTiO_3$, in which the spontaneous strain in the tetragonal phase arises, not because the primary instability is elastic in nature, but because of its coupling with the primary order parameter of the transition (which has the symmetry of a dipole-moment vector). It has been realized over the years that even in improper ferroelastic transitions, the elastic interaction dominates. In the work referred to in the previous paragraph, a theory of multiferroics is sought to be built on the all-important constraining assumption that the integrity of the crystalline lattice must be preserved across all the interfaces separating the various mesophases and ferroic domains in the crystal. Subject to this constraint, strain, polarization (or charge), and magnetization are treated as coupled transition parameters in the Ginzburg–Landau (GL) formalism.

10. Coefficients of the strain tensor are, by definition, derivatives of displacements with respect to space coordinates. As emphasized by Shenoy *et al.* (2003), the true degrees of freedom in an elastic solid continuum are defined by the displacement field, although it is the strain field that enters the GL expansion of the free energy. Instead of treating strains as independent fields, one should treat them as derivatives of a single continuous function defined by the displacements. Any perturbation or short-range ordering occurring locally creates a strain field around itself, which affects the neighbouring regions in the crystal (subject to self-consistent elastic compatibility constraints), which, in turn, affect *their* neighbours, and so on. This deformation at large distances has components coming from non-order-parameter strains also, which thus contribute to the overall *anisotropic long-range* (ALR) interaction. Such models are able to explain a variety of mesoscale hierarchical structures in multiferroics, as also the multiscale complex dynamical phenomena observed in them.

11. Multiferroics offer multiscale, multifunctional, strongly correlated properties which are very easily influenced by, and are therefore tuneable by, external fields. Now this is the stuff *very* smart materials are made of. Strikingly rich technological applications can therefore be envisaged for them.

3.10   Colossal magnetoresistance

CMR is predominantly an effect caused by a field-induced phase transition. It is observed in doped perovskite materials, ABO_3, where A is a rare-earth and B a transition metal; the dopants, which go to the A sites, are typically Ba, Ca, Sr or Pb.

Doped manganite perovskites, particularly $LaMnO_3$, are the best studied CMR materials. They exhibit a metal–insulator and ferromagnetic–paramagnetic phase transition at the *same* temperature (Kaplan and Mahanti 1999; Ramakrishnan *et al.* 2004). In the undoped crystal, manganese is in Mn^{3+} valence state. On doping by Ba, Ca ions, etc., some of the La^{3+} ions get replaced by, say, Ca^{2+}. This makes some of the Mn^{3+} ions change to Mn^{4+} (to maintain charge-neutrality in the unit cell). This mixed-valence system, with a fair concentration of mobile carriers, is well-poised for *complex behaviour*, including magnetic-field-induced phase transitions, resulting in a colossal magnetoresistive effect (Dagotto 2005). For example, $La_{0.67}Ca_{0.33}MnO_x$ shows a 1000-fold change in resistance on the application of just 6 T dc magnetic field (Poole and Owens 2003).

These systems have rich phase diagrams as a function of temperature and composition (Millis 1998). There is even a coexistence of electronically and structurally distinct phases, all based on a common underlying crystalline lattice (Moreo, Yunoki and Dagotto 1999; Renner *et al.* 2002). There are competing interactions (magnetic, electric and elastic), and there is a delicate balance between them for certain compositions; even small perturbations can tip the balance one way or the other (Mathur and Littlewood 2003). This is an ideal situation for nonlinear response, and therefore tuneability, typical of multiferroic materials.

3.11   Shape memory effect in alloys

3.11.1   *One-way shape memory effect*

An important class of alloys (e.g. NiTi, AuCd, CuAlZn, CuAlNi, CuSn), called *shape-memory alloys* (SMAs), exhibit the shape-memory effect (SME), and also *superelasticity*.

The SME is the ability of a material to remember the shape it had above a certain characteristic temperature, even though it has been deformed quite severely at a lower temperature (below the characteristic temperature); the material, after being deformed at the lower temperature, recovers its original shape on being heated to the characteristic temperature.

The characteristic temperature is related to a martensitic phase transition. The terms 'austenite' and 'martensite' are used in a generic sense for the higher-temperature phase and the lower-temperature phase, respectively. The SME arises primarily due to the accommodative reorientation of the austenitic and the martensitic phases (James 1989; Madangopal 1997; Otsuka and Wayman 1998; Bhattacharya 2003).

Superelasticity is the ability of a material to be able to undergo large but *recoverable* strains, provided the deformation is carried out in a characteristic range of temperatures. This property is an example of a purely field-induced phase transition (at a fixed temperature). When the deforming stress is applied at a temperature (within the characteristic temperature range) at which it is in the austenitic phase, the stress sends the material to the martensitic phase, with an accompanying large deformation. When the stress is removed, the material recovers its shape because it reverts back to the austenitic phase.

One of the important and quite prevalent applications of a field-induced phase transition is in the medical industry, using the superelastic behaviour of the shape-memory alloy Ni-Ti (Duerig 1995). Although the SME *per se* is used for deployment purposes in some medical applications, what is really important finally is the superelastic recovery of the shape of the material at body temperature.

In a typical loading–unloading curve of a superelastic alloy, there is a large deformation of the sample, and also a zero or nonzero pseudoplastic strain at the end of the cycle (depending on the sample history). What is important here is the fact that there is at least an inflexion point (if not an extended plateau comprising a whole range of strain values for which the stress remains unchanged) in both the loading and the unloading parts of the curve. This is the range of strain values, at a fixed temperature (the body temperature in a medical device application), for which Hooke's law is violated strongly, and the changes of applied stress (σ) are used up for causing a stress-induced phase transition, rather than for changing the strain (e) the way Hooke's law will demand. The zero slope of the $\sigma(e)$ plot at and around the inflexion point means zero effective stiffness (or infinite effective compliance) of the material. Hence the use of the term superelastic: 'Super' because it is so highly compliant, and 'elastic' because it bounces back when the stress is removed. The bouncing-back action is simply a reminder of the fact that there is a stress-induced, reversible, phase transition from the austenitic to the martensitic phase; no applied stress, no transition. For medical applications the composition chosen for the alloy is such that all this can happen at body temperature.

This action of the material also makes it qualify to be called a smart *material*, in the sense that it passes successfully the Culshaw (1996) criterion described in Chapter 1.

Coming to the SME, a typical *shape-memory cycle* runs as follows. Let us assume that the material is in the martensitic phase at room temperature. It is first heated towards the temperature of the martensite–austenite phase transition. Since this is typically a case of a first-order phase transition, there is a range of temperatures in which the martensitic and the austenitic phases coexist. There is thus a temperature A_s at which the austenitic phase starts forming. Further increase of temperature converts more and more of the martensitic phase to the austenitic phase. Finally there comes a temperature (A_f) at which conversion of the martensitic phase to the austenitic phase is finished. At a temperature above A_f, a deforming load is applied on the specimen material, and, with the deforming load still 'on', the system is cooled gradually towards room temperature. At some stage during the cooling, a temperature M_s is reached, at which the martensitic phase starts forming. On further cooling, a temperature M_f is reached such that conversion of the austenitic phase to the martensitic phase is finished. There are two factors contributing to the phase-transition process: One is the lowering of temperature to values at which the martensitic phase is stable; this is temperature-induced phase transition. In addition, there also occurs a stress-induced phase transition (like in superelastic behaviour described above). The combined effect of these two factors, as also the adjustment of the domain structure of the martensitic phase to the applied stress, leads to an anomalously large deformation of the material. After cooling to room temperature, the applied deforming stress is removed, leaving behind an apparently 'plastically' deformed specimen. When this specimen is heated back to the austenitic phase (thus completing the shape-memory cycle), the domain structure disappears, making the specimen recover the shape it had before the stress was applied.

If the specimen is cooled back to room temperature, it usually does not recover its apparently plastically deformed shape. Since the shape is remembered only on heating to the parent phase, and not also on cooling to the martensitic phase, we speak of a *one-way shape-memory effect.*

It is important for device applications that the performance of the SMA should not deteriorate significantly after only a small number of shape-memory cycles. In any case, it is important to characterize the material thoroughly and determine the 'safe' number of cycles before the deterioration becomes unacceptable. The main cause of degradation in performance due to repeated cyclic use is fatigue, particularly thermal fatigue or *amnesia* (Moumni *et al.* 2005): The characteristic temperatures M_s, M_f, A_s, A_f shift with repeated cycling through the phase transition. The underlying reasons are still not understood very well, and limit a much more widespread use of SMAs in device applications.

Apart from use in actuators, another important application of SMAs stems from their high damping characteristics for vibrations. This property is related to the movement of interfaces between phases and between domains, and also to the absorption of mechanical energy for causing the stress-induced phase transition (Humbeeck and Stalmans 1998; Humbeeck 2001).

3.11.2   *Two-way shape memory effect*

If a material can remember its shape in both the austenitic phase and the martensitic phase, it is said to exhibit the *two-way* shape-memory effect (Guenin 1989; Boyko, Garber and Kossevich 1994; Saburi 1998; Humbeeck and Stalmans 1998; Ren and Otsuka 2000).

The basic requirement for the occurrence of the two-way SME is the creation of some kind of an internal bias or anisotropy in the substructure of the parent phase, with which the martensitic phase transition interacts, resulting in the preferential development of a specific configuration of variants in the martensitic phase. Since every such configuration corresponds to a specific overall shape of the specimen, the specimen remembers its shape in the martensitic phase also. Thus, without the application of any stress, the specimen goes to a particular shape on heating to the austenitic phase, and comes back to another specific shape on cooling to the martensitic phase.

To obtain two-way SME behaviour, a specimen has to be *trained*. A variety of training protocols (thermomechanical treatments) have been described in the literature. Most of them amount to creating a network of dislocations in the specimen. These dislocations have an associated strain field (a *microscale asymmetry*), and that martensitic configuration develops on cooling to the daughter phase for which the overall strain energy (defect energy) is the least. This can happen repeatedly across the heating and cooling cycles.

Training procedures include: repetition of the one-way SME cycle; constrained cycling across the phase transition (i.e. thermal cycling at a constant strain); thermal cycling at constant applied stress; and superelastic cycling.

Creation of complex dislocation patterns is not the only way for making a specimen exhibit the two-way SME. Precipitation of particles of another phase or material can also provide a built-in biasing stress field. For example, the presence of Ti_3Ni_4 particles in Ti-51at% Ni alloy plays a central role in the occurrence of the so-called 'all-round shape memory effect' (a variation on the usual two-way SME) (cf. Saburi 1998). For this Ni-rich composition of Ni-Ti, the occurrence of the R-phase also plays a role. The Ti_3Ni_4 particles are lenticular in shape, and they align themselves in strain-energy-minimizing lattice-coherent ways as the specimen is cycled across the martensitic phase transition.

3.11.3 *Ferromagnetic shape-memory alloys*

Superelasticity, a concomitant feature of shape-memory alloys (SMAs), involves the use of mechanical stress for achieving a large and recoverable shape change (in a specific temperature range around the martensitic phase transition). Mechanical stress is also required for introducing shape changes in the conventional shape-memory effect (SME) exhibited by Ni-Ti and some other alloys. For ferromagnetic SMAs, one can replace mechanical stress by magnetic field, which can be advantageous in certain device-application situations. The first demonstration of the SME by using a magnetic field was by Ullakko *et al.* (1996), and it was for single crystals of the ferromagnetic alloy Ni_2MnGa (cf. Soderberg *et al.* 2005, for a recent review).

The magnetic SME (MSME) occurs because of the reversible rearrangement of ferromagnetic and structural domains by the applied magnetic field. Naturally, magnetocrystalline anisotropy also plays a crucial role (Vassiliev 2002).

3.12 Relaxor ferroelectrics

Relaxor ferroelectrics exhibit a broad and frequency-dependent maximum in their permittivity *vs.* temperature response in the RF regime of frequencies, and this response function falls off rapidly at high frequencies. The permittivity values are very high, as are the electrostriction coefficients (Park and Shrout 1997).

Lead-based relaxor ferroelectrics have the general formula $Pb(B_1B_2)O_3$, where $B_1 = Mg^{2+}, Zn^{2+}, Ni^{2+}, Sc^{3+}$, etc., and $B_2 = Nb^{5+}, Ta^{5+}, W^{6+}$, etc. To understand some of their properties, let us consider the crystal structure of the archetypal relaxor ferroelectric: lead magnesium niobate (PMN), $Pb^{2+}(Mg_{1/3}^{2+}Nb_{2/3}^{5+})O_3^{2-}$. It has the cubic perovskite structure, the same as that of $BaTiO_3$. The Pb^{2+} ions occupy the corners of the cubic unit cell, and the three O^{2-} ions sit at the face-centres. The body centre of the cube is occupied either by Mg^{2+} or by Nb^{5+}, with a relative occupation frequency of 1 to 2, thus ensuring that the average ('global') valence for the occupants of the body-centres is 4 (= $(2 + 5\times2)/3$), the same as that of Ti^{4+} in $Ba^{2+}Ti^{4+}O_3^{2-}$.

Any particular unit cell has either Mg^{2+} or Nb^{5+} at the body centre, and therefore there is a *charge-imbalance* at the unit-cell level, because the needed valence for charge balance is 4+. This unit-cell charge imbalance is the basis of practically all the remarkable properties exhibited by PMN (not all of which are well-understood at present) (Bokov and Ye 2002).

How the structure gets over the local charge-imbalance problem has been a subject of considerable investigation and debate. According to one of the models, there exist *1:1 ordered regions* in the crystal, separated by Nb-rich regions. In this kind of 'chemical' ordering, neighbouring unit cells are occupied by Mg and Nb ions, so that there are $\langle 111 \rangle$ planes occupied alternately by Mg- and Nb-based unit cells. But this cannot go on beyond a certain length-scale ($\sim 2 - 3$ nm) because it corresponds to an average charge of $(2 + 5)/2$ or 3.5 at the body centre, whereas the required value for charge balance is 4. Such negatively charged 1:1 ordered clusters (called *chemical clusters*) are therefore separated by positively charged regions (called *polar clusters*) in which all unit cells have only the Nb ion at the body centre (Chen, Chan and Harmer 1989; Boulesteix *et al.* 1994; Lee and Jang 1998). The chemical clusters and the polar clusters extend over a few nanometres only, and their size distribution is of a dynamically changing nature. At such small sizes they can be readily turned around by thermal fluctuations, and the boundaries between them are constrained to move in a correlated manner. It is envisaged that the 1:1 ordered regions provide a quenched random field, in which the polar clusters undergo reorientations (Westphal, Kleeman and Glinchuk 1992; Pirc and Blinc 1999). Under the influence of an RF probing field, the polar clusters reorient as a whole, thus accounting for the very large permittivity values. At high probing frequencies, the fluctuations in the direction and size of the polar clusters are not able to follow the rapid variations of the probing field. Consequently the dielectric permittivity falls off rapidly with increasing frequency; hence the name *relaxor*.

A reason cited by Chen, Chan and Harmer (1989) in support of the above *space charge model* of ordering in PMN was the observation that the chemical clusters could not be made to grow by annealing procedures: It was argued that the cluster growth does not occur because it would only further exacerbate the charge imbalance problem. Akbas and Davies (1997) advocated the alternative *random site model*.

They worked with PMT ($Pb(Mg_{1/3}Ta_{2/3})O_3$), rather than PMN. By carrying out the annealing at a much higher temperature of $1325°C$, they could make the 1:1 ordered domains grow to sizes as large as 10 nm. In the random-site model, there is no charge imbalance. The 1:1 ordering has a different meaning here. It involves a doubling of the unit cell along all the three pseudo-cubic edges, and the system is believed to adopt an NaCl type fcc structure. Its formula is written as $Pb(B^1_{1/2}B^2_{1/2})O_3$, where $B^2 = Ta^{5+}$, and $B^1 = (Mg^{2+}_{2/3}Ta^{5+}_{1/3})$. If this model is correct, the relaxor behaviour must arise from the random occupancy of B^1 sites within the eight-times-larger unit cell by divalent and pentavalent ions in a 2:1 ratio.

It is conceivable that the real structure of relaxors like PMN involves a *coexistence* of configurations described by the space-charge model and the random-site model. Akbas and Davies (1997), by their very determined annealing protocol for PMT, could achieve only 70% conversion to 1:1 ordering, and the ordered domains did not grow beyond 10 nm.

PMN forms a solid solution with PT (lead titanate) (Noheda *et al.* 2002), and also with PZ (lead zirconate) (Singh *et al.* 2001, 2003). The Ti ion in PT, as also the Zr ion in PZ, has a valence of 4+, and when any of them replaces Mg^{2+} or Nb^{5+} in PMN, the effect is an easing of the charge-imbalance situation in PMN-PT or PMN-PZ, compared to that in pure PMN. An indication of this is that the average size of the mesoscopic clusters in PMN-PT and PMN-PZ is large compared to that in PMN, even when the annealing carried out is not at excessively high temperatures.

3.13   Shape memory effect in ceramics

The shape-memory effect has been observed, not only in alloys and polymers, but also in ceramics. Some examples of such materials are: PLZT (lead lanthanum zirconate titanate) (Wadhawan *et al.* 1981); CeO_2-stabilized tetragonal zirconia (ZrO_2) (Reyes-Morel, Cherng and Chen 1988); Y-Ba-Cu-O (Tiwari and Wadhawan 1991); antiferroelectric compositions based on PZT, like tin-modified PZT (Uchino 1998); $La_{2-x}Sr_xCuO_4$ (Lavrov, Komiya and Ando 2002); and PMN-PT (Pandit, Gupta and Wadhawan 2004, 2006; Wadhawan, Pandit and Gupta 2005). The occurrence of the SME in such insulators offers the possibility of using the electric field as an additional control parameter (apart from mechanical stress), even though the recoverable strain is rather small ($\sim 0.2\%$).

Shape-memory ceramics not only offer controllability of the SME by electric fields, they are also attractive because of the quicker ($\sim$msec) response, the very high actuation stress (tonnes cm^{-2}), and the lower consumption of energy compared to SMAs (Uchino 1998). The ability to operate under harsh environmental conditions, as well as very high resistance to corrosion, have been cited as some additional advantages of shape-memory ceramics (Matsumura *et al.* 2000).

Of particular interest in this context are shape-memory ceramics which are simultaneously ferroelastic and ferroelectric (cf. Wadhawan 2000). The coupling of these two properties offers the possibility of effecting the entire shape-memory cycle by electric (and thermal) means alone, with no need to bring in mechanical loading and unloading. PZT, PLZT and PMN-PT are possible examples of such ceramics.

Because of the rather small recoverable strains, applications of the SME in ceramics involving thermal cycling through the martensitic phase transition are no match for the same effect in SMAs which give much larger recoverable strains. But shape-memory in ceramics involving strain recovery at a fixed temperature, namely *superpiezoelectricity* (cf. the Glossary for the definition of this term), is a very attractive proposition, especially because a trip around the field-induced phase transition need not involve a (cumbersome-to-apply) mechanical load; one can use the electric-field-induced phase transition.

There are several types of oxide materials that may be available for this purpose (Cross 1995). One such type is antiferroelectrics which can undergo an electric-field-induced phase transition to a ferroelectric phase (the 'A-F' transition). Zirconia-rich compositions in the ternary $PbZrO_3$:$PbTiO_3$:$PbSnO_3$ (lead zirconate titanate stannate, or PZTS) constitute a good example of this (Yang and Payne 1992, 1996). A variant of this, namely $Pb_{0.99}Nb_{0.02}[(Zr_{0.6}Sn_{0.4})_{0.94}Ti_y]_{0.98}O_3$, has been discussed by Uchino (1998). The reversible strain caused by the field-induced phase transition is of the order of 0.1%. The field applied is typically 3 kV mm^{-1}.

The strain for this composition of PZTS persists metastably even when the field is removed. The original shape in the antiferroelectric phase can be regained either by applying a small reverse field, or by thermal annealing. The forward and reverse phase transitions are rapid (Pan *et al.* 1989).

The A-F transition is first-order, so there is a range of temperatures in which the two phases coexist. The activation barrier between regions of the two phases is the reason why the field-induced ferroelectric phase remains stranded when the field is removed.

The electric field applied for the A-F, or for the reverse F-A, transition in the mixed-phase system will cause a reversible shape recovery, provided the field needed for domain switching (the coercive field) in the ferroelectric phase is of a higher magnitude than that needed for inducing the phase transition. For PZTS, this is possible only for a narrow range of compositions (Yang and Payne 1992; Uchino 1998; also see Yang *et al.* 2004).

A second type of ceramic materials which can undergo an electric-field-induced phase transition to a normal ferroelectric phase are some of the relaxor ferroelectrics, particularly certain compositions of PLZT (Wadhawan *et al.* 1981; Meng, Kumar and Cross 1985). PLZT $[(Pb^{2+},La^{3+})(Zr,Ti)O_3]$ is a much investigated *transparent* ceramic (Haertling and Land 1971), which is also a ferroelastic–ferroelectric (cf. Wadhawan 2000). Its composition is usually specified by the $x/y/z$ notation, where y/z stands for the Zr/Ti atomic ratio, and x denotes the atomic percentage of La^{3+} ions. The replacement of Pb^{2+} at some of the sites by La^{3+} leads to a charge-imbalance situation, which results in relaxor–ferroelectric behaviour for certain compositions. The first reported study of the effect of electric field on the SME exhibited by PLZT was for the composition 6.5/65/35 (Wadhawan *et al.* 1981).

As mentioned earlier in this chapter, undoped PZT has a morphotropic phase boundary at 52/48. Doping with La^{3+} shifts this boundary progressively towards Zr-rich compositions. For example, PLZT(4.5/65/35) falls close to this boundary. For $x > 4.5$, the x/65/35 composition displays relaxor behaviour. Above the Curie temperature T_c the system behaves like a normal ferroelectric. The phase transition around this temperature is akin to a martensitic transition. The dielectric behaviour of the material immediately below T_c is not like that of a normal ferroelectric. There exist polar distorted microregions shorter than the wavelength of light (Meitzler and O'Bryan 1973). If an electric field is applied at a temperature between T_c and a lower temperature T_p, the microdomains change to macrodomains. In this temperature range the process is reversible: Switching off of the field makes the macrodomains revert back to randomly oriented and distributed microdomains. A manifestation of this is a nearly hysteresis-free or 'thin' polarization-versus-field loop.

For temperatures between T_c and T_p, microdomains of polar short-range order coexist with the paraelectric matrix, just as martensite coexists with austenite in a certain range of temperatures.

If the PLZT specimen is cooled below T_p, it behaves like a 'normal' ferroelectric: There is a 'fat' hysteresis loop, signifying that the macroscopic ferroelectric domains are stable; they do not revert to microdomains when the field is switched off.

Wadhawan *et al.* (1981) argued that T_p and T_c for PLZT(6.5/65/35) can be identified with the 'martensite start' temperature M_s and the 'martensite finish' temperature M_f, respectively. Application of electric field in this temperature range converts the microdomains to macrodomains, resulting in a higher degree of reversible (shape-memory) bending by the applied mechanical stress. A similar interpretation has been given recently for the effect of the electric field on the SME observed in PMN-PT(65/35) ceramic (Pandit, Gupta and Wadhawan 2006).

Meng *et al.* (1985) have demonstrated that PLZT(8/70/30) exhibits superpiezo-electricity. What we have discussed so far may be described as *superpiezoelectricity by transformation*: The specimen is in the austenitic phase to start with, and the electric field causes a large strain through the process of phase transformation to the martensitic phase.

The phenomenon of *superpiezoelectricity by reorientation* is also possible: For this the specimen (a ferroelectric–ferroelastic) is in the martensitic phase to start with, with its attendant ferroelastic domain structure. The electric field is able to deform the specimen by a large amount because it is able to cause ferroelastic domain switching or *reorientation*, making some domains grow at the cost of others. Since the domains grow in a surrounding matrix, a restoring force builds up because of the overall strain caused by the growth of these domains. When the electric field is switched off, the restoring force makes the specimen bounce back to its original macroscopic shape. This superpiezoelectricity by reorientation has its analogy in superelasticity by reorientation, wherein electric field is replaced by uniaxial stress (cf. Wadhawan 2000). Meng *et al.* (1985) have shown that what we are calling superpiezoelectricity by reorientation in this book is possible for certain compositions of PLZT close to the MPB.

3.14   Chapter highlights

- Anything we build requires materials. The crux of smart-structure behaviour is that one or more materials used for making the smart structure should possess the property of nonlinear response to stimuli, and the resultant field tuneability. Ferroic materials are inherently nonlinear in response. No wonder, they are already employed in a large number of smart structures as sensors and actuators (cf. Chapter 9).

- The nonlinear behaviour of ferroic materials arises for two main reasons: (i) domain structure; and/or (ii) vicinity of a ferroic phase transition. The response function corresponding to the order parameter becomes extremely large and non-linear in the vicinity of a phase transition, and for a ferroic phase transition the order parameter can be identified with a *macroscopic* tensor property of the material.

- Field-induced phase transitions in ferroics provide another mechanism for achieving tuneability of properties for smart-structure applications. Such transitions also contribute strongly to the anomalously large or giant property coefficients exhibited by ferroics in certain situations.

- From the vantage point of smart structures, there is an important class of phase transitions, called martensitic phase transitions, the kinetics of which is dominated by strain energy arising from shear-like displacements of atoms. They are usually diffusionless, first-order, structural phase transitions. They are somewhat like ferroelastic phase transitions, but usually of a much more drastic nature, in the sense that they usually involve a change of coordination numbers of the atoms.

- Domain structure and field-induced phase transitions account for the shape-memory effect (SME) exhibited by a number of ferroic materials. The SME is usually associated with a martensitic phase transition, or something analogous to it. The alloy Ni-Ti is the archetypal SME material. As we shall see in Chapter 9, the SME is an important mechanism exploited in a number of actuator applications in smart structures.

- Quartz, the best-known transducer crystal, is a secondary ferroic; it is a ferroelastoelectric. The piezoelectric property of quartz, as also of several other materials (including polymers), provides an important mechanism for actuation and sensor action in smart structures.

- Large single crystals are generally expensive and/or difficult to grow, so one would like to use ferroic and other materials in polycrystalline form wherever possible and sufficient. An as-prepared polycrystalline ceramic like PZT exhibits little or no piezoelectric effect. To get a usable output, one must first pole the material. The poling process gives a preferential direction to the orientation of the grains.

- Only ferroic materials can be poled.

- The poling process can also be used for creating certain desired domain-structure patterns in a ferroic material. Creation of such designer domain patterns can result in an enhancement of certain macroscopic properties of the ferroic.

- Since the various domain types in a ferroic phase are 'equivalent' objects, efforts to create certain domain patterns can, in principle, lead to the emergence of unexpected symmetries sometimes. This can happen because certain types of latent symmetry may become manifest in composite systems under appropriate conditions (cf. the appendix on the symmetry of composite systems).

- Spin glasses are an example of a very special type of magnetic ordering in crystals, in which the spins on embedded magnetic atoms are randomly oriented and located. They are complex systems, characterized by competing interactions, frustration, and a strongly degenerate ground state.

- Orientational glasses are like spin glasses in several aspects. They are crystals, some sites of which are associated randomly with dipole, quadrupole, or higher-order multipole moments which have orientational degrees of freedom, and which interact with one another sufficiently to undergo, at some freezing temperature T_f, a relaxational freezing into a state devoid of spatial long-range correlations.

- In the field of research called spintronics, one tries to exploit the fact that, in addition to the charge that an electron carries, there is also a magnetic spin associated with it, which can have two possible values. Any device application of this fact has a provision for spin-polarizing the electrons, a provision for manipulating this polarized population, and a provision for sensing the difference between the polarization of the input and output populations. One would prefer to do this with magnetic semiconductors because of the already established industry for semiconductor-based electronics. Efforts in this direction have been hampered by the difficulties in injecting spin-polarized electrons into semiconductors. By contrast, spintronic devices using metallic ferromagnets have been already in use in the magnetic recording industry.

- A material in which two or more ferroic properties coexist is called a multiferroic. Multiferroics are the 'creamy layer' among ferroics. Their complex behaviour is far from understood at present, although good progress has been made in recent years. As we understand them better, their applications in smart structures will increase.

- In a multiferroic, two or all three of the electric, magnetic and elastic interactions compete in a delicately balanced manner, and even a very minor local factor can tilt the balance in favour of one or the other. Dominance of any one of these interactions would normally give a fairly definite ground-state configuration for the ferroic. But if there is close ('hairy edge') competition between two or all three, there occur competing ground states. A consequence of this is that, in the same crystal, different portions may order differently. Even the slightest of local perturbations (defects, inclusions, voids, composition variations, etc.) can tilt the balance in favour of ferroelastic, ferroelectric, or ferromagnetic ordering over mesoscopic length-scales.

- Transition-metal oxides are particularly prone to exhibiting multiferroic behaviour involving a strong correlation of spin, charge, and elastic degrees of freedom.

- Very often, there is an underlying competition between short-range and long-range ordering interactions in a multiferroic, resulting in a hierarchy of structures of different length-scales.

- Multiferroics are intrinsically multiscale, not only spatially, but also temporally: There is usually a whole range of relaxation times.

- The term 'glassy behaviour' is used in the context of systems that exhibit non-crystallinity, nonergodicity, hysteresis, long-term memory, history dependence of behaviour, and multiple relaxation rates. Multiferroics usually display a variety of glassy properties.

- Multiferroics offer multiscale, multifunctional, strongly correlated properties which are very easily influenced by, and are therefore tuneable by, external fields. Strikingly rich technological applications can therefore be envisaged for them as *very* smart materials.

- Doped manganite perovskites, particularly $LaMnO_3$, are the best investigated CMR materials. They are mixed-valence systems, with a fair concentration of mobile carriers. They undergo magnetic-field-induced phase transitions, resulting in a colossal magnetoresistive effect. There is also a coexistence of electronically and structurally distinct phases, all based on a common underlying crystalline lattice. There are competing interactions (magnetic, electric and elastic), and there is a delicate balance between them for certain compositions. Even small perturbations can tip the balance one way or the other. This is an ideal situation for nonlinear response, and therefore tuneability, typical of multiferroic materials.

References

Ahluwalia, R., T. Lookman, A. Saxena and S. R. Shenoy (2004). 'Pattern formation in ferroelastic transitions'. *Phase Transitions*, 77: 457.

Aizu, K. (1970). 'Possible species of ferromagnetic, ferroelectric, and ferroelastic crystals'. *Phys. Rev. B*, 2: 754.

Akbas, M. A. and P. K. Davies (1997). 'Domain growth in $Pb(Mg_{1/3}Ta_{2/3})O_3$ perovskite relaxor ferroelectric oxides'. *J. Am. Ceram. Soc.* 80(11): 2933.

Als-Nielsen, J. and R. J. Birgeneau (1977). 'Mean field theory of Ginzburg criterion, and marginal dimensionality of phase transitions'. *Amer. J. Phys.* 45: 554.

Ball. P. (2 December 1999). 'Transitions still to be made'. *Nature*, 402 (Suppl.): C73.

Burgler, D. E. and P. A. Grunberg (2005). 'Magnetoelectronics – magnetism and magnetotransport in layered structures'. In Waser, R. (ed.), *Nanoelectronics and Information Technology: Advanced Electronic Materials and Novel Devices*, 2nd edition, p. 107. KGaA: Wiley-VCH Verlag

Bhattacharya, K. (2003). *Microstructure of Martensite: Why it Forms and How it Gives Rise to the Shape-Memory Effect*. Oxford: Oxford University Press.

Bishop, A. R. (2003). 'Intrinsic inhomogeneity and multiscale functionality in transition metal oxides'. In Bishop, A. R., S. R. Shenoy and S. Sridhar (eds.), *Intrinsic Multiscale Structure and Dynamics in Complex Electronic Oxides*. Singapore: World Scientific.

Bishop, A. R., S. R. Shenoy and S. Sridhar (eds.) (2003). *Intrinsic Multiscale Structure and Dynamics in Complex Electronic Oxides*. Singapore: World Scientific.

Bobnar, V., Z. Kutnjak, R. Pirc, R. Blinc and A. Levstik (2000). 'Crossover from glassy to inhomogeneous-ferroelectric nonlinear dielectric response in relaxor ferroelectrics'. *Phys. Rev. Lett.* 84: 5892.

Bokov, A. A. and Z.-G. Ye (2002). 'Universal relaxor polarization in $Pb(Mg_{1/3}Nb_{1/3})O_3$ and related materials'. *Phys. Rev. B*, 66: 064103-1.

Boulesteix, C., F. Varnier, A. Llebaria and E. Husson (1994). 'Numerical determination of the local ordering of $PbMg_{1/3}Nb_{2/3}O_3$ (PMN)'. *J. Solid State Chem.* 108: 141.

Bowles, J. C. and J. K. Mackenzie (1954). 'The crystallography of martensite transformations. I'. *Acta Metall.* 2: 129.

Boyko, V. S., R. I. Garber and A. M. Kossevich (1994). *Reversible Crystal Plasticity*. New York: American Institute of Physics.

Bratkovsky, A. M., E. K. H. Salje, S. C. Marais and V. Heine (1995). 'Strain coupling as the dominant interaction in structural phase transitions'. *Phase Transitions*, 55: 79.

Castan, T., E. Vives, L. Manosa, A. Planes and A. Saxena (2005). 'Disorder in magnetic and structural transitions: Pretransitional phenomena and kinetics'. In Planes, A., L. Manosa and A. Saxena (eds.), *Magnetism and Structure in Functional Materials*. Berlin: Springer.

Chen, J., H. M. Chan and M. P. Harmer (1989). 'Ordering structure and dielectric properties of undoped and La/Na-doped $Pb(Mg_{1/3}Nb_{2/3})O_3$'. *J. Am. Ceram. Soc.* 72(4): 593.

Courtens, E. (1984). 'Vogel-Fulcher scaling of the susceptibility in a mixed-crystal proton glass'. *Phys. Rev. Lett.* 52: 69.

Cross, L. E. (1995). 'Boundary conditions for shape memory in ceramic material systems'. *J. Intelligent Material Systems and Structures*, 6: 55.

Culshaw, B. (1996). *Smart Structures and Materials*. Boston: Artech House.

Dagotto, E. (8 July 2005). 'Complexity in strongly correlated electronic systems'. *Science*, 309: 257.

Dotsenko, V. (2001). *Introduction to the Replica Theory of Disordered Statistical Systems*. Cambridge, U. K.: Cambridge University Press.

Dragoman, M. and D. Dragoman (2006). *Nanoelectronics: Principles and Devices*. Boston: Artech House.

Duerig, T. W. (1995). 'Present and future applications of shape memory and superelastic materials', in E. P. George, S. Takahashi, S. Trolier-McKinstry, K. Uchino and M. Wun-Fogle (eds.), *Materials for Smart Systems*. MRS Symposium Proceedings, Vol. 360. Pittsburgh, Pennsylvania: Materials Research Society.

Fejer, M. M. (May 1994). 'Nonlinear optical frequency conversion'. *Physics Today*, p. 25.

Fousek, J. (1986). 'Electrooptic effects in crystals induced by phase transitions'. In P. Gunter (ed.), *Electrooptic and Photorefractive Materials* [Proceedings of the International School on Material Science and Technology, Erice, Italy, July 6–17, 1986].

Fousek, J. and L. E. Cross (2001). 'Engineering multidomain ferroic samples'. *Ferroelectrics*, 252: 171.

Fousek, J., D. B. Litvin and L. E. Cross (2001). 'Domain geometry engineering and domain average engineering of ferroics'. *J. Phys.: Condens. Matter*, 13: L33.

Fujimoto, M. (2005). *The Physics of Structural Phase Transitions*. Second edition. Berlin: Springer.

Ginzburg, V. L. (1961). 'Some remarks on phase transitions of the second kind and the microscopic theory of ferroelectric materials'. *Sov. Phys. Solid State*, 2: 1824.

Guenin, G. (1989). 'Martensitic transformation: Phenomenology and the origin of the two-way memory effect'. *Phase Transitions*, 14: 165.

Heinonen, O. (2004). 'Magnetoresistive materials and devices'. In Di Ventra, M., S. Evoy and J. R. Heflin (eds.), *Introduction to Nanoscale Science and Technology*. Dordrecht: Kluwer.

Hill, N. A. (2002). 'Density functional studies of multiferroic magnetoelectrics'. *Ann. Rev. Mater. Res.* 32: 1.

Hochli, U. T., K. Knorr and A. Loidl (1990). 'Orientational glasses'. *Adv. Phys.* 39: 405.

Hochli, U. T., P. Kofel and M. Maglione (1985). 'Dipolar relaxation and limit of ergodicity in $K_{1-x}Li_xTaO_3$'. *Phys. Rev. B*, 32: 4546.

Hu, Z. W., P. A. Thomas, A. Snigirev, I. Snigireva, A. Souvorov, P. G. R. Smith, G. W. Ross and S. Teat (1998). 'Phase mapping of periodically domain-inverted $LiNbO_3$ with coherent X-rays'. *Nature*, 392: 690.

Humbeeck, J. Van (2001). 'Shape memory alloys: A material and a technology'. *Adv. Engg. Mater.* 3: 837.

Humbeeck, J. Van and R. Stalmans (1998). 'Characteristics of shape memory alloys'. In Otsuka, K. and C. M. Wayman (eds.), *Shape Memory Materials*. Cambridge, U. K.: Cambridge University Press.

Izyumov, Yu. A., V. M. Laptev and V. N. Syromyatnikov (1994). 'Phenomenological theory of martensitic and reconstructive phase transitions'. *Phase Transitions*, 49: 1.

James, R. D. (1989). 'Basic principles for the improvement of shape-memory and related materials'. In C. A. Rogers (ed.), *Smart Materials, Structures, and Mathematical Issues*. Lancaster: Technomic Pub. Co.

James, R. D. (2005). 'A way to search for multiferroic materials with "unlikely" combinations of properties'. In Planes, A., L. Manosa and A. Saxena (eds.), *Magnetism and Structure in Functional Materials*. Berlin: Springer.

Jin, Y. M., Y. U. Wang, A. G. Khachaturyan, J. F. Li and D. Viehland (2003). 'Conformal miniaturization of domains with low domain-wall energy: Monoclinic ferroelectric states near the morphotropic phase boundaries'. *Phys. Rev. Lett.* 91: 197601–1.

Joannopoulos, J. D., R. D. Meade and J. N. Winn (1995). *Photonic Crystals: Moulding the Flow of Light*. New Jersey: Princeton University Press.

Kaplan, T. A. and S. D. Mahanti (eds.) (1999). *Physics of Manganites*. New York: Kluwer.

Kartha, S., J. A. Krumhansl, J. P. Sethna and L. K. Wickham (1995). 'Disorder-driven pretransitional tweed pattern in martensitic transformations'. *Phys. Rev. B*, 52: 803.

Kasuya, T. (July 1956). 'A theory of metallic ferro- and antiferromagnetism on Zener's model'. *Prog. Theor. Phys.* 16(1): 45.

Khachaturyan, A. G., S. Semenovskaya and Long-Qing Chen (1994). In M. H. Yoo and M. Wuttig (eds.), *Twinning in Advanced Materials*. Pennsylvania: The Minerals, Metals and Materials Society.

Khachaturyan, A. G., S. M. Shapiro and S. Semenovskaya (1991). 'Adaptive phase formation in martensitic transformation'. *Phys. Rev. B*, 43: 10832.

Khalil-Allafi, J., W. W. Schmahl and T. Reinecke (2005). 'Order parameter evolution and Landau free energy coefficients for the B2 ↔ R-phase transition in a NiTi shape memory alloy'. *Smart Mater. Struct.* 14: S192.

Landau, L. D. (1937). *Phys. Z. Sowjet.* 11: 26 (in Russian). For an English translation, see D. ter Haar (ed.) (1965), *Collected Papers of L. D. Landau*. New York: Gordon and Breach.

Lavrov, A. N., S. Komiya and Y. Ando (2002). 'Magnetic shape-memory effects in a crystal'. *Nature*, 418: 385.

Lee, K.-M. and H. M. Jang (1998). 'A new mechanism of nonstoichiometric 1:1 short-range ordering in NiO-doped Pb(Mg$_{1/3}$Nb$_{2/3}$)O$_3$ relaxor ferroelectrics'. *J. Am. Ceram. Soc.* 81(10): 2586.

Li, S., J. A. Eastman, Z. Li, C. M. Foster, R. E. Newnham and L. E. Cross (1996). 'Size effects in nanostructured ferroelectrics'. *Phys. Lett. A*, 212: 341.

Li, S., J. A. Eastman, R. E. Newnham and L. E. Cross (1 May 1997-II). 'Diffuse phase transition in ferroelectrics with mesoscopic heterogeneity'. *Phys. Rev. B*, 55(18): 12 067.

Litvin, D. B. and V. K. Wadhawan (2001). 'Latent symmetry and its group-theoretical determination'. *Acta Cryst.* A57: 435.

Litvin, D. B. and V. K. Wadhawan (2002). 'Latent symmetry'. *Acta Cryst.* A58: 75.

Litvin, D. B., V. K. Wadhawan and D. M. Hatch (2003). 'Latent symmetry and domain average engineered ferroics'. *Ferroelectrics*, 292: 65.

Lookman, T., S. R. Shenoy, K. O. Rasmussen, A. Saxena and A. R. Bishop (2003). 'Ferroelastic dynamics and strain compatibility'. *Phys. Rev. B*, 67: 024114.

Madangopal, K. (1997). 'The self-accommodating martensitic microstructure of Ni-Ti shape memory alloys'. *Acta Mater.* 45: 5347.

Mathur, N. and P. Littlewood (January 2003). 'Mesoscopic texture in manganites'. *Physics Today*, 56: 25.

Matsumura, T., T. Nakamura, M. Tetsuka, K. Takashina, K. Tajima and Y. Nishi (2000). *Mat. Res. Soc. Symp. Proc.* 604: 161. Editors for this volume: M. Wun-Fogle, K. Uchino, Y. Ito and R. Gotthardt.

Mazumder, S. (2006). 'The phenomenon of dynamical scaling of structure factor – a few open questions'. *Physica B*, 385–386: 7.

Meitzler, A. H. and H. M. O'Bryan (1973). 'Polymorphism and penferroelectricity in PLZT ceramics'. *Proc. IEEE*, 61: 959.

Meng, Z., U. Kumar and L. E. Cross (1985). 'Electrostriction in lead lanthanum zirconate titanate ceramics'. *J. Amer. Ceram. Soc.* 68(8): 335.

Millis, A. J. (1998). 'Lattice effects in magnetoresistive manganese perovskites'. *Nature*, 392: 147.

Moessner, R. and A. P. Ramirez (February 2006). 'Geometrical frustration'. *Physics Today*, 59(2): 24.

Moreo, A., S. Yunoki and E. Dagotto (26 March 1999). 'Phase separation scenario for manganese oxides and related materials'. *Science*, 283: 2034.

Moumni, Z., A. Van Herpen and P. Riberty (2005). 'Fatigue analysis of shape memory alloys: energy approach'. *Smart Mater. Struct.* 14: S287.

Muller, K. A. (2003). 'Essential heterogeneities in hole-doped cuprate superconductors'. In Bishop, A. R., S. R. Shenoy and S. Sridhar (eds.), *Intrinsic Multiscale Structure and Dynamics in Complex Electronic Oxides*. Singapore: World Scientific.

Mydosh, J. A. (1993). *Spin Glasses: An Experimental Introduction*. London: Taylor and Francis.

Neto, A. H. C. (2001). 'Stripes, vibrations, and superconductivity'. *Phys. Rev. B*, 64: 104509–1.

Noheda, B., D. E. Cox, G. Shirane, J. Gao and Z.-G. Ye (2002). 'Phase diagram of the ferroelectric relaxor $(1 - x)PbMg_{1/3}Nb_{2/3}O_3$-$xPbTiO_3$'. *Phys. Rev. B*, 66: 054104–1.

Noheda, B. and D. E. Cox (2006). 'Bridging phases at the morphotropic boundaries of lead oxide solid solutions'. *Phase Transitions*, 79: 5.

Otsuka, K. and C. M. Wayman (eds.) (1998). *Shape Memory Materials*. Cambridge, U. K.: Cambridge University Press.

Pan, W., Q. Zang, A. Bhalla and L. E. Cross (1989). 'Antiferroelectric to ferroelectric switching in modified lead zirconate titanate stannate ceramics'. *J. Amer. Ceram. Soc.* 72(4): 571.

Pandit, P., S. M. Gupta and V. K. Wadhawan (2004). 'Shape-memory effect in PMN-PT(65/35) ceramic'. *Solid State Commun.* 131: 665.

Pandit, P., S. M. Gupta and V. K. Wadhawan (2006). 'Effect of electric field on the shape-memory effect in $Pb[(Mg_{1/3}Nb_{2/3})_{0.70}Ti_{0.30}]O_3$ ceramic'. *Smart Materials and Structures* (2006), 15: 653.

Park, S.-E. and R. Shrout (1997). 'Ultrahigh strain and piezoelectric behaviour in relaxor based ferroelectric single crystals'. *J. Appl. Phys.* 82(4): 1804.

Pirc, R. and R. Blinc (1999). 'Spherical random-bond—random-field model of relaxor ferroelectrics'. *Phys. Rev. B*, 60: 13 470.

Planes, A., L. Manosa and A. Saxena (eds.) (2005). *Magnetism and Structure in Functional Materials*. Berlin: Springer.

Poole, C. P. and F. J. Owens (2003). *Introduction to Nanotechnology*. New Jersey: Wiley.

Ramakrishnan, T. V., H. R. Krishnamurthy, S. R. Hassan and G. V. Pai (2004). 'Theory of insulator metal transition and colossal magnetoresistance in doped manganites'. *Phys. Rev. Lett.* 92: 157203.

Rasmussen, K. O., T. Lookman, A. Saxena, A. R. Bishop, R. C. Albers and S. R. Shenoy (2001). 'Three-dimensional elastic compatibility and varieties of twins in martensites'. *Phys. Rev. Lett.* 87(5): 055704-1.

Ren, X. and K. Otsuka (2000). 'Universal symmetry property of point defects in crystals'. *Phys. Rev. Lett.* 85(5): 1016.

Ren, X. and K. Otsuka (February 2002). 'The interaction of point defects with the martensitic transformation: A prototype of exotic multiscale phenomena'. *MRS Bulletin*, p. 115.

Renner, Ch., G. Aeppli, B.-G. Kim, Yeong-Ah Soh and S.-W. Cheong (2002). 'Atomic-scale images of charge ordering in a mixed-valence manganite'. *Nature*, 416: 518.

Reyes-Morel, P. E., J. S. Cherng and I.-W. Chen (1988). 'Transformation plasticity of CeO_2-stabilized tetragonal zirconia polycrystals: II, Pseudoelasticity and shape memory effect''. *J. Amer. Soc.* 71(8): 648.

Roytburd, A. L. (1993). 'Elastic domains and polydomain phases in solids'. *Phase Transitions*, 45: 1.

Ruderman, M. A. and C. Kittel (1954). 'Indirect exchange coupling of nuclear magnetic moments by conduction electrons'. *Phys. Rev.* 96: 99.

Saburi, T. (1998). 'Ti-Ni shape memory alloys'. In Otsuka, K. and C. M. Wayman (eds.), *Shape Memory Materials*. Cambridge, U. K.: Cambridge University Press.

Sapriel, J. (1975). 'Domain-wall orientations in ferroelastics'. *Phys. Rev. B*, 12: 5128.

Saxena, A., T. Lookman, A. R. Bishop and S. R. Shenoy (2003). 'Nonlinear elasticity, microstructure and complex materials'. In Bishop, A. R., S. R. Shenoy and S. Sridhar (eds.), *Intrinsic Multiscale Structure and Dynamics in Complex Electronic Oxides*. Singapore: World Scientific.

Sethna, J. P. and K. S. Chow (1985). 'Microscopic theory of glassy disordered crystals'. *Phase Transitions*, 5: 317.

Sharma, R. P., S. B. Ogale, Z. H. Zhang, J. R. Liu, W. K. Chu, B. Veal, A. Paulikas, H. Zheng and T. Venkatesan (13 April 2000). 'Phase transitions in incoherent lattice fluctuations in $YBa_2Cu_3O_{1-\delta}$'. *Nature*, 404: 736.

Shenoy, S. R., T. Lookman and A. Saxena (2005). 'Spin, charge, and lattice coupling in multiferroic materials', in A. Planes, L. Manosa and A. Saxena (eds.), *Magnetism and Structure in Functional Materials*. Berlin: Springer.

Shenoy, S. R., V. Subrahmanyam and A. R. Bishop (1997). 'Quantum paraelectric model for layered superconductors'. *Phys. Rev. Lett.* 79: 4657.

Shenoy, S. R., T. Lookman, A. Saxena and A. R. Bishop (1 Nov. 1999-II). 'Martensitic textures: Multiscale consequences of elastic compatibility'. *Phys. Rev. B*, 60(18): R12 537.

Shenoy, S. R., T. Lookman, A. Saxena and A. R. Bishop (2003). 'Composite textured polarons in complex electronic oxides'. In Bishop, A. R., S. R. Shenoy and S. Sridhar (eds.), *Intrinsic Multiscale Structure and Dynamics in Complex Electronic Oxides*. Singapore: World Scientific.

Singh, G., V. S. Tiwari and V. K. Wadhawan (2001). 'Dielectric relaxation in $(1 - x)Pb(Mg_{1/3}Nb_{2/3}O_3) - xPbZrO_3$ ceramic''. *Solid State Comm.* 118: 407.

Singh, G., V. S. Tiwari, Arun Kumar and V. K. Wadhawan (2003). 'Canonical-glass-like behaviour of polycrystalline relaxor ferroelectric $(1 - x)[PbMg_{1/3}Nb_{2/3}O_3]-x[PbZrO_3]$ – A heat capacity study'. *J. Mater. Res.* 18: 531.

Soderberg, O., Y. Ge, A. Sozinov, S.-P. Hannula and V. K. Lindroos (2005). 'Recent breakthrough development of the magnetic memory effect in Ni-Mn-Ga alloys'. *Smart Mater. Struct.* 14: S223.

Stojkovic, B. P., Z. G. Yu, A. L. Chernyshev, A. R. Bishop, A. H. Castro Neto and N. Gronbech-Jensen (15 August 2000-I). 'Charge ordering and long-range interactions in layered transition metal oxides: A quasiclassical continuum study'. *Phys. Rev. B*, 62: 4353.

Strocchi, F. (2005). *Symmetry Breaking*. Berlin: Springer.

Tiwari, R. and V. K. Wadhawan (1991). 'Shape-memory effect in the Y-Ba-Cu-O ceramic'. *Phase Transitions*, 35: 47.

Uchino, K. (1997). 'High electromechanical coupling piezoelectrics – How high energy conversion rate is possible'. In George, E. P., R. Gotthardt, K. Otsuka, S. Trolier-McKinstry and M. Wun-Fogle (eds.), *Materials for Smart Systems II*. MRS Symposium Proceedings, Vol. 459. Pittsburgh, Pennsylvania: The Materials Research Society.

Uchino. K. (1998). 'Shape memory ceramics'. In Otsuka, K. and C. M. Wayman (eds.), *Shape Memory Materials*. Cambridge, U. K.: Cambridge University Press.

Ullakko, K., J. K. Huang, C. Kantner, R. C. O'Handley and V. V. Kokorin (1996). 'Large magnetic-field-induced strains in Ni_2MnGa single crystals'. *Appl. Phys. Lett.* 69: 1966.

Vassiliev, A. (2002). 'Magnetically driven shape-memory alloys'. *J. Magn. Magn. Mater.* 242–245: 66.

Viehland, D. (2000). 'Symmetry-adaptive ferroelectric mesostates in oriented $Pb(BI_{1/3}BII_{2/3})O_3$-$PbTiO_3$ crystals'. *J. Appl. Phys.* 88: 4794.

Wadhawan, V. K. (1988). 'Epitaxy and disorientations in the ferroelastic superconductor $YBa_2Cu_3O_{7-x}$'. *Phys. Rev. B*, 38: 8936.

Wadhawan, V. K. (1998). 'Towards a rigorous definition of ferroic phase transitions'. *Phase Transitions*, 64: 165.

Wadhawan, V. K. (2000). *Introduction to Ferroic Materials*. Amsterdam: Gordon and Breach.

Wadhawan, V. K., M. C. Kernion, T. Kimura and R. E. Newnham (1981). 'The shape-memory effect in PLZT ceramics'. *Ferroelectrics*, 37: 575.

Wadhawan, V. K., P. Pandit and S. M. Gupta (2005). 'PMN-PT based relaxor ferroelectrics as very smart materials'. *Mater. Sci. and Engg. B*, 120: 199.

Wechsler, M. S., D. S. Lieberman and T. A. Read (1953). *Trans. Metall. Soc. AIME*, 197: 1503.

Westphal, V., W. Kleeman and M. D. Glinchuk (1992). 'Diffuse phase transitions and random-field-induced domain states of the "relaxor" ferroelectric $PbMg_{1/3}Nb_{2/3}O_3$'. *Phys. Rev. Lett.* 68(6): 847.

Yang, P. and D. A. Payne (1992). 'Thermal stability of field-forced and field-assisted antiferroelectric–ferroelectric phase transitions in $Pb(Zr, Sn, Ti)O_3$'. *J. Appl. Phys.* 71(3): 1361.

Yang, P. and D. A. Payne (1996). 'The effect of external field symmetry on the antiferroelectric–ferroelectric symmetry phase transition'. *J. Appl. Phys.* 80(7): 4001.

Yang, P., M. A. Rodriguez, G. R. Burns, M. E. Stavig and R. H. Moore (2004). 'Electric field effect on the rhombohedral–rhombohedral phase transformation in tin modified lead zirconate titanate ceramics'. *J. Appl. Phys.* 95(7): 3626.

Yosida, K. (1957). 'Magnetic properties of Cu-Mn alloys'. *Phys. Rev.* 106: 893.

Zheng, H., J. Wang, S. E. Lofland, Z. Ma, L. Mohaddes-Ardabili, T. Zhao, L. Salamanca-Riba, S. R. Shinde, S. B. Ogale, F. Bai, D. Viehland, Y. Jia, D. G. Schlom, M. Wuttig, A. Roytburd and R. Ramesh (30 January 2004). 'Multiferroic $BaTiO_3$-$CoFe_2O_4$ nanostructures '. *Science*, 303: 661.

4

SOFT MATTER

The term 'soft condensed matter' is used for materials which are neither liquids nor solids, but have several intermediate properties (Hamley 2000). Examples include glues, paints, soaps, polymer melts, colloidal dispersions, and liquid crystals.

Biological materials are, by and large, wet and soft. If smart structures are going to mimic them in a major way, it is important to understand soft matter.

Biological systems are able to do their work at moderate temperatures. They are also able to convert chemical free energy directly to work, thus preventing, to a large extent, losses such as heat. They are efficient and therefore economic users of energy. In view of the doubts expressed about the feasibility of mass production of Drexler's assemblers and replicators for making nanomachines (see Chapter 6), the soft-matter route for the same purpose becomes even more attractive. Biological soft matter is Nature's tried, tested, and stable nanomachinery. It makes sense to build on that, rather than depending only on untried inorganic or atom-by-atom routes to nanoassembly and smart structures.

We introduce the basics of soft matter in this chapter, and then continue the discussion in the following chapter where we focus on self-assembly and self-organization, the hallmarks of much of soft matter.

4.1   The hydrophobic interaction

Condensed matter is held together by interatomic and intermolecular forces. Interactions like ionic, covalent and metallic are 'strong' interactions. Van der Waals forces are weak forces, and become effective only at short distances.

The *hydrogen bond* is also not a strong interaction, but it is very ubiquitous in soft materials. Oxygen and nitrogen are some strongly electronegative elements; therefore in O–H and N–H covalent bonds, they hog a larger share of the two electrons comprising the covalent bond. The hydrogen atom is anyway small, and has only a single electron before the formation of the covalent bond. The net result is that its side which is furthest away from the electronegative atom presents a significant amount of positive unshielded charge, which becomes available for bonding with other electronegative atoms. The result is a hydrogen bond, like O–H . . . O or N–H . . . O, with an energy intermediate between a covalent bond and van der Waals interaction. The energy associated with hydrogen bonds is typically 20 kJ mol^{-1}, compared to $\sim$500 kJ mol^{-1} for covalent bonding. C–H . . . O and C–H . . . N bonds are also quite common in biological structures.

Liquid water forms a three-dimensional network of O–H . . . O hydrogen bonds. If a molecule is introduced into this network, and if this foreign molecule is not able to take part in the hydrogen bonding, it is isolated by the network of water molecules, which regroup and form a network of as many hydrogen bonds as possible *around* this intruder molecule. There is a lowering of the possible number of configurations of water molecules in the vicinity of the foreign molecule, which amounts to a lowering of entropy, and a consequent increase in free energy. If a second foreign molecule is available, it is herded towards the first one, so as to keep the total free energy to the minimum. The end result, namely the aggregation, is like an 'attraction' between the two foreign molecules. It is a manifestation of the *hydrophobic interaction*.

The hydrophobic interaction is a key player (apart from entropy) in the self-assembly of both inanimate and animate systems (Discher 2004).

4.2 Colloids

In a colloidal dispersion or suspension there are particles of sizes less than 10 μm, dispersed in a liquid. Examples include paints, inks, mayonnaise, ice cream, blood, and milk.

Colloidal suspensions of insoluble materials (e.g. nanoparticles) in organic liquids are called *organosols*. If the carrier liquid is water, we have *hydrosols*.

If two or more dispersed particles come together, they would generally tend to stick. Various measures are adopted to *stabilize* the colloid against this *aggregation*. One way is to overcome the attractive force through *charge stabilization*. Another is that of *steric stabilization* by coating the particles with a polymer layer. When two such coated particles come close, there is a local increase in the polymer concentration, with an attendant increase in osmotic pressure. This generates a repulsive force.

A characteristic process associated with colloidal media is *micellization*, which will be discussed in the next chapter when we explore self-organization of soft matter.

4.3 Polymers

A polymer is a very long molecule, made up by covalent bonding among a large number of repeat units (*monomers*). There can be variations, either in that the bonding is not covalent everywhere, or in that not all subunits are identical (Rubinstein and Colby 2003). Polymers have acquired additional importance recently because of the progress made in the technology of micromachining them for applications in MEMS (Varadan, Jiang and Varadan 2001).

Examples of polymers and polymer solutions include plastics such as polystyrene and polyethylene, glues, fibres, resins, proteins, and polysaccharides like starch. Complex composites incorporating polymers include wood, tissue, and glass-reinforced plastics. Even DNA is generally regarded as a polymer, although its monomer subunits, namely the bases A, T, G and C, neither repeat in a specific sequence, nor are sequenced randomly.

A *rubber* is a polymer melt in which there is a random cross-linking of adjacent chains of the polymer, forming a macroscopic network.

A generic property of long chain-like molecules is that two such molecules cannot cross each other. This results in their *entanglement* in the polymer, leading to viscoelasticity, etc.

Homopolymers consist of a single type of repeat unit. By contrast, a *copolymer* has more than one type of repeat unit.

A *random copolymer* has a random arrangement of two or more types of repeat units. This results in a state of frozen or quenched disorder, which makes it unlikely for such a copolymer to crystallize.

Sequenced copolymers are different from random copolymers in that, although the sequence of different subunits is not periodic, it is not completely random also. Biopolymers like DNA and proteins are examples of this. Their very specific sequence of subunits, ordained by Nature (through the processes of evolution), results in particular properties, an example being the property of proteins to *fold* and *self-organize* into very specific three-dimensional configurations.

If the synthesis process for a copolymer is such that the different types of constituent monomers get arranged in blocks, one gets a *block copolymer* (Hamley 1998). If the constituent monomers are chemically very different, there is an interplay of conflicting tendencies: The different blocks tend to phase-separate, but are not able to do so because of the covalent bonding. A variety of self-assembly effects like *microphase separation* can occur (Stupp *et al.* 1997; Shenhar, Norsten and Rotello 2004).

When a polymer is formed into a hard and tough object by suitable heat and pressure treatment, it is called a *plastic*.

Vulcanization of a polymer into rubbery material, with good strength and elongation properties, gives what is called an elastic polymer or *elastomer*.

Polymers may be used in liquid form as sealants, adhesives, etc.; they are then said to be used as *liquid resins*. Polymers may also be used as *fibres* after being elongated into filament-like material.

4.3.1 *Cellular polymers*

Cellular materials, or *foams*, have an interconnected network of solid material, forming a space-filling structure of open or closed cells (Gibson and Ashby 1999). Wood is an example. Other examples can be found in cork, sponge, coral, and bone. The open structure gives them a lower density, and yet retains useful properties like high stiffness. Foams can be made from metals, ceramics, polymers, and composites.

Polymers usually have low densities and stiffness (because of the weak van der Walls interaction between the chains). Cellular polymers, or polymeric foams, can be of very low density. Polystyrene is a familiar example.

Let ρ_0 and Y_0 denote the density and Young's modulus for a solid polymer, and ρ and Y the corresponding values for an isotropic open-cell foam made from it. Y is often found to scale as $Y_0(\rho/\rho_0)^2$. *Anisotropic* foams (used in electromechanical transducers) do not obey this simple scaling law. They are soft in the thickness direction, and much stiffer along the length and the breadth.

4.3.2 *Piezoelectric polymers*

The best known piezoelectric polymer, which is also a ferroelectric, is poly(vinylidene fluoride) (PVDF). However, by now there is a long list of known and investigated piezoelectric polymers (cf. Kepler and Anderson 1992; Nalwa 1995): polyamides ('odd' nylons); cyanopolymers; copolymers of vinylidene fluoride and vinyl fluoride; polyureas; polythioureas; and biopolymers like polypeptides and cyanoethyl cellulose. The list should also include ferroelectric liquid-crystal polymers, and several polymer-electroceramic composites.

A variety of cellular polymers exhibit excellent piezoelectric properties. In fact, many of them are not only piezoelectrics, but also ferroelectrics, or rather *ferroelectrets* (Nalwa 1995; Goel 2003; Bauer, Gerhard-Multhaupt and Sessler 2004); cf. the appendix on electrets and ferroelectrets. Piezoelectric action is achieved in them by charging the voids in their cellular structure. In the fabrication process, they are subjected to a high electric field, which generates inside them many microplasma discharges, resulting in the creation of tiny, fairly stable, electric dipoles.

Cellular polypropylene (PP) is one of the earliest polymers investigated for piezoelectric behaviour. When suitable processing is carried out for optimizing the pore size, etc., one can achieve a d_{33} coefficient as high as 600 pC N^{-1} for it. This should be compared with the values 170 for ordinary PZT, and 20 for the polar polymer PVDF.

Why is this piezoelectric response so high? One reason is the high compressibility of the cellular polymer. The other is the inherently large dipole moment associated with the large (micron-sized) cellular dipoles. For these two reasons, there is not only a large relative change of polarization per unit applied stress, but also a large absolute change. Both factors contribute to the immensely large piezoelectric response in cellular PP. The response is related to the high deformation of the large voids. By contrast, the piezoelectric response in a crystalline material is related to the much smaller ionic displacements in the unit cell.

4.3.3   Conducting polymers

The field of conducting polymers started with the discovery that if polyacetylene (PA) is doped by employing charge-transfer reactions, it exhibits a well-defined transition to the metallic regime. Some other well-investigated conducting polymers are: polyaniline (PANI), polypyrrole (PPy), and polythiophene (PT) (cf. Lu *et al.* 2002).

These are π-conjugated polymers, which is the reason why their conductivity is so high; the conductivity of PA, for example, is comparable to that of copper. This is in contrast to the situation in piezoelectric polymers, which have a nonconjugated backbone.

Conducting polymers have found a variety of applications as batteries, capacitors, actuators, and photovoltaic cells. Let us consider actuation. The mechanism involved is primarily that of reversible transport of ions and solvent molecules between the polymer and the electrolyte. Huge stresses are generated by the resulting dimensional changes. It has been demonstrated by Lu *et al.* (2002) that a proper choice of electrolyte plays a crucial role in determining the device performance. In general, the electrolyte can be aqueous, organic, a gel, or a polymer. The overall system should meet the following requirements: high ionic conductivity ($> 10^{-4}$ $m^2V^{-1}s^{-1}$); large electrochemical window (> 1 V) over which the electrolyte is neither reduced nor oxidized at an electrode; fast ionic mobility during redox events ($> 10^{-14}$ $m^2V^{-1}s^{-1}$); low volatility; and environmental stability. Lu *et al.* (2004) have shown that certain room-temperature *ionic-liquid* electrolytes meet all these requirements for the synthesis, fabrication, and operation of π-conjugated-polymer electrochemical actuators, and also of high-performance electrochromic windows and numeric displays.

4.3.4   Shape-memory polymers

Rubber is the most familiar example of a polymer with a shape memory. It can be stretched by several hundred percent, and yet it bounces back to the shape it had before it was stretched. This is like the superelasticity exhibited by shape-memory alloys (Chapter 3). The three-dimensional polymer network makes it possible for rubber to memorize its original shape.

Polyurethane is another example of a shape-memory polymer (SMP). Below a certain phase-transition temperature, it can be deformed into a desired shape, but on heating above that temperature it recovers its original shape spontaneously. It is as if it has a memory of the shape it has in the high-temperature phase (cf. Ashley 2001). Such dramatic behaviour is attributed to two important structural features of the polymer (apart from the fact that there is a thermally induced phase transition): triggering segments that undergo the thermal transition at the desired working temperature, and cross-links that determine the remembered permanent shape.

Depending on the nature of cross-links, SMPs and certain other polymers can be described as either *thermoplastics* or *thermosets*. The former soften when heated and harden on cooling, and this process can be repeated several times. Thermosetting polymers solidify (become set) on going through the heating–cooling cycle, and cannot be melted again. The general term *curing* is used for the process in which the use of heat, radiation and pressure makes a thermosetting polymer an infusible and insoluble mass of solid.

Irie (1998) has described three categories of mechanisms for SMPs. A relevant parameter in this context is the glass transition temperature, T_g (cf. the appendix on the glass transition). It is the temperature above which the elasticity of a polymer changes drastically and a stiff polymer shows rubberlike behaviour. The material is 'glassy' below T_g, and 'rubbery' above T_g. Rubber is elastic at room temperature. If we stretch it at room temperature and then cool it to a temperature below T_g, it does not recover its shape on release of the applied stress. On heating to T_g, the frozen polymer chain segments become mobile, and at higher temperatures the material bounces back to its original shape of zero strain.

Another temperature which can sometimes have relevance similar to that of T_g is the melting temperature T_m. The polymer softens near T_m, and stiffens on cooling to lower temperatures.

Category 1 SMPs exhibit *thermal-responsive SME*. For demonstrating the SME, they are subjected to an SME cycle, just like the shape-memory alloys described in Chapter 3. The shape to be remembered is first created from the polymer powder or pellets by melt-moulding. Cross-linking agents or radiation are used for stabilizing the shape. A deforming stress is then applied at a temperature above T_g, or at a temperature near T_m. The system is cooled below T_g or T_m to fix the deformed shape. The polymer reverts to the original shape on being heated to the high temperature again. Some relevant polymers in this context are (Irie 1998): polynorbornene; polyisoprene; styrene-butadiene copolymer; polyurethane; and polyethylene. Apart from cross-linking, some other possible factors for the stabilization and memory of shapes are chain entanglement, micro-crystals, and glassy-state formation.

The mechanism for *Category 2* SMPs, which exhibit the *photo-responsive SME*, is akin to superelasticity, except that, instead of mechanical stress, one uses photochemical or electrochemical reactions for switching between the deformed and undeformed states. Some examples of *photochromic reactions*, i.e. reactions by which a molecule changes to another isomer under irradiation, and returns to its original configuration either thermally or photochemically, are: *trans–cis* isomerization; zwitter-ion formation; ionic dissociation; and ring-formation and ring-cleavage. Azobenzene is an example of an azo-dye which can act as the cross-linking agent. When incorporated in poly(ethylacrylate), it is responsible for photostimulated reversible shape changes in the material.

Category 3 (*chemo-responsive* SME, or '*chemoelasticity*') is like Category 2, except that it is chemical reactions which are responsible for the deformed and undeformed states. This mechanism has been known for a long time (Kuhn *et al.* 1950). The shape of ionizable polymeric molecules (e.g. polyacids and polybases) is a sensitive function of their degree of ionization. For example, ionizing the carboxyl groups of a polyacid like polymethacrylic acid generates electrostatic repulsion along the chain, leading to a stretching of the coiled molecule. Even 50% ionization results in a near-full-length stretching of the molecule. And the stretched molecules can be made to contract reversibly by adding suitable acid or alkali in water.

SMPs offer the advantages of light weight, low cost, and easy control of recovery temperature. Photo-responsive SMPs are particularly attractive for applications in smart structures as actuators. Their main drawback is the low recovery stress, which is typically only 1% of that possible for shape-memory alloys.

4.4   Polymer gels

A gel is a three-dimensional network of bonded subunits, which may have a substantial amount of solvent trapped in it.

Polymer gels are composites comprising a polymer and a liquid. The tangled mess of the polymer molecules prevents the liquid from flowing, and the liquid makes the gel soft and resilient, and prevents the polymer network from collapsing into a compact mass.

Let us first see how a gel forms. Consider a solution containing some long-chain molecules like polystyrene. At high dilutions the polymer molecules do not interact much with one another, and we have a *sol*. But as their concentration is increased, their interactions hinder flow, thus increasing the viscosity of the solution.

At still higher concentrations the molecules get tangled like spaghetti. We now have a material with properties intermediate between those of a liquid and a solid; it is *viscoelastic*, with properties of both a viscous liquid and an elastic solid.

If the right conditions are created (e.g. by hydrolysis), the molecules do not just entangle; they actually form chemical bonds, so that we end up with one meg-amolecule filling the entire container, with a certain amount of porosity, i.e. pockets filled with solvent. The cross-linking bonds can be covalent bonds, hydrogen bonds, hydrophobic interaction, etc. The bonding is strongly influenced by the nature of interactions with the solvent, and is sensitive to influences such as temperature, pH value, presence of ions, etc. This is a *gel*.

One speaks of a *sol–gel transition* in this context. It can be viewed as a *connectivity transition* in the language of composite materials (cf. the appendix on composites). The gel has 3–0 connectivity: the polymer phase is connected to itself in all three dimensions, and the solvent phase is isolated to small pores or pockets and is not self-connected in any dimension. By contrast, in the sol, with which we started, the solvent was self-connected in all three dimensions, and it was the polymer molecules which were isolated and thus zero-connected to themselves. The connectivity of the composite was 0–3 before the sol–gel transition, and 3–0 after the transition. *The sol–gel transition is a 0–3 to 3–0 connectivity transition.*

Variation of temperature, as also of the chemical composition and proportion of the two components, can lead to a rich variation of the properties of a gel, with scope for smart-structure applications. Following Tanaka (1981), we consider a specific gel to illustrate some of these properties. The example considered is that of a polymer matrix of cross-linked polyacrylamide.

After forming the polymer gel from monomeric units, a process of hydrolysis was carried out for various periods of time: the gel was immersed in a basic solution of pH value $\sim$12. In the native polymer, every second carbon atom has a $-CONH_2$ side chain. Hydrolysis changes some of these to $-COOH$. About 60 days of immersion are needed for maximum hydrolysis, which changes about one-fourth of the side chains. Lower immersion periods mean less conversion.

In solution, some $-COOH$ groups ionize to yield H^+ and COO^- ions. The H^+ ions enter the interstitial fluid, leaving a net negative charge on the polymer network.

A batch of identical gels was prepared, and each specimen was hydrolysed for a different amount of time. These were then swollen with water. The fully hydrolysed gel was found to swell to 30 times its original volume.

The swollen gels were then put in different mixtures of acetone and water and observed for several days. The behaviour observed was found to depend on the degree of hydrolysis and the concentration of acetone (and on temperature).

Consider first the unhydrolysed gel. It remains swollen if the concentration of acetone is low. For a higher acetone concentration (20%) the gel shrinks a bit. For 60% acetone the shrinking is by a factor of $\sim$10. The shrinking is a monotonous function of acetone concentration in this case.

For a gel that was hydrolysed for two days, the swelling curve (acetone content *vs.* swelling ratio) develops an inflexion point with a brief horizontal slope, and the shrinking is larger. For gels with a higher degree of hydrolysis, the inflexion point broadens into a discontinuity.

Around the discontinuity an infinitesimal increase in acetone concentration causes a large reduction of volume; the gel collapses into a condensed phase and a more rarefied one. For the fully hydrolysed gel, the volume collapse is by a factor of $\sim$350. This is highly nonlinear response, reminiscent of a phase transition.

Further, the whole process is reversible. And its dynamics is slow, as it is controlled by the diffusion of the acetone and water molecules into the gel.

Temperature has an effect similar to that of acetone concentration: for a fixed acetone concentration, the gel expands at high temperatures, and shrinks at low temperatures.

To understand the reasons for this behaviour, one has to look at the forces acting on the gel. There are three of them: the rubber elasticity; the polymer–polymer affinity; and the hydrogen-ion pressure. Together they determine the osmotic pressure of the gel (Tanaka 1981). By convention, the osmotic pressure is positive if it tends to expand the gel, and negative if it makes it shrink.

The rubber elasticity arises from the resistance the individual polymer strands offer to stretching or compression. Its strength depends on how actively the polymer segments are moving, and therefore on temperature. An increase in temperature makes a collapsed gel to expand.

Polymer–polymer affinity is related to what the solvent is doing. If the polymer–solvent interaction is more attractive than the polymer–polymer interaction, the total free energy is reduced if the polymer surrounds itself by the solvent. In the opposite case, the solvent is expelled and the polymer tends to coagulate (i.e. negative osmotic pressure). The latter is the case for the example discussed above. And between acetone and water, acrylamide is less soluble in acetone than in water.

The third contributor to the osmotic pressure of the gel, namely the hydrogen-ion pressure, involves a delicate balance with the negative charges on the polymer network.

There is competition among the three components, leading to phase transitions and critical phenomena (Tanaka 1981). The critical fluctuations in gels are the local variations in the density of the polymer network. The gel exhibits a critical point, at which the gel makes large excursions into density variations.

Gels offer tremendous tuneability for smart-structure applications (Amato 1992; Hoffman 2004). In the example considered here, one could play with other solvents, change pH value, apply electric field, etc. The response time can be speeded up by making the gel small or finely divided. A cylindrical gel 1 μm in diameter will swell or shrink in a few milliseconds.

4.5   Liquid crystals

Liquid crystals are equilibrium phases of soft matter in which the degree of ordering in the arrangement of molecules is intermediate between that existing in a solid crystal and a disordered liquid or glass. Such phases are found in certain organic compounds with highly anisotropic (rod-like or disc-like) molecular shapes, polymers having units with a high degree of rigidity, and polymers or molecular aggregates with rigid rod-like structures in solution.

The least ordered liquid-crystalline phase is the *nematic* phase: there is no positional order, but the molecules tend to align around a direction called the *director*.

Chiral nematics (better known as *cholesterics*) are liquid crystals composed of molecules which lack inversion symmetry or mirror-reflection symmetry. In such systems, there is a slight tendency for neighbouring molecules to align at a small angle to each other, making the director to form a helix in space. The pitch of the helix is often comparable to optical wavelengths, giving rise to a variety of optical effects. The highly nonlinear and large response of liquid crystals to external fields has made them very popular in display devices (Zorn and Wu 2005) and other applications.

Smectics have a higher degree of order than nematics. In them the molecules are arranged in sheets or layers. Within a layer the molecules point roughly along the same direction, but there is no positional order. Two variants of this ordering are the smectic A phase and the smectic C phase.

In a smectic A phase the director is along the normal to the layer, whereas in a smectic C phase there is a small angle between the two. Thus the smectic C phase consists of layers of tilted molecules.

Finally, if the molecules making up a liquid crystal are disc-shaped, rather than rod-shaped, they constitute a *columnar* or *discotic* phase. The molecules stack themselves along long columns, and these columns can pack themselves in a variety of ways, giving that many different liquid-crystalline phases.

Several liquid crystals, particularly the smectic-C type, exhibit ferroelectric properties. Therefore, like other ferroic materials, they are obvious candidates for possible applications in smart structures. Optical switches based on them are an example (Lines and Glass 1977).

4.6 Ferrofluids

Ferrofluids (ferromagnetic fluids), or *magnetofluids*, are colloidal suspensions of ~ 10 nm-sized particles of a ferromagnetic material (e.g. magnetite, Fe_3O_4) in a liquid like kerosene or transformer oil. To prevent the nanoparticles from segregating, a surface-active agent (*surfactant*) is used, which forms a coating over them (Berkovsky 1982).

The particle size is so small that the material cannot introduce domain walls for splitting into domains. Each particle is therefore a single domain. This is an example of *superparamagnetism*. In the absence of an external magnetic field, there is no net magnetization of the fluid as thermal fluctuations are able to flip the individual magnetic moments around, randomly. When a magnetic field is present, the single-domain magnets align preferentially along the direction of the field. The degree of alignment (as manifested in the total magnetization of the ferrofluid) increases with the external field, and the phenomenon is reversible and practically hysteresis-free.

The magnetic-field-induced alignment in the soft magnetic material brings in anisotropy of optical and other properties, just like the effect of electric field on liquid crystals. This induced and reversible birefringence of ferrofluids can be used for making optical switches.

High enough magnetic fields can make the magnetic particles align like strings or columns, which pack up into a hexagonal lattice, rather like vortex lattices in Type II superconductors. This can serve as a magnetic-field-tuneable diffraction grating.

The field-induced alignment of the magnetic nanoparticles also increases the viscosity of the colloidal suspension, and this fact is exploited in vacuum seals for high-speed high-vacuum motorized spindles. Such seals are used in the computer industry, gas lasers, motors, blowers, clean-room robotics, etc. There are several other applications also, commercial as well as in Nature (Poole and Owens 2003).

4.7 Chapter highlights

- Biological matter is soft matter. Therefore, artificial smart structures based on soft matter may enable us to mimic Nature in the most efficient and versatile manner.
- The hydrophobic interaction is ubiquitous in the mesoscopic structure of soft matter. It is a rather weak interaction, and thus even moderate changes of pH or concentration, etc. can invoke large changes in the structure and dynamics of soft matter.

- Biological systems are able to do their work at moderate temperatures. They are also able to convert chemical free energy directly to work, thus preventing, to a large extent, losses such as heat. They are efficient users of energy. Therefore the soft-matter route to making nanomachines and other smart structures is very attractive. Biological soft matter is Nature's tried, tested, and stable nanomachinery. It makes sense to build on that, rather than depending only on untried inorganic or atom-by-atom routes to nanoassembly and smart structures.

- Polymers have acquired additional importance recently because of the progress made in the technology of micromachining them for applications in MEMS.

- Cellular polymers offer some unique advantages for applications in smart structures as transducers: Large piezoelectric response; flexibility of design for shape and size; low cost; low density; good matching of acoustic impedance with aqueous systems; and low or zero toxicity.

- Conducting polymers find a variety of applications as batteries, capacitors, actuators, and photovoltaic cells. Their use as actuators is particularly important because the involved mechanism of reversible transport of ions and solvent molecules between the polymer and the electrolyte can generate very large stresses.

- Shape-memory polymers (especially the photoresponsive ones) are particularly attractive for use as actuators in smart structures, as they offer the advantages of light weight, low cost, and easy control of recovery temperature.

- The sol–gel transition can be defined succinctly in the language of connectivity of composites: It is a 0–3 to 3–0 connectivity transition.

- Gels offer tremendous tuneability for smart-structure applications.

- Although the most common application of liquid crystals is in display devices, ferroelectric liquid crystals are versatile stuff for possible use in smart structures.

- Ferrofluids already find uses in clean-room robotics. Many more applications are envisaged.

- Smart structures of the future will certainly involve a crucial role for soft matter. The main weakness of soft matter for device applications is that it is not usable in harsh conditions like high temperatures. Although life as we know it is based on soft matter, other kinds of artificial life are conceivable which will be based on silicon, steel, and other inorganic matter.

- The ultimate smart structures of the distant future are likely to be an optimum blend of soft and 'hard' matter.

References

Amato, I. (17 March 1992). 'Animating the material world'. *Science* 255: 284.

Ashley, S. (May 2001). 'Shape-shifters'. *Scientific American,* 284(5): 15.

Bauer, S., R. Gerhard-Multhaupt and G. M. Sessler (February 2004). 'Ferroelectrets: Soft electroactive foams for transducers'. *Physics Today*, p. 37.

Berkovsky, B. (ed.) (1982). *Thermomechanics of Magnetic Fluids*. Washington: Hemisphere Pub. Co.

Discher, D. E. (2004). 'Biomimetic nanostructures'. In Di Ventra, M., S. Evoy and J. R. Heflin (eds.), *Introduction to Nanoscale Science and Technology*. Dordrecht: Kluwer.

Gibson, L. J. and M. F. Ashby (1999). *Cellular Solids: Structure and Properties*. New York: Cambridge University Press.

Goel, M. (25 August 2003). 'Electret sensors, filters and MEMS devices: New challenges in materials research'. *Current Science*, 85(4): 443.

Hamley, I. W. (1998). *The Physics of Block Copolymers*. Oxford: Oxford University Press.

Hamley, I. W. (2000). *Introduction to Soft Matter*. Chichester: Wiley.

Hoffman, A. S. (2004). 'Applications of "smart polymers" as biomaterials'. In B. D. Ratner, A. S. Hoffman, F. J. Schoen and J. E. Lemons (eds.), *Biomaterials Science: An Introduction to Materials in Medicine*, 2nd edn. Amsterdam: Elsevier.

Irie, M. (1998). 'Shape memory polymers'. In Otsuka, K. and C. M. Wayman (eds.) (1998), *Shape Memory Materials*. Cambridge: Cambridge University Press.

Kepler, R. G. and R. A. Anderson (1992). 'Ferroelectric polymers'. *Adv. Phys.* 41: 2.

Kuhn, W., B. Hargitay, A. Katchalsky and H. Eisenberg (1950). 'Reversible dilation and contraction by changing the state of ionization of high-polymer acid networks'. *Nature*, 165: 514.

Lines, M. E. and A. M. Glass (1977). *Principles and Applications of Ferroelectrics and Related Materials*. Oxford: Clarendon Press.

Lu, W., A. G. Fadeev, B. Qi, E. Smela, B. R. Mattes, J. Ding, G. M. Spinks, J. Mazurkiewicz, D. Zhou, G. G. Wallace, D. R. MacFarlane, S. A. Forsyth and M. Forsyth (9 August 2002). 'Use of ionic liquids for π-conjugated polymer electrochemical devices'. *Science*, 297: 983.

Nalwa, H. S. (ed.) (1995). *Ferroelectric Polymers: Chemistry, Physics, and Applications*. New York: Marcel Dekker.

Poole, C. P. and F. J. Owens (2003). *Introduction to Nanotechnology*. New Jersey: Wiley.

Rubinstein, M. and R. H. Colby (2003). *Polymer Physics*. Oxford: Oxford University Press.

Shenhar, R., T. B. Norsten and V. M. Rotello (2004). 'Self-assembly and self-organization'. In Di Ventra, M., S. Evoy and J. R. Heflin (eds.), *Introduction to Nanoscale Science and Technology*. Dordrecht: Kluwer.

Stupp, S. I., V. LeBonheur, K. Walker, L. S. Li, K. E. Huggins, M. Keser and A. Amstutz (1997). 'Supramolecular materials: self-organized nanostructures'. *Science*, 276: 384.

Tanaka, T. (1981). 'Gels'. *Scientific American,* 244(1): 110.

Varadan, V. K., X. Jiang and V. V. Varadan (2001). *Microstreolithogrphy and Other Fabrication Techniques for 3D MEMS*. New York: Wiley.

Zorn, R. and S.-T. Wu (2005). 'Liquid crystal displays'. In Waser, R. (ed.), *Nanoelectronics and Information Technology: Advanced Electronic Materials and Novel Devices*, 2nd edition, p. 887. KGaA: Wiley-VCH Verlag

5

SELF-ASSEMBLY AND SELF-ORGANIZATION OF MATTER

As the wind of time blows into the sails of space,
the unfolding of the universe nurtures the evolution of matter
under the pressure of information.
From divided to condensed
and on to organized, living, and thinking matter,
the path is toward an increase in complexity through self-organization.
– Jean-Marie Lehn (2002)

5.1 From elementary particles to thinking organisms

According to the second law of thermodynamics, phenomena occur because their occurrence results in a lowering of the global free energy. In the beginning, i.e. immediately after the big bang, there were elementary particles and the interactions among them. Their very nature implied the presence of certain built-in and characteristic *information*, which made them spontaneously evolve into condensed matter of a certain specific kind, and then, in due course, some of this condensed matter further self-organized into complex, living and thinking entities. How did all this happen? How did complexity arise in matter as we know it? What other forms of complex matter are also possible? In supramolecular chemistry, we attempt to answer such questions (Lehn 1988). The understanding we gain can help us design smart structures modelled on the present living entities. What is more, discoveries of other possible types of complexity may enable the designing and evolution of new types of smart structures, not seen in Nature at present.

5.2 Supramolecular chemistry

Supramolecular means 'beyond the molecular'. Modern chemistry is defined as the science of structure and transformation of matter. Molecules as entities have a predominantly covalently bonded structure. Supramolecular aggregates, usually formed under near-ambient conditions, involve mainly noncovalent or 'secondary' interactions (van der Waals and Coulomb interactions; hydrogen bonding; hydrophobic interaction; etc.). This is true for crystal growth also. Crystals of soft matter can be regarded as supramolecular aggregates (Desiraju 1995). Only high-temperature crystal growth (e.g. of Si crystals from the melt), or growth conditions conducive to high chemical reactivity, can give covalently bonded crystals.

The aim of supramolecular chemistry is to understand and develop complex chemical systems from components interacting via noncovalent forces (Atwood *et al.* 1996; Cragg 2005). There is an enormous difference in the information content of merely condensed matter on the one hand, and highly organized supramolecular or complex matter on the other. This information is contained in the *shapes* of the component molecules, and in the well-defined *interaction patterns* among them. The build-up of information content, as the degree of supramolecular complexity increases, involves a succession of stages (Lehn 1995, 1999): molecular recognition; self-assembly; self-organization; and adaptation and evolution.

In this chapter we shall discuss each of these, as also some other aspects of complexity. Information storage and information processing is involved at every stage. Supramolecular diversity of structures is achieved by reversible noncovalent self-assembly of basic components, the various potential combinations in the number and nature of which represent a *virtual combinatorial library* (VCL) (Huc and Lehn 1997).

5.3 Molecular recognition

Molecules must have the requisite structure, shape, and bonding proclivities if they are to self-assemble into specific functional entities. This requires preorganization of molecular design, with specific receptor sites *which can be mutually recognized*, so that the requisite spontaneous evolution towards complexity can take place.

Selective molecular recognition in the presence of a plethora of other molecules is the essence of much of molecular biology (cf. Section 2.7).

Since the intermolecular interactions involved in supramolecular assembly are noncovalent and weak, the bonds can get readily broken and re-formed, in a time-reversible manner, till the most stable supramolecular configuration has been achieved. *Reversibility of bonding is the hallmark of self-assembly through molecular recognition.*

5.4   Self-assembly

Molecular recognition, and related processes like supramolecular reactivity, catalysis, and transport, result in self-assembly. Molecular self-assembly may be defined as a spontaneous aggregation of component molecules into a confined, ordered entity, which can exist in a stable state (Hosokawa, Shimoyama and Miura 1995). It is a combinatorial process, with a search sequence determined by the kinetic and thermodynamic parameters governed by the shapes and interactions of the building blocks.

Self-assembly is a much more ubiquitous phenomenon than just molecular self-assembly. In general terms, it is defined as the 'autonomous organization of components into patterns or structures without human intervention' (Whitesides and Grzybowski 2002). Many soft materials are able to self-assemble into a variety of morphologies, and at a variety of length-scales. They usually contain some amount of water, and the hydrophobic interaction is the most important mechanism for the self-assembly. Other examples of self-assembly include crystals, liquid crystals, bacterial colonies, beehives, ant colonies, schools of fish, weather patterns, and galaxies.

There are two mains types of self-assembly: static, and dynamic. Static self-assembly occurs in systems which are at equilibrium (global or local), and which do not dissipate energy. Examples include crystals, as also most kinds of folded globular proteins.

The hallmark of dynamic self-assembly is *dissipation of energy*. Examples include oscillating and reaction–diffusion reactions, weather patterns, and galaxies. The interactions responsible for pattern formation in these systems occur only if there is energy dissipation. It is still not well understood as to how dissipation of energy leads to the emergence of ordered structures from disordered components (however, cf. appendix on chaos).

Whitesides and Grzybowski (2002) have taken note of two more types of self-assembly: templated self-assembly, and biological self-assembly. The former involves the presence of regular features (templates) in the environment. Crystallization (e.g. epitaxial growth) on a substrate is an example.

Biological self-assembly is a class by itself because of the complexity and diversity of the resultant structures and functions. The myriad intra- and intercellular processes are the very essence of life. They involve dynamic self-assembly, requiring the flow of energy through the cell. Stop this flow and the cell dies.

Returning to molecular self-assembly, *amphiphiles* are molecules with a hydrophilic end and a hydrophobic end. The latter is usually a saturated hydrocarbon chain, which cannot take part in hydrogen bonding, and is therefore shunned by water. The hydrophilic end is usually ionic, and can readily form hydrogen bonds.

Phospholipids, which are major components of biological membranes, are important examples of amphiphiles with *two* hydrocarbon chains (instead of one) attached to the hydrophilic end.

When amphiphilic or surfactant molecules are put in water, they aggregate into *micelles* so as to optimally meet the requirements of the hydrophobic interaction.

For example, they may form spherical aggregates, with only the hydrophilic heads on the surface, and all the hydrophobic tails tucked inside so as to avoid contact with water.

Micellization is a process of *concentration-induced phase transition* at nanoscales. It occurs when the surfactant concentration exceeds a certain value called CMC (critical micelle concentration) (Stupp *et al.* 1997).

Apart from spherical micelles, a variety of other shapes are also possible (cylinders, bilayers, vesicles, etc.), depending on the concentration of the solute and presence of other chemical species in the solution.

These aggregates can further self-assemble (undergo phase transitions) into secondary configurations, at critical concentrations of the primary component micelles.

There are several reasons why self-assembly is an important subject of investigation (Whitesides and Boncheva 2002). Life depends on self-assembly. A large number of complex structures in the biological cell form by self-assembly. DNA is one of the most programmable assemblers, not only in Nature, but also for chemical synthesis in the laboratory (Storhoff and Mirkin 1999; Seeman 2004; Baker *et al.* 2006). As mentioned above, growth of molecular crystals is really a process of self-assembly. Some other systems with a regular structure, which form by self-assembly, are liquid crystals, as also semicrystalline and phase-separated polymers. Self-assembly is not limited to molecules (Whitesides and Grzybowski 2002). Even galaxies can self-assemble.

Structures at mesoscopic length-scales (colloids, nanowires, nanospheres, quantum dots) are being created by self-assembly (Sun *et al.* 2000; Klimov *et al.* 2001; Whitesides and Love 2001). Structures built at still larger length-scales include those used in microelectronics (Sirringhaus *et al.* 2000), and photonics (Jenekhe and Chen 1999).

Whitesides and Boncheva (2002) have listed five factors responsible for a successful self-assembly of matter.

- *Components.* The components should have the pre-organized structure and interaction capability conducive to the desired self-assembly into a more ordered state (e.g. a crystal or a folded macromolecule). Complementarity in the shapes of the different interacting molecules, or in different parts of the same molecular species, is also important.
- *Interactions.* The interactions conducive to self-assembly are generally weak and noncovalent (the so-called secondary interactions). Relatively weak covalent or coordination bonding may also be operational sometimes.
- *Reversibility (or adjustability).* Weak interactions, with energies comparable to thermal energies ($\sim$20 kJ mol^{-1} for hydrogen bonding, compared to $\sim$2.5 kJ mol^{-1} for thermal fluctuations), ensure that the bonds can be made and unmade reversibly, until the lowest-energy ordered configuration has been reached. Growth of molecular crystals is an example of this. Reversibility implies that the growing (self-assembling) system is close to equilibrium at all times. Reversibility or easy adjustability is also the crux of smart behaviour.

- *Environment.* The interacting molecules have to *move* to their final positions, usually through a solution and across an interface. In crystal growth, for example, the height of the activation barrier, as also the mobility of the molecules, depends sensitively on the nature of the environment in which the building blocks assemble into a large crystal.

- *Mass transport and agitation.* For the self-assembly of molecules, Brownian motion is often enough to ensure that the molecules are able to reach their final sites. As the size of the building blocks increases, additional problems like low mobility, friction, and gravity have to be tackled.

A variety of approaches have been adopted for designing and creating self-assembled structures in the laboratory. Analogies have been drawn from the self-assembly of molecular systems in Nature to create millimetre-sized structures artificially. We consider a few of them here.

5.4.1 *Self-assembly controlled by capillary interactions*

The components to be self-assembled either float at a fluid–fluid interface (Bowden *et al.* 1997), or are suspended in a nearly isodense liquid (Jackman *et al.* 1998). Both two-dimensional and three-dimensional macroscopic structures have been created. Capillary action provides the necessary motility to the millimetre-sized components, the shapes of which are designed to mimic crystallization of molecular crystals. For example, polyhedral components, with surfaces patterned with solder dots, wires, and LEDs, self-assembled into three-dimensional electrical networks (Gracias *et al.* 2000; also see Breen *et al.* 1999).

5.4.2 *Fluidic self-assembly*

The technique of fluidic self-assembly was developed by Yeh and Smith (1994) for the quasi-monolithic integration of GaAs LEDs on Si substrates. GaAs LEDs, having the shape of trapezoidal blocks, are suspended in a carrier fluid, which flows over a *templating surface*, namely the Si substrate. The substrate has a series of pits etched in it, of a shape complementary to that of the flowing polyhedral components. When a component falls into a pit, it gets stuck there if it happens to have the correct orientation, and thus is able to resist getting dislodged by the force of the flow. Otherwise, it gets removed from the pit, washed away by the current. This is exactly how crystals grow, albeit molecule by molecule.

5.4.3 *Templated self-assembly*

The fluidic self-assembly case considered above is also an example of templated self-assembly. The etch-pit pattern in the Si substrate acts as a template for effecting the desired self-assembly. An advantage of such an approach is that it limits the occurrence of defects by superimposing geometrical constraints (Velev, Furusawa and Nagayama 1996). Choi *et al.* (2000) have demonstrated the use of templated self-assembly based on capillary interactions for creating an ordered array of millimetre-sized poly-(dimethylsiloxane) (PDMS) plates suspended at a water–perfluorodecalin interface. The hydrophobic interaction is mainly responsible for the self-assembly in this case. The use of templates forces the system towards a particular kind of pattern formation.

5.4.4   *Constrained self-assembly*

Self-assembly of identical objects leads to a high degree of symmetry in the aggregate. There is at least the translational symmetry, akin to a crystal. Memory or display devices come under this category, in which there is a repetition of a memory unit or a display unit. In contrast to this, other microelectronic devices, which comprise of distinct blocks, are *asymmetric*. Self-assembly of the latter type of devices requires special measures. Once again, it has been instructive to see how Nature does it. Protein folding provides an example worth imitating. How a chain of amino acids carries the requisite information that makes it self-fold into a very specific globular protein is still a mystery. Drexler (1992) pointed out that *de novo* protein-folding *design* should be easier than protein-folding *prediction*. In the same spirit, Boncheva *et al.* (2002) have described the creation of a linear chain of millimetre-sized polyhedra, patterned with logic devices, wires, and solder dots, connected by a flexible wire. The *de novo* design of the shapes and bonding tendencies of the polyhedra was such that the linear chain self-folded into two domains: a ring oscillator and a shift register.

5.5   Self-organization

Self-organization is the 'spontaneous but information-directed generation of organized functional structures in equilibrium conditions' (Lehn 2002).

5.5.1   *Chemical coding for self-organization*

The information for self-organization is programmed into the molecular-recognition and self-assembly tendencies of the component molecules (e.g. through the nature of hydrogen bonding, or through the nature of coordination architecture of metal ions), and also into how the self-assembled edifice organizes itself into a functional structure in equilibrium. There also has to be chemical coding for telling the system when to *stop* the process of assembly and organization (Yates 1987).

The chemical algorithm, depending on its nature, may give rise to either robust and strong self-organization, or fragile and unstable self-organization.

- *Robust coding.* Robust self-assembly and self-organization behaviour means insensitivity to perturbations like the presence of foreign particles, or variations in concentrations, or other interfering interactions. This imparts stability to the overall process of assembly and organization. It also makes possible phenomena like 'self-selection with self-recognition', necessary for the correct pairing in strands of DNA, etc. (Kramer *et al.* 1993).

- *Fragile coding.* An inherently unstable process of organization can occur in a certain range of controlling parameters. While such situations may be undesirable for the commercial success of some fabrication ventures, they also bear the germ of innovation through branching, adaptation and evolution in Nature. The poor stability of the system implies strongly nonlinear behaviour: Even minor changes in the environment may lead to phenomena like *bifurcation*, resulting in the organization and evolution of totally new structures (cf. the appendix on nonlinear dynamical systems).

- *Coding involving multiplicity plus information.* Chemical coding incorporating both robust and fragile instructions can result in fascinating organization and adaptation possibilities (Kramer *et al.* 1993). Biological processes, including evolution, are examples of such algorithms. As in the biological cell, one aims at creating in the laboratory mixtures of 'instructed components', enabling the multiple assembly and organization of the desired interacting and interdependent supramolecular species. The chemical algorithms involved are species-specific, as well as interactional and adaptive.

- *Coding for multiple self-organization.* Self-organization can occur either through a single chemical code, or through multiple codes (subroutines) operating under the same umbrella code (Lehn 1999, 2002). The latter arrangement offers scope for chemical adaptation and evolution. There can, for example, be a differential expression of the hydrogen-bonding information coded in a molecule, defined by the recognition code of the bonding molecules, resulting in different aggregation and organization scenarios.

  The coexistence of multiple chemical codes in a single algorithm offers potentially powerful possibilities for computation, based on molecular self-assembly and organization (Adleman 1994, 1998; Chen and Wood 2000; Benenson *et al.* 2001).

- *Self-organization by selection.* What we have described so far in this section is self-organization by *design*. In this knowledge-based approach, one starts with a library of preorganized or designed molecules, and the rest follows as the information content in the designed component molecules, with their attendant intermolecular interactions, unfolds itself. It is also possible to have supramolecular self-organization by *selection*, so that, in essence, the dynamic edifice *designs itself*. This is reminiscent of natural selection in Darwinian evolution.

  The two approaches are not mutually exclusive, and can be combined. One starts with molecules with a designed information and instruction content. The molecules and the interactions among them, however, are not completely unchangeable. This provides adequate scope for the evolution of self-designed supramolecular structures, through the processes of function-driven selection of the fittest molecules and pathways (Lehn and Eliseev 2001; Oh, Jeong and Moore 2001).

5.5.2 *Self-organized criticality*

Bak, Tang and Wiesenfeld (1987, 1988) introduced and analysed the notion of self-organized criticality (SOC), an important fundamental notion, applicable to a large number of branches of science and other disciplines. As subsequent work has shown (see Bak 1996, and references therein), a surprisingly large number of phenomena, processes and behaviours can be understood in terms of this very basic property of complex systems.

Imagine a horizontal tabletop, on which grains of dry sand are drizzling vertically downwards (very slowly, one grain at a time). In the beginning, the sand pile just remains flat and grows thicker with time, and the sand grains remain close to where they land. A stage comes when, even as fresh sand keeps drizzling in, the old sand starts cascading down the sides of the table. The pile becomes steeper with time, and there are more and more sand-slides. At a certain stage the total weight of the sand, as well as the average slope, become constant with time; this is described as a *stationary state*.

With time the sand-slides (*avalanches* or *catastrophes*; Gilmore 1981) become bigger and bigger. Eventually, some of the sand-slides may span all or most of the pile. The system is now far removed from equilibrium, and its behaviour has become *collective*, in the sense that it can no longer be understood in terms of the dynamics of individual grains. Falling of just one more grain on the pile *may* cause a huge avalanche (or it may not). The sand pile is then said to have reached a *self-organized critical state*, which can be reached from the other side also by starting with a very big pile: the sides would just collapse until all the excess sand has fallen off.

The state is *self-organized* because it is a property of the system and no outside influence has been brought in. It is *critical* because the grains are critically poised: The edges and surfaces of the grains are interlocked in a very intricate pattern, and are just on the verge of giving way. Even the smallest perturbation can lead to a chain reaction (avalanche), which has no relationship to the smallness of the perturbation; the response is unpredictable, except in a statistical-average sense. The period between two avalanches is a period of tranquillity (*stasis*) or equilibrium; *punctuated equilibrium* is a more appropriate description of the entire sequence of events.

Big avalanches are rare, and small ones frequent. And all sizes are possible. There is a *power law behaviour*: the average frequency of occurrence, $N(s)$, of any particular size, s, of an avalanche is inversely proportional to some power τ of its size:

$$N(s) = s^{-\tau} \tag{5.1}$$

Taking the logarithm of both sides of this equation gives

$$\log N(s) = -\tau \log s \tag{5.2}$$

So, a log–log plot of the power-law equation gives a straight line, with a negative slope determined by the value of the exponent τ. The system is *scale-invariant*: the same straight line holds for all values of s (however, see Held *et al*. 1990). Large catastrophic events (corresponding to large values of s) are consequences of the same dynamics which causes small events.

This law is an experimental observation, and it spans a very wide variety of phenomena in Nature, each exhibiting a characteristic value of τ. Because of the complexity and unpredictability of the phenomena involved, it has not been possible to build realistic theoretical models which can predict the value of the critical exponent τ for a system. It has been possible, however, to calculate the number of possible sandpile configurations that may exist in the critical state (Dhar 1990).

This is complex behaviour, and according to Bak (1996), SOC is so far the only known general mechanism to generate complexity. Bak defined a complex system as that which consists of a large number of components, has large variability, and the variability exists on a wide range of length-scales. Large avalanches, not gradual change, can lead to qualitative changes of behaviour, and may form the basis for emergent phenomena and complexity.

In his book *How Nature Works*, Bak (1996) explains that Nature operates at the SOC state. As in the sandpile experiment described above, large systems may start from either end, but they tend to approach the SOC state. Even biological evolution is an SOC phenomenon (Bak and Sneppen 1993; Bak, Flyvbjerg and Sneppen 1994).

How do systems reach the SOC state, and then tend to stay there? The sandpile experiment provides an answer. Just like the constant input drizzle of sand in that system, a steady input of energy, or water, or electrons, can drive systems towards criticality, and they self-organize into criticality by *repeated spontaneous pullbacks from supercriticality*, so that they are always poised at or near the edge between chaos and order. The openness of the system (to the inputs) is a crucial factor. A closed system (and there are not many around) need not have this tendency to move towards an SOC state.

The intensity and frequency of occurrence of earthquakes provides an example of SOC, wherein the system (the earth) started with a supercritical state and has climbed down towards SOC by shedding the excess energy of its fault zones in the initial stages of its history.

Complex systems are neither fully ordered like crystals, nor completely disordered like gases. There is a window of partial order in which complexity may manifest itself.

Crystals which undergo continuous or second-order phase transitions exhibit critical fluctuations in the vicinity of the critical temperature T_c. This complex behaviour in a closed system at equilibrium occurs only at a particular temperature (the critical temperature), and not at other temperatures. This is therefore not a very typical example of SOC or complexity. Nature is operated by a '*blind watchmaker*' (Dawkins 1986) who is unable to do continuous fine-tuning.

Chaos is another such example. According to Bak (1996), chaos theory cannot explain complexity. Chaos theory explains how simple, deterministic systems can sometimes exhibit unpredictable behaviour, but complex-looking behaviour occurs in such systems only for a specific value of the control parameter; i.e. the complexity is not *robust*. There is no general power-law behaviour, which is a signature of complex critical systems.

A third example of criticality which is not self-organized, and therefore not robust, is that of a nuclear-fission reactor. The reactor is kept critical (neither subcritical nor supercritical) by the operator, and cannot, on its own, tend towards (and be stable around) a state of self-organized criticality.

The SOC idea, along with its all-important scale-invariance feature, has been applied successfully to a large number of diverse problems (Peng *et al.* 1993; Bak 1996; Stanley *et al.* 1998; Boettcher and Percus 2000):

1. The sandpile model, and related models.
2. Real sandpiles and landscape formation.
3. The origin of fractal coastlines.
4. Sediment deposition.
5. The $1/f$ noise problem.
6. The intensity and frequency of earthquakes.
7. Starquakes and pulsar glitches.
8. Black holes.
9. Solar flares.
10. The emergence and evolution of life on earth.
11. Mass extinctions and punctuated equilibria.
12. The functioning of the brain.
13. Models of the economy.
14. Traffic jams.
15. The question of coding and noncoding DNA.
16. Inflation of a degassed lung.
17. The sequence of interbeat intervals in the beating of a heart.
18. The foraging behaviour of the wandering albatross.
19. Urban growth patterns.

5.6   Supramolecular assemblies as smart structures

When a giant vesicle, which happens to have a smaller vesicle inside it, is exposed to octyl glucoside, the smaller vesicle can pass through the outer membrane into the external medium ('birthing'). The resulting injury to the membrane of the host vesicle heals immediately. Addition of cholic acid, on the other hand, induces a feeding frenzy in which a vesicle grows rapidly as it consumes its smaller neighbours. After the food is gone, the giant vesicle then self-destructs (a case of 'birth, growth, and death'). Such lifelike morphological changes were obtained by using commercially available chemicals; thus these processes should be assigned to organic chemistry, and not to biology or even biochemistry.

– Menger and Gabrielson (1995)

(a)

(b)

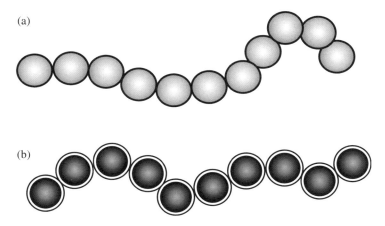

Fig. 5.1   A covalently bonded polymer (a), and a supramolecular polymer (b). In the monomer constituting a supramolecular polymer, there is usually an outer shell, such that the interaction between two contiguous monomers is directional and *reversible*. Such polymers can therefore respond to external stimuli in a manifestly adaptive way reminiscent of biological systems.

Vesicles are spherical supramolecular assemblies separating an aqueous interior volume from the external solvent by means of lipid bilayers (Fig. 5.1). They are also called *liposomes*, and are quite similar to micelles. Given the right conditions, lipids can self-assemble into giant vesicles the size of biological cells. The basic driving force for their self-assembly is the hydrophobic interaction. As vividly described above by Menger and Gabrielson (1995), vesicles can mimic the living cell in many ways, and the underlying processes form the subject of *cytomimetic chemistry*.

In Chapter 1, we defined very smart structures as those which possess many properties normally attributed to living entities. Giant vesicles are very smart structures indeed. They can exhibit features such as birth, sensing, actuation, feeding, growth, healing, and death.

The essence of such smart behaviour in supramolecular structures is the coming together of components having reactive functional groups (Lehn 1995, 1999, 2002). The weak (noncovalent or secondary) and reversible nature of the intermolecular bonding in them gives them the all-important *dynamic* character and lability: Their structure can readily undergo assembly/disassembly processes in response to a changing environment, and this *adaptive behaviour* is the very essence of smartness.

Supramolecular chemistry offers several examples of smart materials/structures. Like giant vesicles, another case in point is liquid-crystalline *supramolecular polymers*. These are defined as monomeric units held together by specific directional secondary interactions (Brunsveld *et al.* 2001). By contrast, a normal polymer has an unchanging covalently bonded backbone (Fig. 5.1). The information contained in the complementary and labile molecular subunits makes them recognize one another, and (reversibly) self-assemble into supramolecular polymers. Such polymers exhibit biomimetic behaviour. By processes such as growing or shortening, rearranging or exchanging components, they show sophisticated abilities like healing, growth, adaptability, and sensory and actuator action.

5.7   Chemical adaptation and evolution

Biological evolution was preceded by chemical evolution (cf. Section 2.2 on biological evolution). The 'molecules of life' like DNA evolved from lower-level molecules and aggregates of molecules. It must have been a case of self-assembly and self-organization, gradually bringing in elements of chemical adaptation and evolution. As discussed above, supramolecular assemblies can react to a changing environment in a dynamic and time-reversible way. This means that they can *adapt* to the environment. Adaptability is the crux of evolution, whether chemical or biological; it leads to the emergence of new features and complexity (Yates 1987). 'Adaptation becomes evolution when acquired features are conserved and passed on' (Lehn 2002).

In the sequence of events called evolution, two broad features stand out:

1. The unidirectional flow of time brings in irreversibility at the global level (even when there is scope for reversibility and self-correction at a local, molecular level). The macroscopic irreversibility implies that such systems are *dissipative*, and therefore evolve towards an *attractor* in phase space (cf. the appendix on nonlinear dynamical systems).
2. As the level of complexity increases, the ability to *remember* and *learn* from the processed information evolves.

5.8   Systems at the edge of order and chaos

Information processing is at the heart of chemical and biological evolution. *Automata* are devices that convert information from one form to another according to a definite procedure.

Cellular automata (CA) represent discrete dynamical systems, the evolution of which is dictated by *local* rules. They are usually realized on a lattice of cells (in a computer program), with a finite number of discrete states associated with each cell, and with a local rule specifying how the state of each cell is to be updated in discrete time steps.

CA have several features of nonlinear dynamical physical systems. Wolfram (1984, 1985, 1986, 2002) argued that, broadly speaking, all CA rules fall into just four universality classes:

Class I has rules such that, no matter what pattern of living or dead cells one starts with, the system reaches a single equilibrium state ('dies') in just one or two time steps.

Class II is similar to Class I, except that there are more than one (but just a few) stable configurations towards which the system moves quickly.

Class III is just the opposite of Class I and Class II, in that it goes to the opposite extreme: Such automata are too lively. There is no stable configuration, and no predictability; a case of total disorder or chaos.

Class IV rules are the most intriguing. They do not produce the dead frozenness of Classes I and II, nor do they lead to the total chaos characteristic of Class III. They produce coherent structures that propagate, grow, disintegrate, and recombine. There is no equilibrium.

Confirmation of these ideas has been described by Heudin (1999).

Langton (1989), working on *artificial life*, noted an overlap of these properties with the properties of nonlinear dynamical physical systems (also see Kauffman 1991). Class I was seen as corresponding to nonlinear systems with a single *point* attractor (cf. the appendix on chaos); Class II to *periodic* attractors; and Class III to *strange* attractors.

Class IV systems are the most fascinating of them all. They never settle down. There is *perpetual novelty*, just like in living systems (cf. Section 2.3.5). It is not possible to predict the rules that would result in Class IV behaviour. One has to just try various rules to see what happens. They are complex systems. Thus:

$$[(I, II) \rightarrow IV \rightarrow III] \text{ corresponds to } [Order \rightarrow Complexity \rightarrow Chaos].$$

Langton noted a similarity between this correspondence and a second-order phase transition (with the attendant critical fluctuations) in a physical system (cf. Section 3.1 for a discussion of phase transitions). At the critical point, the physical system is precariously perched between chaos and order. Just above the critical temperature, T_c, there is chaos (disorder, and higher symmetry), but there are short-lived regions of various sizes in which order (lower symmetry) prevails (because of the large excursions that the order parameter can make without costing free energy). And, just below T_c the ordered regions are larger and longer-lived; at low-enough temperatures below T_c, complete order prevails.

The critical point for a crystal in the vicinity of a phase transition thus represents the meeting point of order and chaos, or *the edge of order and chaos*. One gets complex behaviour at this point. In the language of Bak (1996), this is self-organized criticality (SOC), albeit only at one temperature, namely the critical temperature.

Langton carried this analogy further and argued that, in computation science, Classes I and II correspond to *halting behaviour*, Class III to *nonhalting behaviour*, and Class IV to *undecidable behaviour* (see Waldrop 1992). Most of the Class III and IV systems can only be modelled by mathematical algorithms which are *computationally irreducible* (CIR).

It has been believed that one cannot predict the future evolution of a process described by a CIR algorithm. Recently, Israeli and Goldenfeld (2004) have shown that, at least for some systems, the complexity of which stems from very simple rules, useful approximations can be made that enable prediction of their future behaviour.

5.9   Origin of life

Till recently, the standard picture about the origin of life was that, in prehistoric times, molecules of DNA, RNA, proteins, polysaccharides, etc., somehow got synthesized from smaller building blocks like amino acids, sugars, and nucleic acids, in some primordial soup. Once DNA and RNA got formed, self-replication followed, and natural selection did the rest.

The problem with this picture was that, if everything depended on random chance events, it is extremely unlikely that large molecules like DNA would get formed in reasonable time (Cairns-Smith 1985; Zimmer 2004). In any case, as Kauffman (1991, 1993, 1995, 2000) asked, where is the evidence that the origin of life on earth should be equated with the appearance of DNA? We describe some salient features of Kauffman's work here, as the scenario established by him has direct relevance to the processes of technological change and evolution.

Kauffman (1993) had established, by the cellular-automata approach, that the regulatory genetic networks arose spontaneously by self-organization. But the question still remained as to how an extremely large molecule like DNA, which is the bearer of the genetic information, came into existence in the first place. Could there be other mechanisms whereby large protein molecules could form spontaneously, without the intervention of DNA? How did those proteins arise which help the double-helix DNA molecule to uncoil itself and split into two strands for replication purposes?

There must have been a *nonrandom* origin of life. There must have been another way, involving simple chemical reactions, and independent of the need to involve DNA molecules, for self-replicating molecular systems to have got started. Kauffman propounded his *autocatalysis theory* to explain how this must have happened.

Imagine the primordial soup, either just a fluid, or a fluid in contact with rocks of various types, in which the basic building blocks, namely the relatively small molecules of amino acids, sugars, etc., just flitted around. It is impossible that, given enough time, some of them would not have undergone some random polymerization reactions with one another. It is certainly possible that at least some of these end-products, with some side chains and branches hanging around, will act as *catalysts* for facilitating the production of other molecules which may also be catalysts for another chemical reaction. Thus: A facilitates the production of B, and B does the same job for C, and so on. Given enough time, and a large enough pool containing all sorts of molecules, it is probable that, at some stage a molecule Z will get formed (aided by catalytic reactions of various types), which would be a catalyst for the formation of the catalyst molecule A we started with.

Once the loop closes on itself, it would head towards what we now call *self-organized criticality* (and order). There will be more production of A, which will lead to more production of B, and so on. In this scenario, there is no need to wait for random reactions. And once a threshold has been crossed, the system inches towards the 'edge of chaos and order', and acquires robustness against destabilizing agencies.

This spontaneous order, emerging out of molecular chaos, was akin to life: The system could consume (metabolize) raw materials, and grow into more and more complex molecules. It progressed into a situation where the forebears of DNA started appearing, with potential for replication. The rest, as they say, is history.

In this model, life originated *before* the advent of DNA. The model can incorporate features like reproduction, and competition and cooperation for survival and evolution (including coevolution).

Life just had to appear, to satisfy the tendency of open natural systems to move towards self-organized order and complexity.

5.10   Chapter highlights

• Self-assembly and self-organization are the hallmarks of natural biological phenomena. It makes sense to develop smart structures by the same approach.

• There is an enormous difference in the information content of merely condensed matter on the one hand, and highly organized supramolecular or complex matter on the other. The latter is qualitatively different from the former, and the information is contained in the shapes of the component molecules, and in the interaction patterns among them.

• Life depends on the hydrogen bond. This bond is much weaker than the covalent bond, and yet strong enough to sustain self-assembled biological structures, enabling them to withstand the disintegrating influences of thermal fluctuations and other perturbations. Hydrogen bonding, and the associated hydrophobic interaction, have the right kind of strength to enable superstructures to self-assemble without the need for irreversible chemical reactions. There is a strong element of reversibility associated with these weak interactions, enabling the spontaneous making and breaking of assemblies until the lowest free-energy configuration has been attained.

• The noncovalent and reversible nature of the intermolecular bonding in supramolecular structures gives them the all-important *dynamic* character and lability. They can readily undergo assembly/disassembly processes in response to a changing environment, and this adaptive behaviour is the very essence of smartness.

• To develop artificial smart structures by self-assembly, we have to begin with molecules having the requisite structure, shape, and bonding proclivities. This requires preorganization of molecular design, with specific receptor sites which can be mutually recognized, so that the desired evolution towards complexity can take place. Reversibility of bonding is the key feature for self-assembly through molecular recognition.

• Artificial structures at mesoscopic length-scales (colloids, nanowires, nanospheres, quantum dots) are being created in the laboratory by self-assembly. Structures built at still larger length-scales include those used in microelectronics and photonics.

• Supramolecular self-organization in the laboratory can be either by design, or by selection. The former is a knowledge-based approach, in which one starts with a library of preorganized or designed molecules, which then organize themselves as the information content in their designed shapes and interactions unfolds itself. In the latter approach, the dynamic edifice designs itself. This is like natural selection in Darwinian evolution. In a combination of these two approaches, one starts with molecules with a designed information and instruction content. The molecules and the interactions among them, however, are not completely unchangeable. This provides adequate scope for the evolution of self-designed supramolecular structures, through the processes of function-driven selection of the fittest molecules and pathways.

- Bak's very fundamental notion of self-organized criticality has not received the attention it deserves. Large avalanches, rather than gradual change, can lead to qualitative changes of behaviour, and can form the basis for emergent phenomena and complexity.

- Giant vesicles can mimic the smart action of the living cell in many ways, even though their behaviour is purely in the realm of organic chemistry, and not biology, or even biochemistry. They qualify to be called *very* smart structures. They can exhibit features such as birth, sensing, actuation, feeding, growth, healing, and death. Liquid-crystalline supramolecular polymers are another example of this type.

- Supramolecular assemblies can react to a changing environment in a dynamic and time-reversible manner. They can adapt to the environment, and adaptability is the crux of evolution, both chemical and biological. Adaptability leads to the emergence of new features and complexity. Adaptation becomes evolution when acquired features are conserved and passed on.

- To satisfy the tendency of open natural systems to move towards self-organized order and complexity, the appearance of life was inevitable. The same can be said about the evolution of 'artificial' smart structures.

References

Adleman, L. M. (11 Nov. 1994). 'Molecular computation of solutions to combinatorial problems'. *Science,* 266: 1021.

Adleman, L. M. (Aug. 1998). 'Computing with DNA'. *Scientific American,* 279: 34.

Atwood, J. L., J. E. D. Davies, D. D. MacNicol, F. Vogtle and J.-M. Lehn (eds.) (1996). *Comprehensive Supramolecular Chemistry.* Oxford: Pergamon.

Bak, P. (1996). *How Nature Works: The Science of Self-Organized Criticality.* New York: Springer.

Bak, P., H. Flyvbjerg and K. Sneppen (1994). 'Can we model Darwin ?'. *New Scientist,* 12: 36.

Bak, P. and K. Sneppen (1993). 'Punctuated equilibrium and criticality in a simple model of evolution'. *Phys. Rev. Lett.* 24: 4083.

Bak, P., C. Tang and K. Wiesenfeld (1987). 'Self-organized criticality: An explanation of $1/f$ noise'. *Phys. Rev. Lett.* 59: 381.

Bak, P., C. Tang and K. Wiesenfeld (1988). 'Self-organized criticality'. *Phys. Rev. A,* 38: 364.

Baker, D., G. Church, J. Collins, D. Endy, J. Jacobson, J. Keasling, P. Modrich, C. Smolke and R. Weiss (June 2006). 'Engineering life: Building a FAB for biology'. *Scientific American,* 294(6): 44.

Benenson, Y., T. Paz-Elizur, R. Adar, E. Keinan, Z. Livneh and E. Shapiro (2001). 'Programmable and autonomous computing machine made of biomolecules'. *Nature,* 414: 430.

Boettcher, S. and A. Percus (2000). 'Nature's way of optimizing'. *Artificial Intelligence*, 119: 275.

Boncheva, M., D. H. Gracias, H. O. Jacobs and G. W. Whitesides (2002). 'Biomimetic self-assembly of a functional asymmetrical electronic device'. *PNAS, USA*, 99(8): 4937.

Bowden, N., A. Terfort, J. Carbeck and G. M. Whitesides (1997). 'Self-assembly of mesoscale objects into ordered two-dimensional arrays'. *Science*, 276: 233.

Breen, T. L., J. Tien, S. R. J. Oliver, T. Hadzic and G. M. Whitesides (1999). 'Design and self-assembly of open, regular, 3D mesostructures'. *Science*, 284: 948.

Brunsveld, L., B. J. B. Folmer, E. W. Meijer and R. P. Sijbesma (2001). 'Supramolecular polymers'. *Chem. Rev.* 101: 4071.

Cairns-Smith, A. G. (June 1985). 'The first organisms'. *Scientific American,* 253: 90.

Chen, J. and D. H. Wood (2000). 'Computation with biomolecules'. *PNAS, USA*, 97: 1328.

Choi, I. S., M. Weck, B. Xu, N. L. Jeon and G. M. Whitesides (2000). 'Mesoscopic, templated self-assembly at the fluid–fluid interface'. *Langmuir*, 16: 2997.

Cragg, P. J. (2005). *A Practical Guide to Supramolecular Chemistry*. West Sussex: Wiley.

Dawkins, R. (1986). *The Blind Watchmaker: Why the Evidence of Evolution Reveals a Universe Without Design*. New York: Norton.

Desiraju, G. R. (1995). *The Crystal as a Supramolecular Entity*. [*Perspectives in Supramolecular Chemistry*, Vol. 2]. Chichester: Wiley.

Dhar, D. (1990). 'Self-organized critical state of sandpile automata models'. *Phys. Rev. Lett.* 64: 1613.

Drexler, K. E. (1992). *Nanosystems: Molecular Machinery, Manufacturing and Computation*. New York: Wiley.

Gilmore, R. (1981). *Catastrophe Theory for Scientists and Engineers*. New York: Wiley.

Gracias, D. H., J. Tien, T. L. Breen, C. Hsu and G. M. Whitesides (2000). 'Forming electrical networks in three dimensions by self-assembly'. *Science*, 289: 1170.

Held, G. A., D. H. Solina II, D. T. Keane, W. J. Haag, P. M. Horn and G. Grinstein (1990). 'Experimental study of critical-mass fluctuations in an evolving sandpile'. *Phys. Rev. Lett.* 65: 1120.

Heudin, J.-C. (1999) (ed.). *Virtual Worlds: Synthetic Universes, Digital Life, and Complexity*. Reading, Massachusetts: Perseus Books.

Hosokawa, K., I. Shimoyama and H. Miura (1995). 'Dynamics of self-assembling systems: Analogy with chemical kinetics'. *Artif. Life*, 1: 413.

Huc, I. and J.-M. Lehn (1997). 'Virtual combinatorial libraries: Dynamic generation of molecular and supramolecular diversity by self-assembly'. *PNAS, USA*, 94: 2106.

Israeli, N. and N. Goldenfeld (2004). 'Computational irreducibility and the predictability of complex physical systems'. *Phys. Rev. Lett.* 92(7): 074105-1.

Jackman, R. J., S. T. Brittain, A. Adams, M. G. Prentiss and G. M. Whitesides (1998). 'Design and fabrication of topologically complex three-dimensional microstructures'. *Science* 280: 2089.

Jenekhe, S. A. and L. X. Chen (1999). 'Self-assembly of ordered microporous materials from rod-coil block polymers'. *Science*, 283: 372.

Langton, C. G. (ed.) (1989). *Artificial Life*. Santa Fe Institute Studies in the Sciences of Complexity, Proceedings Vol. 6. Redwood City, CA: Addison Wesley.

Lehn, J.-M. (1988). 'Supramolecular chemistry – scope and perspectives. Molecules, supermolecules, and molecular devices' (Nobel lecture). *Angew. Chem. Int. Ed. Engl.*, 27: 90.

Lehn, J.-M. (1995). *Supramolecular Chemistry: Concepts and Perspectives*. New York: VCH.

Lehn, J.-M. (1999). In R. Ungaro and E. Dalcanale (eds.), *Supramolecular Science: Where It Is and Where It Is Going*. Dordrecht: Kluwer.

Lehn, J.-M. (16 April 2002). 'Toward complex matter: Supramolecular chemistry and self-organization'. *PNA, USA*, 99(8): 4763.

Lehn, J.-M. and A. Eliseev (2001). 'Dynamic combinatorial chemistry'. *Science*, 291: 2331.

Kauffman, S. A. (August 1991). 'Antichaos and adaptation'. *Scientific American,* 265: 64.

Kauffman, S. A. (1993). *The Origins of Order*. Oxford: Oxford University Press.

Kauffman, S. A. (1995). *At Home in the Universe*. Oxford: Oxford University Press.

Kauffman, S. A. (2000). *Investigations*. Oxford: Oxford University Press.

Kramer, R., J.-M. Lehn and A. Marquis-Rigault (1993). 'Self-recognition in helicate self-assembly: Spontaneous formation of helical metal complexes from mixtures of ligand sans metal ions (programmed supramolecular systems/polynuclear metal complexes/instructed mixture paradigm)'. *PNAS, USA*, 90: 5394.

Menger, F. M. and K. D. Gabrielson (1995). 'Cytomimetic organic chemistry: Early developments'. *Angew. Chem. Int. Ed. Engl.* 34: 2091.

Oh, K., K.-S. Jeong and J. S. Moore (2001). 'Folding-driven synthesis of oligomers'. *Nature*, 414: 889.

Peng, C.-K., J. Mietus, J. M. Hausdorff, S. Havlin, H. E. Stanley and A. L. Goldberger (1993). 'Long-range anticorrelations and non-Gaussian behaviour of the heartbeat'. *Phys. Rev. Lett.*, 70: 1343.

Seeman, N. C. (June 2004). 'Nanotechnology and the double helix'. *Scientific American,*, p. 64.

Sirringhaus, H., T. Kawase, R. H. Friend, T. Shimoda, M. Inbasekaran, W. Wu and E. P. Woo (2000). 'High-resolution inkjet printing of all-polymer transistor circuits'. *Science*, 290: 2123.

Stanley, H. E., L. A. N. Amaral, J. S. Andrade, S. V. Buldyrev, S. Havlin, H. A. Makse, C.-K. Peng, B. Suki and G. Viswanathan (1998). 'Scale-invariant correlations in the biological and social sciences'. *Phil. Mag. B*, 77: 1373.

Storhoff, J. J. and C. A. Mirkin (1999). 'Programmed materials synthesis with DNA'. *Chem. Rev.* 99: 1849.

Stupp, S. I., V. LeBonheur, K. Walker, L. S. Li, K. E. Huggins, M. Keser and A. Amstutz (1997). 'Supramolecular materials: self-organized nanostructures'. *Science*, 276: 384.

Sun, S. H., C. B. Murray, D. Weller, L. Folks and A. Moser (2000). 'Monodisperse FePT nanoparticles and ferromagnetic FePT nanocrystal superlattices'. *Science*, 287: 1989.

Velev, O. D., K. Furusawa and K. Nagayama (1996). 'Assembly of latex particles by using emulsion droplets as templates. 1. Microstructured hollow spheres'. *Langmuir*, 12: 2374.

Waldrop, M. M. (1992). *Complexity: The Emerging Science at the Edge of Order and Chaos*. New York: Simon and Schuster.

Whitesides, G. M. and M. Boncheva (16 April 2002). 'Beyond molecules: Self-assembly of mesoscopic and macroscopic components'. *PNA, USA*, 99(8): 4769.

Whitesides, G. M. and B. Grzybowski (29 March 2002). 'Self-assembly at all scales'. *Science*, 295: 2418.

Whitesides, G. M. and J. C. Love (September 2001). 'The art of building small'. *Scientific American*, 285: 33.

Wolfram, S. (September 1984). 'Computer software in science and mathematics'. *Scientific American*, p. 140.

Wolfram, S. (1985). 'Undecidability and intractability in theoretical physics'. *Phys. Rev. Lett.* 54: 735.

Wolfram, S. (ed.) (1986). *Theory and Applications of Cellular Automata*. Singapore: World Scientific.

Wolfram, S. (2002). *A New Kind of Science*. Wolfram Media Inc. (for more details about the book, cf. www.wolframscience.com).

Yates, F. E. (Ed.) (1987). *Self-Organizing Systems*. New York: Plenum.

Yeh, H.-J. J. and J. S. Smith (1994). 'Fluidic self-assembly for the integration of GaAs light-emitting diodes on Si substrates'. *Photon. Technol. Lett. IEEE*, 6(6): 706.

Zimmer, C. (June 2004). 'What came before DNA?'. *Discover* 25: 34.

6

NANOSTRUCTURES

What the computer revolution did for manipulating data,
the nanotechnology revolution will do for manipulating matter,
juggling atoms like bits.

– R. Merkle

How soon will we see the nanometer-scale robots envisaged by
K. Eric Drexler and other nanotechnologists?
The simple answer is never.

– R. E. Smalley

6.1   Introduction

In the great Indian epic *Ramayan*, Hanuman, the supreme devotee of Lord Ram, had the power to change his size at will. At one stage in the story, he decides to become invisible to his adversaries by shrinking into a miniscule ('*sukshama*') state. Nearer to our times, we have had books by George Gamow on the adventures of Mr. Tompkins in Wonderland, who could change his size to suit the thought processes: For example, he would shrink to nuclear dimensions when he was 'looking' at fundamental particles. A modern version of this scenario has been bandied around from the world of future nanomedicine (Alivisatos 2001; Kakade 2003). Ever since Drexler (1986) wrote his book *Engines of Creation*, one of the fond hopes of some people has been that problems like cancer, old age, and blocked arteries would become a thing of the past because, one day, it would be possible to make cell-repair kits so small that they will be sent into the human body like little submarines. There is a long way to go before such feats become possible (Schummer and Baird 2006).

Nanoscience is science at nanometre length-scales. Dimensions from 0.1 nm to 1000 nm (i.e. from 1 Angstrom to 1 micron) can be taken as falling under the purview of nanoscience. Qualitatively new effects can arise at such small dimensions. Quantum mechanical effects become important, and there is a huge surface to volume ratio. Totally new size-dependent phenomena can arise.

Nanotechnology is technology based on the manipulation or self-assembly of individual atoms and molecules, leading to the formation of structures with complex specifications at the atomic level. It is the technology of the future, that would hopefully give us thorough and inexpensive control on the desired (and feasible) structure of matter (Kelsall, Hamley and Geoghegan 2005). However, not many manmade nanoscale machines exist at present.

Nanostructured materials include atomic clusters, films, filamentary structures, and bulk materials having at least one component with one or more dimensions below 100 nm (Chandross and Miller 1999).

A recent (trivial) application of nanomaterials, which has been already commercialized, is the use of nano-sized particles of ZnO in sunscreen creams. This has made the previously white-coloured creams *transparent* because the ZnO particles are too small to scatter visible light substantially.

The central thesis of nanotechnology is that almost any chemically stable structure that can be specified can be built (Rieth 2003; Balzani, Venturi and Credi 2003; Ozin and Arsenault 2005; Minoli 2006). One hopes that, in the near future, it would be possible to build and programme *nanomachines* or *molecular machines* that build structures (smart structures included!) atom by atom, or molecule by molecule (Drexler 1992). These machines will thus act as *assemblers*. If we can also develop *replicators*, it would be possible to have copies of these assemblers in large enough numbers, so that even large and complex structures and systems will be assembled in reasonable time, and at low cost (Drexler 1981, 1986, 1992, 2001; Drexler, Peterson and Pergamit 1991).

What will the nanomachines build? Here is a partial wish-list:

- More powerful, faster and cheaper computers.
- Biological nanorobots that can enter the human body and perform disease-curing operations.
- Structures that can heal or repair themselves.
- Smart buildings, aircraft, automobiles, etc.
- All kinds of smart structures with a cortex-like brain.

'The age of nanofabrication is here, and the age of nanoscience has dawned, but the age of nanotechnology – finding practical uses of nanostructures – has not really started yet' (Whitesides and Love 2001). By contrast, in living cells, Nature employs extraordinarily sophisticated mechanisms for the production and assembly of nanostructures. The nanoscale biomachines manipulate genetic material and supply energy. Cell multiplication and differentiation are some of the marvels of Nature which we hope to emulate for mass-producing our smart nanostructured materials and systems (Seeman and Belcher 2002).

By and large, what we have at present is the *top-down* approach, wherein material is removed or etched away from larger blocks or layers of materials, usually by lithography (Wallraff and Hinsberg 1999; Harriott and Hull 2004; Okazaki and Moers 2005). However, substantial successes in the directed self-assembly of molecules offer hope for the *bottom-up* route so common in Nature. As discussed in Chapter 5, molecular self-assembly is the spontaneous coming together of molecules into stable, structurally well-defined, aggregates (Wilbur and Whitesides 1999; Shenhar, Norsten and Rotello 2004; Kohler and Fritzsche 2004; Ozin and Arsenault 2005). Self-assembly is an important approach for the cost-effective production of nanostructured materials in large amounts (Whitesides and Boncheva 2002). There have also been some efforts at exploiting the folding of tailor-made DNA molecules for creating nanoscale shapes and patterns (Storhoff and Mirkin 1999; Seeman 2004; Rothemund 2006). Microfabrication and nanofabrication techniques (see, e.g. Brodie and Muray 1992) can seldom be low-cost techniques for such purposes.

Cost is, however, not the only consideration. Self-assembly also results in a high degree of thermodynamic equilibrium, leading to a high degree of stability and perfection (Wilbur and Whitesides 1999).

The published literature on nanoscience and nanotechnology is already huge. In this chapter, we have a very limited agenda: To introduce the subject, and to delineate its overlap with the subject of smart structures. This includes the following:

- *Size-dependent effects*, with an eye for nonlinear variation of properties, so that *size-tuneability* can be used for achieving a 'smart' purpose. Like phase transitions in macroscopic systems, phase transitions in small systems can offer good scope for smart innovation. In fact, self-assembly itself is a process of phase transition, growth of crystals being its most familiar manifestation.
- The top-down and bottom-up fabrication of nanostructures which may function as sensors, actuators and 'brains' in smart structures.

6.2   Molecular building blocks for nanostructures

We introduce a few important *molecular building blocks* (MBBs) which can be exploited for building nanostructured materials by the bottom-up or self-assembly route.

Diamondoids.   These are polycyclic, saturated, cage-like hydrocarbons, in which the carbon skeleton has the same spatial structure as that in diamond crystals (cf. Drexler 1992). The smallest diamondoid molecule is adamantane. Its homologous molecules are: diamantane, triamantane, tetramantane, and hexamantane (Dahl, Liu and Carlson 2003). Diamondoids are regarded as a very important class of MBBs in nanotechnology. They may find applications in the fabrication of nanorobots and other molecular machines.

Fullerenes.   The four allotropes of carbon are: graphite, diamond, fullerenes, and nanotubes. Graphite and diamond are well known. Fullerenes and nanotubes are more recent discoveries (cf. Dresselhaus, Dresselhaus and Eklund 1996; Moriarty 2001).

The fullerene molecules were discovered by Kroto *et al.* (1985). The most commonly occurring fullerene is C_{60} (Crespi 2004; Dorn and Duchamp 2004). It has the structure of a classical football: A truncated icosahedron, i.e. a polygon with 60 vertices and 32 faces, 12 of which are pentagonal and 20 are hexagonal. The carbon atoms sit at the 60 vertices.

Still larger fullerenes, like C_{70}, C_{76}, C_{80}, C_{240}, also exist. They can be put to a variety of uses, either as single molecules, or as self-assembled aggregates.

Carbon nanotubes.   Carbon nanotubes (CNTs) were discovered by Iijima (1991). Graphite has a hexagonal-sheet structure. Carbon nanotubes can be regarded as rolled up versions of graphite, the direction of rolling determining the nature of the hollow nanotube (Ajayan 1999; Smith and Luzzi 2004; Evoy, Duemling and Jaruhar 2004). Their diameters range from ~ 1 nm to several nm. Like the fullerenes, CNTs also find a variety of applications in nanotechnology (Baughman, Zakhidov and de Heer 2002; Dragoman and Dragoman 2006; O'Connell 2006).

A variety of other MBBs for nanotechnology have also been developed and used. These include nanowires, nanocrystals, peptides, nucleotides, and a number of other organic structures (cf. Mansoori 2005).

6.3   Phase transitions in small systems

When we discussed phase transitions in Section 3.1, it was assumed that we are dealing with 'large' systems. That is, we worked in the *macroscopic limit*, or the *thermodynamic limit*, defined as: $N \rightarrow \infty; V \rightarrow \infty; N/V =$ finite. Here N is the number of atoms or molecules in the system, and V is the volume occupied by it. The number N can be considered to be large if it is of the order of the Avogadro number. The largeness, or otherwise, of the volume V, or of the spatial dimensions, is somewhat tricky to define. The dimensions may be very small in one direction, and very large in the other two (as in thin films). In any case, one has to consider it with reference to the *range* of the interaction responsible for a phase transition. The nucleus of an atom is indeed a small system; its size is of the same order as the range of the nuclear interaction (Bondorf *et al.* 1995). Even an astrophysical system can be a 'small' system because of the infinite range of the gravitational interaction.

A solid has a surface (or an interface with another material or phase). The atomic structure and bonding at a surface or interface are different from those in the bulk of the solid. For small surface-to-volume ratios, the surface effects can be usually ignored. But the situation changes for nanomaterials: They are more surface than bulk. Naturally, their properties depend in a sensitive way on their size, i.e. on the surface-to-volume ratio.

In bulk materials, phase transitions can occur as a function of temperature, pressure, composition, electric field, magnetic field, or uniaxial stress. As we go down to mesoscales, size becomes another important parameter for determining the most stable phase at a given size. Let us take the bottom-up route to get a feel for what happens to the structure as its size *increases*, starting from a single molecule.

6.3.1 *The cluster-to-crystal transition*

When a few molecules bond together, we either have a *cluster* or a *supermolecule*. The term 'clusters' has come to mean aggregates of atoms or molecules that, unlike supermolecules, are not found in appreciable numbers in vapours at equilibrium.

As a very small cluster grows in size, it undergoes a process of *reconstruction* at every event of growth. Every time a unit (atom or molecule) of the growing species attaches itself to the cluster, the units rearrange themselves completely, so as to minimize the sum total of surface and volume free energies. Naturally, the symmetry possessed by the cluster also changes every time there is a reconstruction of the structure. In other words, there are size-driven symmetry-changing transitions in the structure as it grows, unit by unit.

A stage comes in the increasing size of the cluster when it no longer has to reconstruct itself drastically on the attachment of additional units. We then have what is called a *microcrystal* or *microcrystallite*, which has the same symmetry as that of the bulk crystal: The symmetry does not change on further increase of size. One speaks of a *cluster-to-crystal transition* in this context (cf. Multani and Wadhawan 1990; Reinhard and Suraud 2004).

What is the symmetry possessed by clusters when they are not large enough to be at the cluster-to-crystal transition stage? A striking fact is that noncrystallographic symmetries are possible (simply because there is no translational periodicity in the structure). In particular, icosahedral symmetry is often a favoured symmetry for small clusters. Gold clusters, for example, have icosahedral symmetry in the 4–15 nm size range. As their size is increased beyond 15 nm, they make a transition to a crystallographic symmetry (fcc), using multiple twinning as an adjustment mechanism (Ajayan and Marks 1990).

6.3.2 *Thermodynamics of small systems*

A crystal has an underlying lattice. The lattice, by definition, has all its points equivalent. This is possible only if the lattice is infinite. Real crystals, however, are never infinite. Their finiteness implies the presence of a surface, which introduces a surface-energy term in the total free energy of the crystal. As we go down to nanometre sizes, the surface term becomes very important in the thermodynamic behaviour of the system.

Phase transitions in small systems are poorly understood; the experiments are difficult to carry out. Moreover, small systems may not be in equilibrium in many real-life conditions. In such situations, computer simulations, particularly Monte Carlo (MC) (Challa, Landau and Binder 1990) and molecular dynamics (MD) calculations, can provide valuable insights into the nature of the system (Mansoori 2005).

Nanostructured materials are not only small systems that are practically all surface and very little volume, their surface interfaces with a surrounding material with a different density. Therefore, phase transitions in small systems cannot, in general, be discussed without reference to the surroundings. For similar reasons, the *geometry* of the small system is also a relevant parameter, which can strongly influence the phase transition. So can the size (Chacko *et al.* 2004).

In the thermodynamics of large systems, a state parameter is said to be *intensive* if it is independent of the size of the system; for example, temperature and pressure. An *extensive* state parameter, by contrast, is (linearly) proportional to the size of the system. Internal energy is an example. So is entropy. Such concepts break down for small systems. Entropy, for example, becomes nonadditive, and therefore nonextensive, for small systems (Tsallis, Mendes and Plastino 1998). The appendix on nonextensive thermostatistics discusses such systems.

Following Berry (1990a), let us focus attention on clusters of 4–200 atoms or molecules. It turns out that, for certain clusters, there can be a coexistence of solid-like and liquid-like states over a finite range of temperatures (Berry 1990a, 1998; Schmidt *et al.* 1998). This means that the melting point and the freezing point may not always coincide for small systems (contrary to the situation for large systems) (Berry 1990b). Why does this happen?

As already stated, a very substantial fraction of a cluster is surface, rather than volume. For example, in a 55-atom argon cluster, 42 atoms are surface atoms. Such atoms have less tight binding than those in the interior; in fact, the binding of atoms in the interior is akin to that in bulk material. The tighter the binding in the interior, the deeper is the potential well, and the more widely spaced are the energy levels involved in the transitions from one energy state to another.

The surface atoms, on the other hand, undergo vibrations and other excitations in a shallower and variable potential, having more closely spaced and more unevenly spaced energy levels (Berry 1990a).

One can generally associate a melting point T_m and a freezing point T_f with a cluster. By definition, for $T < T_f$ the entire cluster is solid, and for $T > T_m$ the entire cluster is liquid. T_f is the lower limit for the thermodynamic stability of the liquid-like form, and T_m is the upper limit of the thermodynamic stability of the solid-like form. The interesting thing about these small systems is that T_m and T_f need not coincide: For $T_f < T < T_m$, atoms in the interior of the cluster are solid-like, and the surface atoms are liquid-like (Berry 1990b). This very important conclusion is best understood in terms of the free-energy density $F (= U - TS)$:

All systems tend to minimize the free energy under all conditions. For low temperatures $(T < T_f)$, the entropy-term contribution (TS) to the free energy is low, and mainly the lower-energy levels are occupied. The system is rigid, corresponding to the fully solid state of the cluster.

As the temperature is raised to a value just above T_f, a substantial fraction of atoms are excited to the higher energy states, which are the more closely spaced states of the shallow potential. The entropy term is now large enough to force the creation of a second minimum in the plot of free energy *vs.* temperature and pressure. This implies the coexistence of a solid-like phase and a liquid-like phase. Atoms in the liquid-like phase are easily able to undergo thermal fluctuations among the closely spaced energy states of the shallow potential, in keeping with the 'soft' and nonrigid nature of the liquid-like phase.

With increasing temperature, the proportion of the solid-like phase decreases, till it disappears at T_m.

The situation in large systems in no different, except that the fraction of surface atoms is so small that the difference between T_f and T_m is practically indiscernible.

So far we have implicitly assumed that we are dealing with clusters of a particular size at a fixed pressure, and that temperature is the control parameter. We may as well keep the temperature and pressure fixed, and look at a certain range of cluster sizes. It is found that, for a suitable temperature, larger clusters are solid-like and smaller ones are liquid-like; in between, there is a range of sizes for which the solid-like and liquid-like states coexist (Berry 1990a).

Although there can be exceptions (Berry 1990b), by and large the solid–liquid transition in small systems has the following characteristics (Schmidt *et al.* 1997; Bertsch 1997):

1. There may (but need not) be a range of temperatures over which the two phases coexist. *This is stable coexistence, different from the metastable coexistence of phases in a supercooled or superheated large system.*
2. The melting point decreases with decreasing cluster size.
3. The latent heat decreases with decreasing cluster size.

Experimental proof for these conclusions was first provided by Schmidt *et al.* (1997) for a free cluster of 139 sodium atoms. It has also been observed that the variation of melting point with size is not always monotonic (Berry 1998; Schmidt *et al.* 1998).

Small clusters of atoms like argon adopt a *shell structure*, rather like the shell structure of electrons in an atom. Atomic clusters of icosahedral symmetry get filled up at cluster sizes (N) of 13, 55, 147, 309 for argon. One might expect that binding energies, and therefore melting points, of clusters should show irregular variation with size because of this shell effect. However, certain things still remain unexplained (Berry 1990b; Schmidt *et al.* 1998).

Berry (1990b, 1999) has formulated thermodynamic criteria for the stability of a system *of any size*. According to him, the essential condition for the *local* stability (or metastability) of any system is the occurrence of a local minimum in the free energy, F, with respect to some suitable control parameter (for fixed external variables). A parameter γ is introduced to represent quantitatively the degree of local nonrigidity of a cluster. $F(T, \gamma)$ must be a minimum everywhere in the cluster. Coexistence of solid-like and liquid-like forms means that $F(T, \gamma)$ has two minima as a function of γ at a given T. The theory shows that, in a temperature range ΔT_c ($= T_m - T_f$), the two forms may coexist in thermodynamic equilibrium in a ratio fixed by the chemical equilibrium constant K:

$$K \equiv \frac{[\text{solid}]}{[\text{liquid}]} = \exp\left(-\frac{\Delta F}{N k_B T}\right) \tag{6.1}$$

Here, $\Delta F = F(solid) - F(liquid)$ at temperature T.

$K = \infty$ for $T < T_f$ because the entire system is solid-like at low temperatures. Similarly, $K = 0$ for $T > T_m$. It has a nonzero finite value between T_f and T_m. Thus, the chemical equilibrium constant exhibits discontinuities at T_f and T_m. These discontinuities are not signatures of either a first-order or a second-order phase transition, but something unique to finite systems.

In the thermodynamic limit defining a large system, the two discontinuities merge to a single discontinuity characteristic of a regular first-order phase transition.

We can write $\Delta F = N \Delta \mu$, where $\Delta \mu$ is the difference in chemical potential, i.e. the difference in free energy per constituent atom or molecule of the solid-like and liquid-like forms. The local coexistence equilibrium of the two forms is thus described by $\Delta \mu = 0$ (a condition applicable for systems of any size).

Berry's theory provides a seamless thermodynamic description of transitions in small, intermediate, and large systems. A solid is fully rigid, and a liquid is fully nonrigid. Anything in between, as quantified by the local nonrigidity parameter γ for a cluster, is a different form of matter, unique to nanoclusters. What is more, the local degree of nonrigidity has been observed to vary strongly with the addition or subtraction of just a few atoms or molecules constituting the cluster (Schmidt *et al.* 1998). *This offers the possibility of designing nanoclusters with size-dependent properties that we select* (Berry 1998).

6.3.3   *Fragmentation*

Small or finite systems can exhibit fragmentation, a phenomenon not possible in large (infinite) systems. A finite many-body system can develop global density fluctuations, culminating in its break-up into several fragments. This is somewhat like boiling, except that relatively large chunks of matter can be identified in the aftermath.

Fragmentation ('multifragmentation') was first observed in nuclear systems (cf. Bondorf *et al.* 1995; Cole 2004). It is also a rather ubiquitous phenomenon in polymer degradation, break-up of liquid droplets, etc. (Courbin, Engl and Panizza 2004). Fragmentation of water confined in a CNT has been observed by Stupp *et al.* (1997).

6.4   Size dependence of macroscopic properties

6.4.1   *Size effects in ferroic clusters and crystallites*

Ferromagnetics

For fairly comprehensive reviews of this topic, the articles by Multani *et al.* (1990) and Haneda and Morrish (1990) should be consulted, as also the book by Kelsall, Hamley and Geoghegan (2005).

Four successive stages of properties are encountered as the size of a ferromagnetic crystallite is reduced (Newnham, Trolier-McKinstry and Ikawa 1990): (1) polydomain; (2) single domain; (3) superparamagnetic; and (4) paramagnetic. A similar succession of four size-dependent stages also occurs for ferroelectrics.

The polydomain stage is typical of any bulk ferroic phase; it prevails for sizes above $\sim$1000 nm. And Stage 4, namely the paramagnetic stage, comes about when the size is so small ($\sim$10 nm) that no significant long-range cooperative ordering, characteristic of a ferromagnetic phase, is possible.

Stage 2 (occurring in the size range 1000–50 nm) is a consequence of the fact that creation of a domain wall costs energy, and above a certain surface-to-volume ratio the trade-off between this cost and the gain of decreasing the demagnetizing field gets reversed. Fe particles suspended in mercury provide an example of this effect. The crystallite size below which a single-domain configuration is favoured for them is 23 nm. And for particles of $Fe_{0.4}Co_{0.6}$ this size is $\sim$28 nm.

Stage 3 sets in, typically, at dimensions below $\sim$50 nm. The material, being ferromagnetic, has an inherent tendency for cooperative ordering, but the ordering energy scales with volume. The volume is so small that the ordering energy, as also the energy barriers for a collective spin flip (for the entire crystallite) from one easy direction of magnetization to another, become comparable to thermal fluctuation energies. Thus, although the instantaneous value of the spontaneous magnetization is large (because all spins point the same way), the time-averaged value is zero. This is a superparamagnetic state because the entire giant magnetization of the crystallite responds collectively to a probing magnetic field, implying a very high magnetic permeability. And yet, because of the thermal flipping around of the spontaneous magnetization, the net magnetization of the single-domain particle is zero (with an attendant high time-averaged symmetry) (Haneda and Morrish 1990; Moriarty 2001).

At mesoscale dimensions the spontaneous magnetization M_s is found to decrease with particle size. This has been observed for γ-Fe_2O_3, CrO_2, $NiFe_2O_4$, and $CoFe_2O_4$. It is also true for particles of Fe, FeNi, and FeCo, although the underlying mechanism is different. Similarly, the temperature for the transition to the ferromagnetic phase, T_c, is found to decrease with decreasing particle size. $BaFe_{12}O_{19}$, a ferrite, is an example of this (Haneda and Morrish 1990). This material finds applications, not only in permanent magnets, but also as a recording medium and a magneto-optic medium. Smaller particle sizes mean greater information-storage densities.

Ferroelectrics

Ferroelectric phase transitions are *structural* phase transitions, implying a strong coupling between electric and elastic degrees of freedom. By contrast, ferromagnetic phase transitions need not always be structural phase transitions, and magneto-elastic coupling effects are, as a rule, much weaker than electro-elastic coupling effects. Therefore, extraneous factors like sample history and defect structure have a strong influence on size-dependent effects in small ferroelectric particles.

In contrast to this, it is also true that the exchange interaction responsible for spin–spin coupling in ferromagnetics is much stronger than the dipole–dipole interaction in ferroelectrics. This makes the intrinsic size effects in ferroelectrics considerably weaker than in ferromagnetics (Gruverman and Kholkin 2006).

Bulk ferroelectrics have an optimum domain structure. Below a certain size, a single-domain configuration is favoured. Because of the strong electro-elastic coupling mentioned above, even the identity of the phase stable at a given temperature depends on the size of the specimen. For example, for $BaTiO_3$ powder the critical size is 120 nm. For sizes smaller than this, the normally stable tetragonal phase at room temperature reverts back to the cubic phase, although the behaviour depends strongly on residual strains (Newnham and Trolier-McKinstry 1990b). Similarly, for $PbTiO_3$ powder, the tetragonal phase is stable down to 20 nm sizes.

Even the method of preparation of ultra-fine powders of ferroelectric materials has an effect on the phase stability, as exemplified by $BaTiO_3$. Its nanoparticles prepared by the sol–gel method have a critical size ($\sim$130 nm), below which the occurrence of the tetragonal–cubic phase transition is suppressed. By contrast, nanoparticles prepared by the 'steric acid gel' method remain cubic for all the sizes studied (cf. Wadhawan 2000). This behaviour has been attributed to intergrowth effects like alternating polytypic layers of cubic and hexagonal forms of $BaTiO_3$, and to vacancies, and charged and uncharged dopants.

The ferroelectric transition temperature, T_c, for $BaTiO_3$ is found to decrease with decreasing particle size below 4200 nm, down to 110 nm. The corresponding size limits for $PbTiO_3$ are 90 nm and 25 nm.

The effect of small size on particles of PZT has been investigated extensively (cf. Multani *et al.* 1990). For particle sizes as small as 16 nm for the morphotropic phase boundary (MPB) composition, the MPB itself seems to disappear. What we have instead is a near-amorphous phase.

Reduced dimensions of ferroelectric materials, including those brought about by the formation of a thin film, have the effect of smearing out the ferroelectric phase transition, i.e. making it a diffuse phase transition (DPT) (Li *et al.* 1996). This feature can be advantageous in some device applications, because it implies a less steep temperature dependence of certain properties.

Ferroelastics

Ferroelastics are different from ferromagnets and ferroelectrics in one important aspect. For a *free* particle of a purely ferroelastic material, there is no equivalent of the demagnetization field of a ferromagnet, or the depolarization field of a ferroelectric. Therefore a free particle of a purely ferroelastic material need not have a domain structure (Arlt 1990). But one can still envisage four regimes of the size dependence of the ferroelastic properties of crystalline materials, just like for ferromagnets and ferroelectrics (described above):

For purely ferroelastic crystals larger than $\sim$1000 nm, the characteristic domain structure (transformation twinning) may still occur (because of certain surface effects), particularly if it does not cost much energy to create a domain wall. If the ferroelastic is also a ferroelectric or a ferromagnet, then the depolarization or the demagnetization field provides the reason for the occurrence of the ferroelastic domain structure.

For particle sizes in the 1000 nm to 100 nm range, the surface-energy term competes strongly with the volume-energy term. The surface-energy term includes contributions from domain walls also. For a particle of size d, the volume term scales as d^3, and the surface term scales as d^2. The size, d_{crit}, at which the two terms match (they have opposite signs) is the size below which an absence of domain walls is favoured (Arlt 1990).

At sizes below 100 nm, a *superparaelastic* phase may occur. By analogy with superparamagnetism, superparaelasticity implies very large elastic compliance, and yet a near-zero macroscopic spontaneous strain (Newnham and Trolier-McKinstry 1990a, b).

For still smaller particle sizes, the concept of long-range ferroelastic ordering loses its meaning, and the system may revert to a higher-symmetry paraelastic phase.

6.4.2 *Size effects in polycrystalline ferroics*

The powder of a ferroic material can be sintered to obtain a polycrystalline material, with a certain average grain size. The grain size can have a very substantial effect on the properties of the polycrystalline ferroic. For example, certain phase transitions are arrested if the grain size is in the sub-micron range (cf. Wadhawan 2000).

For grain sizes approaching mesoscopic length-scales, the grain boundaries occupy a significant fraction of the total volume. A grain boundary is a region of the poly-crystal that acts as a transition layer between two grains, and thus has a structure, composition, and purity level quite different from that of the grains separated by it. A preponderance of the grain-boundary fraction in the total volume of a polycrystal can have a dominating influence at nanoscale grain sizes.

Polycrystalline ferromagnetic materials have been reviewed by Valenzuela (1994). The reader is referred to this excellent text for detailed information.

All ferroelectrics are piezoelectrics as well. Nanostructured ferroelectrics have been reviewed by Xu (1991), Mayo (1993), and Li *et al.* (1996). The most used polycrystalline piezoelectric is PZT. Moreover, its doping with La gives another remarkable electroceramic, namely PLZT, which can be made transparent and has extensive optical applications (cf. Wadhawan 2000).

Although a free particle of a purely ferroelastic material need not split into domains, this is no longer the case when we are dealing with a polycrystalline purely fer-roelastic material. Any grain is now in an anisotropic environment provided by the grains touching it. Its spontaneous deformation at a ferroelastic phase transi-tion creates a strain field in the surrounding material, and we have what may be called a *mechanical depolarizing field*, by analogy with the electric depolarizing field (Arlt 1990).

Shape-memory alloys (SMAs) provide the best-investigated examples of purely ferroelastic polycrystalline materials. Another example of a polycrystalline ferroe-lastic, which is not simultaneously ferroelectric or ferromagnetic, is the high-T_c

superconductor Y-Ba-Cu-O (Wadhawan 1988). It exhibits the shape-memory effect (Tiwari and Wadhawan 1991).

A consequence of mesoscale grain sizes is that the fraction of the material residing in the grain boundaries, rather than in the grains, increases drastically. For example, whereas there is only 3% material in the grain boundaries for an average grain size of 100 nm, this rises to 50% if the grain size is as small as 5 nm. The atomic structure in the grain boundaries is very different from that in the grains. It may even tend to be amorphous (Eastman and Siegel 1989). The macroscopic manifestations of this situation include better hardness and fracture characteristics (through more efficient crack dissipation), and higher ductility.

6.4.3 *Nanocomposites*

Composites offer immense scope for design and optimization. *Smart composites* are among the earliest of materials envisaged for incorporating sensors, actuators and microprocessor controllers into a single integrated smart entity (Grossman *et al.* 1989; Thursby *et al.* 1989; Davidson 1992). Typically, optical fibres (sensors) and shape-memory wires of Nitinol (actuators) can be embedded in the composite at the fabrication stage itself, and the whole assembly can be integrated to an underlying IC chip, or to a distribution of chips, for microprocessor control. The distributed optical-fibre sensors can also play the dual role of carrying laser energy for heating the SMA wires for the actuation action.

Ferrofluids are examples of ferromagnetic nanocomposites (cf. Section 4.6).

Another major class of ferromagnetic nanocomposites is that of γ-Fe_2O_3 nanoparticles dispersed in a binder and coated on a tape or disc, for use in the recording industry (including computer memories). By going to nanometre sizes, one is able to increase the density of information storage. Like ferrofluids, this class of nanocomposites is also based on 0–3 connectivity (cf. the appendix on composites). Some other connectivity patterns for improving the performance of magnetic recording media have been discussed by Newnham, Trolier-McKinstry and Ikawa (1990).

Spin glasses can be regarded as self-assembled, or natural, magnetic nanocomposites. Similarly, relaxor ferroelectrics are natural nanocomposites involving electrically ordered nanoclusters.

The electrical analogues of ferrofluids, called ferroelectric fluids, also exist. Poled nanoparticles of $PbTiO_3$, of size $\sim$20 nm, are dispersed in a polymer matrix. The composite exhibits piezoelectricity. Similarly, milled particles of $BaTiO_3$, having a size of $\sim$10 nm, are dispersed in a mixture of heptane and oleic acid, and the composite exhibits a permanent dipole moment.

The best-known example of a ferroelastic nanocomposite is alumina impregnated with sub-micron sized particles of zirconia. Such a composite has a very high fracture-toughness. A material is said to be tough if it resists fracture by resisting the propagation of cracks. This is realized in this nanocomposite by the fact that at sub-nanometre sizes there is a suppression of the phase transition in zirconia particles from tetragonal to monoclinic symmetry. When a crack tends to propagate in the nanocomposite, the stress field associated with the crack causes the size-suppressed phase transition to occur, which has the effect of dissipating the energy of the crack, thus preventing it from propagating (cf. Wadhawan 2000).

Nanocomposites based on CNTs promise to be among the wonder materials of the future. CNTs have the best known stiffness and strength. They are thermally stable in vacuum up to 2800°C, and their thermal conductivity is better than that of diamond. CNT-reinforced nanocomposites are therefore expected to have exceptionally good mechanical properties, and yet a low value for the density.

6.5   Stability and optimum performance of artificial nanostructures

Two approaches are possible for obtaining nanostructured materials by the bottom-up approach. One is the 'natural' self-assembly approach, which we shall discuss later in this chapter. The other is the 'artificial' nanofabrication and nanomanipulation approach, including what was advocated by Drexler (1992) for creating nanostructures atom-by-atom or molecule-by-molecule by manipulating their movement and positioning. This latter approach makes use of various MBBs, as also special techniques like near-field scanning microscopy. Nanostructures made by this approach include electrical devices based on the use of carbon nanotubes (Huang *et al.* 2001; Bachtold *et al.* 2001; Service 2001), as also devices like optical sieves (Ebbesen *et al.* 1998).

Not all artificially created or conceivable nanostructures may be stable. The laws of thermodynamics of small systems must be obeyed. This is where the advantage of self-assembled nanostructures becomes obvious: Since they assemble spontaneously, their stability can be taken for granted (to a large extent).

For nanostructures which are going to be put together atom by atom, it is necessary to ascertain the stability, not only of the end product, but also of the intervening stages. Computational physics calculations can be put to good use for this purpose (Mansoori 2005). The problem is more tractable for metallic structures, compared to those with covalent bonding.

Rieth (2003) has reported extensive molecular dynamics (MD) calculations for examining the stability and therefore the feasibility of fabricating nanostructures of complex shapes and interactions. For nanomachines particularly, the MD calculations must even make a provision for interactions which change with moving parts. Certain surfaces and edges get ruled out at nanodimensions.

A possible anticipated complication is that, at very small sizes, when two parts in a nanomachine move past each other, their interaction may distort their shapes, temporarily or permanently (Landman *et al.* 1990). The choice of materials and geometries becomes important in this context. At still smaller sizes, nothing much can be done to prevent nanoparticles from undergoing spontaneous changes of shape (fluctuations). For example, gold clusters of 1–10 nm radius (containing ~460 atoms) undergo fluctuation-induced changes of shape, as observed by electron microscopy (Sugano and Koizumi 1998).

At present, the fabrication of most of the devices coming under the heading 'MEMS' (micro-electro-mechanical systems), e.g. accelerometers and other similar sensors, is carried out by photolithography and related techniques coming under semiconductor microtechnology. The term 'micromachining' is also used in the context of fabricating MEMS. It is essentially a set of etching processes (Gaitan 2004), and we shall discuss it shortly. Extension of such approaches to NEMS (nano-electro-mechanical systems) is not at all straightforward.

Although the matter of stability is more relevant for NEMS than for MEMS, *optimal performance* is crucial for both. Therefore, a large volume of modelling and simulation studies are carried out by engineers for designing MEMS, NEMS and other smart structures properly (Macucci *et al.* 2001). The finite element method (FEM) is particularly popular, although it has its limitations. Like in any composite structure or system, the performance may depend critically on relative sizes, positions, and connectivities of the various components. One first tries to optimize the configuration on the computer, rather than by actually building the structure and testing it. The presence of functional or smart components in the composite structure makes the task of modelling even more challenging. The response functions of such components, and the coupling of this response to the rest of the structure, are more difficult to model than when only structural components are present (Tzou 1993).

6.6   Two-dimensional microfabrication procedures

Planar technology is a key concept in the fabrication of IC chips, which is being also used for making three-dimensional microdevices, and even some nanodevices. One makes *patterned layers*, one on top of the other, with materials of different properties. This sandwich of patterned layers is used for making circuit elements like resistors, capacitors, transistors, rectifiers, which are suitably interconnected by a patterned conducting overlayer.

Various techniques are used for creating the different layers. For example, silicon has the unique property that it forms an amorphous SiO_2 layer on itself in an oxidizing atmosphere, a layer that is impervious and strongly bonded. This layer can be used both as a protective coating and as a dielectric insulation between circuit functions. Another way of modifying the silicon substrate is by doping it. Layers can also be deposited from an external source by evaporation or sputtering.

Patterning is normally effected by lithography (optical, X-ray, or electron-beam lithography) (Cerrina 1997; Wallraff and Hinsberg 1999).

Conventional planar technology is essentially a three-step process: (i) deposition; (ii) lithography; and (iii) etching. While it has met outstanding commercial success at the micron-level scale, its extension to nanometre scales and the fabrication of NEMS is fraught with difficulties (Lyshevski 2002, 2005), and economical batch production has not been possible to a substantial extent. Photolithography does not give high-enough resolution; and electron and X-ray lithographies are not commercially economical. One promising, high-throughput alternative is that of *nanoimprint lithography* (Chu, Krauss and Renstrom 1996). In it, a resist is patterned by first taking it to a high temperature where it softens, and then physically deforming its shape with a mould having the nanostructure pattern on it. This creates a contrast pattern in the resist. The compressed portions are then etched away.

6.7   Fabrication of small three-dimensional structures

There have been a few successful attempts at making small three-dimensional structures directly, i.e. by going beyond the essentially planar technology of photolithography (cf. Amato 1998). We shall discuss fabrication of three-dimensional MEMS in Chapter 9. A glimpse of what can be done by an assortment of techniques is given here.

6.7.1   *Variable-thickness electroplating*

This is a variant of the photolithography technique. A photosensitized gel is exposed through a grey-scale mask. This results in varying amounts of cross-linking in the gel, in proportion to the exposure at various points. This, in turn, determines the resistance offered to ion-transport through the gel. Therefore, when electroplating is carried out through such a gel, the original grey-scaling gets transferred to thickness variations in the final deposit (Angus *et al.* 1986).

6.7.2   *Soft lithography onto curved surfaces*

A soft lithography technique has been used by Jackman *et al.* (1998) in which two-dimensional patterns are first transferred to cylindrical substrates. Uniaxial straining is then resorted to for transforming this pseudo three-dimensional structure to a truly three-dimensional structure of different symmetry. The patterns are then connected on intersecting surfaces to obtain free-standing three-dimensional structures. A microelectrodepostion step provides further strengthening, as well as welding of the nonconnected structures.

For a general description of soft lithography, as also 'dip-pen lithography', see Whitesides and Love (2001).

6.7.3   *Arrays of optically trapped structures*

A tightly focussed laser beam exerts forces which can be used for trapping and manipulating small transparent objects. MacDonald *et al.* (2002) have used this 'optical tweezing' technique for creating vertical arrays of micron-sized dielectric particles. An interferometric pattern between two annular laser beams was used for this purpose. The 'angular Doppler effect' was used for achieving continuous rotation of the trapped structure. A large number of particles can be loaded, and then coaxed into self-assembling into stable configurations.

6.8   Self-assembled nanostructures

We have discussed self-assembly and self-organization of structures in Sections 5.4 and 5.5. In the field of nanostructures, a substantial amount of activity has been going on for inducing the desired structures to self-assemble from the constituent molecules (see, for example, Stupp *et al.* 1997). This approach 'strives for self-fabrication by the controlled assembly of ordered, fully integrated, and connected operational systems by hierarchical growth' (Lehn 2002). Some examples of self-assembled nanostructures are those for microelectronics and nanoelectronics (Gau *et al.* 1999; Sirringhaus *et al.* 2000; Dragoman and Dragoman 2006), photonics (Jenekhe and Chen 1999), and nanocrystal superlattices (Sun *et al.* 2000; Black *et al.* 2000).

The dynamic, time-reversible nature of supramolecular assembly and organization processes makes this approach ideal for imbuing such nanostructures with a high degree of smartness. They can self-adjust in response to a changing environment, and even have capabilities like healing, adaptation, evolution, and learning.

The self-assembly approaches discussed in a general way in Section 5.4 were mostly for millimetre-sized components. More such examples can be found in Davis *et al.* (1998), Xia *et al.* (1999), Jager, Smela and Inganas (2000), and Schon, Meng and Bao (2001). Do these approaches have the potential to be scaled down to mesoscale structures? This is not easy, because it amounts to being able to design *molecules* (rather than millimetre-sized components) which have the requisite shape and bonding tendencies for molecular recognition, followed by self-assembly and self-organization. The predominantly covalent nature of the bonding in molecules makes this a highly nontrivial task. This is why most of the existing techniques for three-dimensional micro- and nanofabrication are based on photolithography. Nevertheless, attempts continue to be made for creating nanostructures by self-assembly (Whitesides, Mathias and Seto 1991).

We mention here the work of Lopes and Jaeger (2001) as an example of the potential that the so-called *hierarchical self-assembly* approach has for creating nanostructures. They demonstrated the assembly of organic–inorganic copolymer–metal nanostructures, such that one level of self-assembly guided the next level. As a first step, ultra-thin diblock copolymer films were made to self-split into a regular scaffold of highly anisotropic stripe-like domains. In the next step, differential wetting guided the diffusion of metal atoms such that they aggregated selectively along the scaffold into highly organized nanostructures. Depending on the kinetics of the process, two distinct modes of assembly of the metal atoms, namely chains of separate nanoparticles and continuous nanowires, were observed.

Semiconductor *quantum dots* (QDs) constitute an important class of self-assembled nanostructures. They are nanocrystals having typical dimensions of a few nanometres across (Denison *et al.* 2004). When they are prepared by a technique like MBE (molecular beam epitaxy), they self-assemble as *islands* on a substrate (Weinberg *et al.* 2000; Kelsall, Hamley and Geoghegan 2005). Although some tricks can be used to make them of nearly the same size, there is bound to be some size variation.

The electronic structure of QDs depends very sensitively on the size. This is an example of size-dependent variation of properties of nanostructures. In fact, it is not always unwelcome. QDs of CdSe or CdTe are used in biological research as colour-coded fluorescent markers: When radiation of a short-enough wavelength falls on them, electron–hole pairs are created; when these pairs recombine, visible light is emitted (Alivisatos 2001; Whitesides and Love 2001).

Other possible applications for QDs include those in semiconductor electronics (discussed in the next section), optical devices (Poole and Owens 2003; Fafard 2004; Heflin 2004; Xia, Kamata and Lu 2004), quantum communication, and quantum computing (Flatte 2004).

Semiconductors can also be produced as nanoclusters (cf. Alivisatos 1996; Kelsall, Hamley and Geoghegan 2005). Since they have a tendency to coagulate, versatile coating strategies have been evolved to prevent this from happening (Ishii *et al.* 2003).

Molecules like C_{60} (fullerene) and C_{∞} (nanotubes) are nanostructures with the important feature that there are no errors when they are formed; they are identical (notwithstanding isotopic variation). However, errors do begin to appear when we deal with larger and larger self-assembled structures. Such occasional errors notwithstanding, self-assembly *can* give long-range ordering. *Self-assembled monolayers* (SAMs) are an example of this. In these systems, there is usually a surface-active head group (surfactant) that attaches to its corresponding substrate via a chemical adsorption process. As a result of this process, ultra-thin monolayers are obtained (Wilbur and Whitesides 1999; Mansoori 2005).

Layer-by-layer (LBL) deposition is an extension of SAM procedures for obtaining thick, even spherical, functional nanostructures (Shenhar, Norsten and Rotello 2004).

6.9   Quantum dots and other 'island' structures for nanoelectronics

We introduced QDs in the above section. They are an example of *island structures*, in which electrons are *confined* over nanometre dimensions. Other such examples include resonant tunnelling devices (RTDs) (Maezawa and Forster 2005), and single-electron transistors (SETs) (Uchida 2005). The drive towards more and more miniaturization in computing devices for achieving ultra-dense circuitry has led to a surge of investigations on such systems (Asai and Wada 1997; Dupas, Houdy and Lahmani 2006). Similarly, in molecular computers, a molecule with predominantly covalent bonding serves as the island in which the valence electrons are confined (Mayor, Weber and Waser 2005).

These devices exploit the wavelike nature of electrons. When confined to such small dimensions, the electron waves set up interference patterns. These devices should work better and better as we build them smaller and smaller. At present, their operations have been demonstrated only at low temperatures, because even thermal fluctuations at higher temperatures are enough to bump the electrons from one crest to another in the interference pattern. The separation between the crests widens as the dimensions of the system are reduced, and for confinements extending over ∼10 nm, even room-temperature operation of the devices should be possible.

In IC chips or other digital computing devices, the two basic functions performed are switching (between the states 0 and 1), and amplification (Waser 2005). In conventional transistors, these operations involve the movement of large numbers of electrons in bulk matter. By contrast, as stated above, the new-generation nanoelectronic devices are based on quantum interference effects at nanometre scales. Two broad categories can be identified for the latter:

1. solid-state quantum-effect devices and single-electron devices; and
2. molecular-electronic devices.

The former set of nanoelectronic configurations requires the presence of an island of semiconductor or metal, in which there is a quantum confinement of electrons. They can be classified in terms of the number of degrees of freedom for these electrons (Goldhaber-Gordon et al. 1997): (i) QDs (zero degrees of freedom); (ii) RTDs (one or two degrees of freedom); and (iii) SETs (three degrees of freedom).

Confinement of electrons in the island or well has two consequences: Discretization of the energy states in which electrons can exist; and quantum tunnelling of the electrons through the island boundary. These two factors strongly influence the flow of charge across the nanodevice.

The smaller the size of the island, the greater is the separation between two quantized energy states. By contrast, in bulk transistors the energy levels are very closely spaced; they form 'bands' of a range of energies.

The island is surrounded by a 'source' and a 'drain' for electron flow (cf. Goldhaber-Gordon 1997). It is possible to adjust the energies of the discrete states in the potential well with respect to the energy bands in the source and the drain. Application of a bias voltage is one such way of doing so.

The island in the case of RTDs is usually long and narrow; e.g. a 'quantum wire' or a 'pancake', with shortest dimensions typically in the 5–10 nm range.

QDs involve quantization in all three dimensions (Lent and Tougaw 1997).

SETs are three-dimensional devices, comprising of a gate, a source, and a drain. Small changes in the charge on the gate result in a switching on or switching off of the source-to-drain current. The change in charge may be as small as a single electron.

It is expected that computers with nanometre-sized components will be thousands of times faster than the present computers, and will also pack thousands of times higher density of components. They will consume much less power because only a few electrons will have to be moved across quantum crests, rather than moving them in large numbers through resistive material. Declining costs will be a natural fallout.

Patterned magnetic nanostructures and quantized magnetic discs offer a still higher density of nanocircuits for computing (Chou 1997; Solaughter, DeHerrera and Durr 2005).

In molecular electronics, the molecule itself is the confining device, which offers several unique features, yet to be fully exploited: The molecules are naturally occurring nanostructures. They are identical. And they can be manufactured cheaply in very large numbers. Switching and memory cells made of single molecules have been demonstrated. They have the potential of having a billion times the packing density of present-day circuits for computing.

The ultimate in computing power may be the quantum computers, in which the whole computer acts in a wavelike manner. The content of any memory cell of such a computer is a superposition of both 0 and 1, and the whole computer is a superposition of all possible combinations of the memory states. The contents of the memory cells get modified by a sequence of logical transformations as the computation proceeds. At the end of a job, the final solution is extracted from quantum interference of all the possibilities (Moravec 1998; Flatte 2004; Mooji 2004; Le Bellac 2006; Collins 2006).

6.10   Nanotechnology with DNA

The nanoscale is also the scale at which biological systems build their structures by self-assembly and self-organization. Therefore it is only natural that we should think of exploiting such systems for building our artificial nanostructures (Wagner 2005). Of special interest in this context is the ubiquitous role played by DNA in the creation of nanostructures (Barbu, Morf and Barbu 2005).

DNA has the famous double helix structure (Dennis and Campbell 2003). Each of the two backbone helices consists of a chain of phosphate and deoxyribose sugar molecules, to which are attached the bases adenine (A), thymine (T), cytosine (C), and guanine (G), in a certain sequence. A phosphate molecule and the attached sugar molecule and base molecule constitute a *nucleotide*. A strand of DNA is a *polynu-cleotide*. The two helices in the double-helix structure of DNA are bonded to each other, along their length, through hydrogen bonds between complementary base pairs: A bonds to T, and C bonds to G.

DNA usually exists as B-DNA, which has the right-handed screw conformation. However, conditions can be created such that the left-handed conformation (called Z-DNA) is the more stable of the two.

DNA has outstanding properties for building structures by self-assembly and self-organization. The two strands of DNA have a strong affinity for each other. What is more, they form *predictable* structures when they bond to each other along their lengths. Modern biotechnology has made it possible to synthesize long or short DNA molecules having a sequence of base pairs chosen at will (cf. Shih, Quispe and Joyce 2004; Baker *et al.* 2006). This is an important enabling factor for designing even those DNA configurations which are not found in Nature.

DNA is highly and consistently selective in its interactions with other molecules, depending strongly on the sequence of its base pairs. This is important for molecular recognition (cf. Section 5.3), a precursor to self-assembly (Section 5.4).

Although linear DNA double helices are probably of limited utility, it has been possible to design synthetic DNA molecules having stable *branched* structures. This has made possible the creation of highly complex nanostructures by self-assembly. The branching of DNA occurs, for example, when the double helix partially unravels into its component strands, and each of the branches then seeks and gets complementary bases for pairing from another DNA strand. In biological processes, the branching occurs during, for example, replication and recombination. Replication involves the pairing of the bases, all along the length of the strand. Recombination of DNA is a familiar process in genetics. In the well-known *crossover* process, a portion of DNA from a strand from one parent is exchanged with that from the other parent.

The strong molecular-recognition feature of DNA is further enhanced by the possible occurrence of *sticky ends* in it, particularly those in branched structures. Imagine a DNA strand with a certain number, N, of bases along the chain. If all N bases are paired to N complementary bases of another strand having the same length, then there are no sticky ends. But if one strand is longer or shorter than the other, then one or both ends of the double helix would have some unpaired bases, hungry for sticking (bonding) to bases complementary to them (Fig. 6.1). This fact has been exploited for long in genetic engineering, and is now being put to great use in nanotechnology (see, for example, Friend and Stoughton 2002; Seeman and Belcher 2002).

DNA is a particularly suitable molecule for building nanostructures. It scores over, for example, proteins for this purpose, even though the latter also have, just like the nucleotide chains of DNA, the polypeptide chains having a sequence of amino acids. This sequence of amino acids determines the way a protein *folds* into a complicated (though unique) globular shape. Protein folding is still not a properly understood process. By contrast, DNA simply has a double-helical conformation.

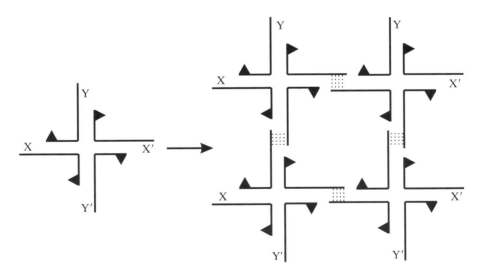

Fig. 6.1   Formation of a two-dimensional 'crystal' from a DNA building block (or 'monomer') with sticky ends. X and Y are sticky ends, and X′ and Y′ are their complements. Four of the units on the left are shown as having self-assembled into the complex on the right. [After Seeman and Belcher (2002). Copyright 2002, National Academy of Sciences, USA. Figure redrawn with the permission of the publisher.]

We mention here a few of the uses to which DNA has been put for creating nanostructures by self-assembly.

A variety of branched DNA configurations with sticky ends can be created in the laboratory by designing the sequence of bases attached to the backbone of the DNA strand in a specific way. A proper sequence can make them self-assemble into specific branched configurations (building blocks) with sticky ends (Seeman and Belcher 2002). Then follows the next level of self-assembly and self-organization, such that the building blocks form orderly repetitive structures (like in a crystal). Building blocks having shapes such as cubes, octahedra, and truncated octahedra have been produced (Seeman 2003, 2004; Turberfield 2003; Shih, Quispe and Joyce 2004).

The crystal-like repetitive structure into which the building blocks self-assemble can act as a scaffolding or template for holding *guest molecules* in a definite orientation at every lattice point, in one, two, and (eventually) three dimensions (Braun *et al.* 1998; Seeman 2004). X-ray diffraction studies can be carried out on such 'crystals' to obtain information about the structure of the guest molecules, which may be difficult or impossible to crystallize otherwise.

Another application of DNA is for fabricating nano-sized or molecular machines. A conformational transition between the B and Z forms of DNA involves movement of large segments of the molecule, making it useable for nanomecahnical devices like tweezers or rotating shafts (Yurke *et al.* 2000; Seeman 2004). B-DNA is the stable configuration in typical aqueous solutions. Introduction of chemicals like cobalt hexamine in the solution makes it possible for the Z-DNA configuration to form, provided it has a stretch of alternating C and G bases. Thus microactuator or even nanoactuator action can be achieved by effecting the B-Z conformational transition by altering the nature of the surrounding liquid.

The invention of planar technology in 1957 for making IC chips (cf. Section 6.6 above) heralded a revolution in electronics, which has still not abated. Before this remarkable combination of technology and methodology was introduced, electronic circuits were manufactured by literally wiring together the individual transistors one by one, with the attendant irreproducible results and high costs. A similar revolution is now emerging for the flexible and reliable fabrication of the biological equivalent of ICs, with genes taking the place of transistors (Baker *et al.* 2006). Genes are long strands of DNA, having a specific sequence of nucleotides. Mass fabrication of nanodevices based on genetic circuits would need a planar-technology-like approach for the assembly-line production of long stretches of DNA in a reproducible, dependable, and low-cost manner.

We would like such mass production to be error-free. This has proved to be a tough task. We discussed in Section 2.7.6 how Nature has evolved processes like kinetic proofreading for minimizing errors in the production of biomolecules involved in the complex biological circuitry. In Nature, enzymes such as polymerase can manufacture DNA molecules at incredibly high speeds, and with extremely low chances of mistakes. Multiple polymerases, working in parallel, can manufacture five million DNA bases in 20 minutes. The error rate is as low as one base in a billion. By contrast, the best that humans have been able to achieve so far in the laboratory is an error rate of one base in 10,000 in the production of synthetic DNA (Baker *et al.* 2006).

6.11   Chapter highlights

- The nanoscale is where most of the action takes place in natural biological phenomena. The fields of nanotechnology and smart structures will therefore grow hand in hand.
- Diamondoids are an important class of molecular building blocks in nanotechnology. They may be used in the fabrication of nanorobots and other molecular machines.
- Phase transitions in mesoscale systems, particularly those occurring when their size is changed even slightly (for example due to changes in the surroundings), offer the possibility of designing smart nanoclusters with properties that we can select and control.
- The melting point and the freezing point may not always coincide in very small systems, contrary to the situation for large systems. This happens because for atoms in the interior of the cluster the chemical binding is solid-like, and for surface atoms it is liquid-like.
- Berry's theory provides a seamless thermodynamic description of transitions in small, intermediate, and large systems. A solid is fully rigid, and a liquid is fully nonrigid. Anything in between, as quantified by the local nonrigidity parameter γ for a cluster, is a different form of matter, unique to nanoclusters. The local degree of nonrigidity varies strongly with the addition or subtraction of just a few atoms or molecules constituting the cluster. This offers the possibility of designing nanoclusters with size-dependent properties.
- Size effects in nanostructured ferroic materials include properties like superparamagnetism, superparaelectricity, and superparaelasticity, with good potential for applications in smart structures.
- Ferroic and other nanocomposites offer immense scope for design and optimization of properties for applications in smart structures. Smart composites are among the earliest of materials envisaged for incorporating sensors, actuators and microprocessor controllers into a single integrated smart entity.
- Modelling and simulation studies are an integral part of designing MEMS, NEMS, and other smart structures optimally (Baier and Dongi 1999). Like in any composite structure or system, the performance of a smart structure may depend critically on relative sizes, positions, and connectivities of the various components. The presence of functional or smart components in the composite structure makes the modelling task more challenging. The response functions of such components, and the coupling of this response to the rest of the structure, are more difficult to model than when only structural components are present (Lenhard 2004).
- The dynamic, time-reversible nature of supramolecular assembly and organization processes makes the self-assembly approach of making nanostructures ideal for

imbuing such structures with a high degree of smartness. They can self-adjust in response to a changing environment, and can even have capabilities like healing, adaptation, evolution, and learning.

- The hierarchical self-assembly approach has good potential for creating nanostructures. As demonstrated for the assembly of organic–inorganic copolymer–metal nanostructures, one level of self-assembly guides the next level.

- The drive towards more and more miniaturization in computing devices for achieving ultra-dense circuitry has led to a surge of investigations on 'island structures', in which the electrons exhibit quantum-mechanical confinement effects. Similarly, in molecular computers, a molecule with predominantly covalent bonding serves as the island in which the valence electrons are confined. When confined to such small dimensions, the electron waves set up interference patterns. These devices should work better and better (i.e. become more stable) as we build them smaller and smaller.

- Computers with nanometre-sized components are expected to be thousands of times faster than present computers, and they will also pack thousands of times higher density of components. They will consume much less power because only a few electrons will have to be moved across quantum crests, rather than moving them in large numbers through resistive material.

- DNA has outstanding properties for building nanostructures by self-assembly and self-organization. The two strands of DNA have a strong affinity for each other, and, unlike polypeptides, they form *predictable* structures on bonding along their lengths. Modern biotechnology has made it possible to synthesize DNA molecules having a sequence of base pairs chosen at will. This is an important enabling factor for designing even those DNA configurations which are not found in Nature.

- The strong molecular-recognition feature of DNA (required for self-assembly and self-organization of nanostructures) is further enhanced by the possible occurrence of sticky ends in it, particularly those in branched structures. A variety of branched DNA configurations with sticky ends can be created in the laboratory by designing the sequence of bases attached to the backbone of the DNA strand in a specific way. A proper sequence can make them self-assemble into specific branched configurations (building blocks) with sticky ends. The building blocks then self-assemble into orderly repetitive structures.

- Microactuator or nanoactuator action can be achieved by effecting the B-Z conformational transition in DNA. This has possibilities for creating molecular machines.

References

Ajayan, P. M. (1999). 'Nanotubes from carbon'. *Chem. Rev.* 99: 1787.

Ajayan, P. M. and L. D. Marks (1990). 'Phase instabilities in small particles'. *Phase Transitions*, 24–26: 229.

Alivisatos, A. P. (1996). 'Semiconductor clusters, nanocrystals, and quantum dots'. *Science*, 271: 933.

Alivisatos, A. P. (September 2001). 'Less is more in medicine'. *Scientific American,*, 285: 59.

Amato, I. (16 October 1998). 'Fomenting a revolution, in miniature'. *Science*, 282: 402.

Angus, J. C., U. Landau, S. H. Liao and M. C. Yang (1986). 'Controlled electroplating through gelatine films'. *J. Electrochem. Soc.* 133: 1152.

Arlt, G. (1990). 'Twinning in ferroelectric and ferroelastic ceramics: stress relief'. *J. Mater. Sci.* 25: 2655.

Asai, S. and Y. Wada (1997). 'Technology challenges for integration near and below 0.1 μm'. *Proc. IEEE*, 85(4): 505.

Bachtold, A., P. Hadley, T. Nakanishi and C. Dekker (2001). 'Logic circuits with carbon nanotube transistors'. *Science*, 294: 1317.

Baier, H. and F. Dongi (1999). 'Simulation of adaptronic systems', in Janocha, H. (1999) (ed.). *Adaptronics and Smart Structures: Basics, Materials, Design, and Applications.* Berlin: Springer-Verlag.

Baker, D., G. Church, J. Collins, D. Endy, J. Jacobson, J. Keasling, P. Modrich, C. Smolke and R. Weiss (June 2006). 'Engineering life: Building a FAB for biology'. *Scientific American,* 294(6): 44.

Balzani, V., M. Venturi and A. Credi (2003). *Molecular Devices and Machines: A Journey into the Nanoworld.* Wiley-VCH.

Barbu, S., M. Morf and A. E. Barbu (2005). 'Nanoarchitectures, nanocomputing, nanotechnologies and the DNA structure'. In R. S. Greco, F. B. Prinz and R. L. Smith (eds.), *Nanoscale Technology in Biological Systems.* New York: CRC Press.

Baughman, R. H., A. A. Zakhidov and W. A. de Heer (2 August 2002). 'Carbon nanotubes – the route toward applications'. *Science*, 297: 787.

Berry, R. S. (August 1990a). 'When the melting and the freezing points are not the same'. *Scientific American,*, p. 68.

Berry, R. S. (1990b). 'Melting, freezing and other peculiarities in small systems'. *Phase Transitions*, 24–26: 259.

Berry, R. S. (21 May 1998). 'Size is everything'. *Nature*, 393: 212.

Berry, R. S. (1999). In J. Jellinek (ed.), *Theory of Atomic and Molecular Clusters.* Berlin: Springer-Verlag.

Bertsch, G. (1997). 'Melting in clusters'. *Science*, 277: 1619.

Black, C. T., C. B. Murray, R. L. Sandstrom and S. Sun (2000). 'Spin-dependent tunnelling in self-assembled cobalt-nanocrystal superlattices'. *Science*, 290: 1131.

Bondorf, J. P., A. S. Botvina, A. S. Iljinov, I. N. Mishustin and K. Snepen (1995). 'Statistical multifragmentation of nuclei'. *Phys. Rep.* 257: 133.

Braun, E., Y. Eichen, U. Sivan and G. Ben-Yoseph (19 February 1998). 'DNA-templated assembly and electrode attachment of a conducting silver wire'. *Nature*, 391: 775.

Brodie, I. and J. M. Muray (1992). *The Physics of Micro/Nano-Fabrication.* New York: Plenum Press.

Cerrina, F. (1997). 'Application of X-rays to nanolithography'. *Proc. IEEE*, 85(4): 644.

Chacko, S., K. Joshi, D. G. Kanhere and S. A. Blundell (2004). 'Why do gallium clusters have a higher melting point than the bulk?'. *Phys. Rev. Lett.* 92(13): 135506-1.

Challa, M. S. S., D. P. Landau and K. Binder (1990). 'Monte Carlo studies of finite-size effects at first-order transitions', 24–26: 343.

Chandross, E. A. and R. D. Miller (eds.) (1999). 'Nanostructures: Introduction'. *Chem. Rev.* 99: 1641.

Chou, S. Y. (1997). 'Patterned magnetic nanostructures and quantized magnetic disks'. *Proc. IEEE*, 85(4): 652.

Chou, S. Y., P. R. Krauss and P. J. Renstrom (1996). 'Imprint lithography with 25-nanometer resolution'. *Science* 272: 85.

Cole, A. J. (2004). 'Analytical treatment of constraints in fragmentation'. *Phys. Rev. C*, 69: 054613-1.

Collins, G. P. (April 2006). 'Computing with quantum knots'. *Scientific American, India*, 1(11): 40.

Courbin, L., W. Engl and P. Panizza (2004). 'Can a droplet break up under flow without elongating? Fragmentation of smectic monodisperse droplets'. *Phys. Rev. E*, 69: 061508-1.

Crespi, V. (2004). 'The geometry of nanoscale carbon'. In Di Ventra, M., S. Evoy and J. R. Heflin (eds.), *Introduction to Nanoscale Science and Technology.* Dordrecht: Kluwer.

Dahl, J. E., S. G. Liu and R. M. K. Carlson (2003). 'Isolation and structure of higher diamondoids, nanometer-sized diamond molecules'. *Science*, 299: 96.

Davidson, R. (April 1992). 'Smart composites: where are they going?.' *Materials and Design*, 13: 87.

Davis, W. B., W. A. Svec, M. A. Ratner and M. R. Wasielewski (1998). 'Molecular-wire behaviour in p-phenylenevinylene oligomers'. *Nature*, 396: 60.

Denison, A. B., L. J. Hope-Weeks, R. W. Meulenberg and L. J. Terminello (2004). 'Quantum dots'. In Di Ventra, M., S. Evoy and J. R. Heflin (eds.), *Introduction to Nanoscale Science and Technology*. Dordrecht: Kluwer.

Dennis, C. and P. Campbell (eds.) (23 January 2003). 'The double helix – 50 years'. Special feature in *Nature*, 421: 395.

Dorn, H. C. and J. C. Duchamp (2004). 'Fullerenes'. In Di Ventra, M., S. Evoy and J. R. Heflin (eds.), *Introduction to Nanoscale Science and Technology*. Dordrecht: Kluwer.

Dragoman, M. and D. Dragoman (2006). *Nanoelectronics: Principles and Devices*. Boston: Artech House.

Dresselhaus, M. S., G. Dresselhaus and P. C. Eklund (1996). *Science of Fullerenes and Carbon Nanotubes*. New York: Academic Press.

Drexler, K. E. (1981). 'Molecular engineering: An approach to the development of general capabilities for molecular manipulation'. *Proc. Natl. Acad. Sci. USA*, 78: 5275.

Drexler, K. E. (1986). *Engines of Creation*. New York: Anchor Books.

Drexler, K. E. (1992). *Nanosystems: Molecular Machinery, Manufacturing and Computation*. New York: Wiley.

Drexler, K. E. (Sept. 2001). 'Machine-phase nanotechnology'. *Scientific American,* 285: 66.

Drexler, K. E., C. Peterson and G. Pergamit (1991). 'Unbounding the future: The nanotechnology revolution'. New York: William Morrow and Co.

Dupas, C., P. Houdy and M. Lahmani (eds.) (2006). *Nanoscience: Nanotechnologies and Nanophysics*. Berlin: Springer.

Eastman, J. and R. Siegel (January 1989). 'Nanophase synthesis assembles materials from atomic clusters'. *Research and Development*, p. 56.

Ebbesen, T. W., H. Lezec, H. F. Ghaemi, T. Thio and P. A. Wolff (1998). 'Extraordinary optical transmission through sub-wavelength hole arrays'. *Nature*, 391: 667.

Evoy, S., M. Duemling and T. Jaruhar (2004). 'Nanoelectromechanical systems'. In Di Ventra, M., S. Evoy and J. R. Heflin (eds.), *Introduction to Nanoscale Science and Technology*. Dordrecht: Kluwer.

Fafard, S. (2004). 'Quantum-confined optoelectronic systems'. In Di Ventra, M., S. Evoy and J. R. Heflin (eds.), *Introduction to Nanoscale Science and Technology*. Dordrecht: Kluwer.

Flatte, M. E. (2004). 'Semiconductor nanostructures for quantum computation'. In Di Ventra, M., S. Evoy and J. R. Heflin (eds.), *Introduction to Nanoscale Science and Technology*. Dordrecht: Kluwer.

Friend, S. H. and R. B. Stoughton (February 2002). 'The magic of microarrays'. *Scientific American,* p. 34.

Gaitan, M. (2004). 'Introduction to integrative systems'. In Di Ventra, M., S. Evoy and J. R. Heflin (eds.), *Introduction to Nanoscale Science and Technology*. Dordrecht: Kluwer.

Gau, H., S. Herminghaus, P. Lenz and R. Lipowsky (1999). 'Liquid morphologies on structures surfaces: from microchannels to microchips'. *Science*, 283: 46.

Goldhaber-Gordon, D., M. S. Montemerlo, J. C. Love, G. J. Opiteck, and J. C. Ellenbogen (1997). 'Overview of nanoelectronic devices'. *Proc. IEEE*, 85(4): 521.

Grossman, B., T. Alavie, F. Ham, J. Franke and M. Thursby (1989). 'Fibre-optic sensor and smart structures research at Florida Institute of Technology', in E. Udd (ed.), *Proc. SPIE (Fibre Optic Smart Structures and Skins II)*, 1170: 123.

Gruverman, A. and A. Kholkin (2006). 'Nanoscale ferroelectrics: processing, characterization, and future trends', *Rep. Prog. Phys.*, 69: 2443.

Haneda, K. and A. H. Morrish (1990). 'Mossbauer spectroscopy of magnetic small particles with emphasis on barium ferrite'. *Phase Transitions*, 24–26: 661.

Harriott, L. R. and R. Hull (2004). 'Nanolithography'. In Di Ventra, M., S. Evoy and J. R. Heflin (eds.), *Introduction to Nanoscale Science and Technology*. Dordrecht: Kluwer.

Heflin, J. R. (2004). 'Organic optoelectronic nanostructures'. In Di Ventra, M., S. Evoy and J. R. Heflin (eds.), *Introduction to Nanoscale Science and Technology*. Dordrecht: Kluwer.

Huang, Y., X. Duan, Y. Cui, L. J. Lauhon, K.-H. Kim and C. M. Lieber (2001). 'Logic gates and computation from assembled nanowires building blocks'. *Science*, 294: 1313.

Iijima, S. (1991). 'Helical microtubes of graphitic carbon'. *Nature*, 345: 56.

Ishii, D., K. Kinbara, Y. Ishida, N. Ishii, M. Okochi, M. Yohda and T. Aida (5 June 2003). 'Chaperonin-mediated stabilization and ATP-triggered release of semiconductor nanoparticles'. *Nature*, 423: 628.

Jackman, R. J., S. T. Brittain, A. Adams, M. G. Prentiss and G. M. Whitesides (1998). 'Design and fabrication of topologically complex three-dimensional microstructures'. *Science* 280: 2089.

Jager, E. W. H., E. Smela and O. Inganas (2000). 'Microfabricating conjugated polymer actuators'. *Science*, 290: 1540.

Jenekhe, S. A. and L. X. Chen (1999). 'Self-assembly of ordered microporous materials from rod-coil block polymers'. *Science*, 283: 372.

Kakade, N. (April 2003). 'Nanotechnology: The progress so far'. *Electronics for You*, p. 47.

Kelsall, R., I. Hamley and Geoghegan (eds.) (2005). *Nanoscale Science and Technology*. London: Wiley.

Kohler, M. and W. Fritzsche (2004). *Nanotechnology: An Introduction to Nanostructuring Techniques*. Wiley-VCH Verlag GmbH and Co., KGaA.

Kroto, H. W., J. R. Heath, S. C. O'Brien, R. F. Curl and R. E. Smalley (14 November 1985). 'C_{60}: Buckminsterfullerene'. *Nature*, 318: 165.

Landman, U., W. D. Luedtke, N. A. Burnham and R. J. Colton (1990). 'Atomistic mechanisms and dynamics of adhesion, nanoindentation, and fracture'. *Science*, 248: 454.

Le Bellac, M. (2006). *A Short Introduction to Quantum Information and Quantum Computation*. Cambridge: Cambridge University Press.

Lehn, J.-M. (16 April 2002). 'Toward complex matter: Supramolecular chemistry and self-organization'. *PNA, USA*, 99(8): 4763.

Lenhard, J. (2004). 'Nanoscience and the Janus-faced character of simulations', in D. Baird, A. Nordmann and J. Schummer (eds.), *Discovering the Nanoscale*. Amsterdam: IOS Press.

Lent, C. S. and P. D. Tougaw (1997). 'A device architecture for computing with quantum dots'. *Proc. IEEE*, 85(4): 541.

Li, S., J. A. Eastman, Z. Li, C. M. Foster, R. E. Newnham and L. E. Cross (1996). 'Size effects in nanostructured ferroelectrics'. *Phys. Lett. A*, 212: 341.

Lopes, W. A. and H. M. Jaeger (2001). 'Hierarchical self-assembly of metal nanostructures on diblock scaffolds'. *Nature*, 414: 735.

Lyshevski, S. E. (2002). *MEMS and NEMS: Systems, Devices, and Structures*. New York: CRC Press.

Lyshevski, S.E. (2005). *Nano- and Micro-Electromechanical Systems: Fundamentals of Nano- and Microengineering*, 2nd edn. London: CRC Press.

Macucci, M., G. Iannaccone, J. Greer, J. Martorell, D. W. L. Sprung, A. Schenk, I. T. Yakimenko, K. F. Berggren, K. Stokbro and N. Gippius (2001). 'Status and perspectives of nanoscale device modelling'. *Nanotechnology*, 12: 136.

MacDonald, M. P., L. Paterson, K. Volke-Sepulveda, J. Arlt, W. Sibbett and K. Dholakia (2002). 'Creation and manipulation of three-dimensional optically trapped structures'. *Science* 296: 1101.

Maezawa, K. and A. Forster (2005). 'Quantum transport devices based on resonant tunnelling'. In Waser, R. (ed.), *Nanoelectronics and Information Technology: Advanced Electronic Materials and Novel Devices*, 2nd edition, p. 405. KGaA: Wiley-VCH Verlag

Mansoori, G. A. (2005). *Principles of Nanotechnology: Molecular-Based Study of Condensed Matter in Small Systems*. Singapore: World Scientific.

Mayo, M. J. (1993). 'Synthesis and applications of nanocrystalline ceramics'. *Materials and Design*, 14: 323.

Mayor, M., H. B. Weber and R. Waser (2005). 'Molecular electronics'. In Waser, R. (ed.), *Nanoelectronics and Information Technology: Advanced Electronic Materials and Novel Devices*, 2nd edition, p. 499. KGaA: Wiley-VCH Verlag

Minoli, D. (2006). *Nanotechnology Applications to Telecommunications and Networking*. New Jersey: Wiley Interscience.

Mooji, H. (December 2004). 'Superconducting quantum bits'. *Phys. World*, 17: 29.

Moravec, H. (1998). 'When will computer hardware match the human brain?' *J. Evolution and Technology*, 1: 1.

Moriarty, P. (2001). 'Nanostructured materials'. *Rep. Prog. Phys.*, 64: 297.

Multani, M. S. and V. K. Wadhawan (eds.) (1990). *Physics of Clusters and Nanophase Materials*. Special issue of *Phase Transitions*, 24–26: 1–834.

Multani, M. S., P. Ayyub, V. Palkar and P. Guptasarma (1990). 'Limiting long-range ordered solids to finite sizes in condensed-matter physics'. *Phase Transitions*, 24–26: 91.

Newnham, R. E. and S. E. Trolier-McKinstry (1991a). 'Crystals and composites' *J. Appl. Cryst.*, 23: 447.

Newnham, R. E. and S. E. Trolier-McKinstry (1990b). 'Structure-property relationships in ferroic nanocomposites'. *Ceramic Transactions*, 8: 235.

Newnham, R. E., S. E. Trolier-McKinstry and H. Ikawa (1990). 'Multifunctional ferroic nanocomposites'. *J. Appl. Cryst.* 23: 447.

O'Connell, M. J. (2006) (ed.). *Carbon Nanotubes: Properties and Applications*. London: CRC Press.

Okazaki, S. and J. Moers (2005). 'Lithography'. In Waser, R., *Nanoelectronics and Information Technology: Advanced Electronic Materials and Novel Devices*, 2nd edition, p. 221. KGaA: Wiley-VCH Verlag

Ozin, G. A. and A. C. Arsenault (2005). *Nanochemistry: A Chemical Approach to Nanomaterials*. Cambridge, U. K.: The Royal Society of Chemistry (RSC) Publishing.

Poole, C. P. and F. J. Owens (2003). *Introduction to Nanotechnology*. New Jersey: Wiley.

Reinhard, P.-G. and E. Suraud (2004). *Introduction to Cluster Dynamics*. KGaA, Weinheim: Wiley-VCH Verlag GmBH and Co.

Rieth, M. (2003). *Nano-Engineering in Science and Technology: An Introduction to the World of Nano-Design*. Singapore: World Scientific.

Rothemund, P. W. K. (16 March 2006). 'Folding DNA to create nanoscale shapes and patterns'. *Nature*, 440: 297.

Schmidt, M., R. Kusche, W. Kronmuller, B. von Issendorff and H. Haberland (1997). 'Experimental determination of the melting point and heat capacity for a free cluster of 139 sodium atoms'. *Phys. Rev. Lett.* 79(1): 99.

Schmidt, M., R. Kusche, B. von Issendorff and H. Haberland (21 May 1998). 'Irregular variations in the melting point of size-selected atomic clusters'. *Nature*, 393: 238.

Schon, J. H., H. Meng and Z. Bao (2001). 'Self-assembled monolayer organic field-effect transistors'. *Nature*, 413: 713.

Schummer, J. and D. Baird (eds.) (2006). *Nanotechnology Challenges: Implications for Philosophy, Ethics and Society*. New Jersey: World Scientific.

Seeman, N. C. (23 January 2003). 'DNA in a material world'. *Nature*, 421: 427.

Seeman, N. C. (June 2004). 'Nanotechnology and the double helix'. *Scientific American,*, p. 64.

Seeman, N. C. and A. M. Belcher (30 April 2002). 'Emulating biology: building nanostructures from the bottom up'. *PNAS USA*, 99(2): 6451.

Service, R. F. (2001). 'Assembling nanocircuits from the bottom up'. *Science*, 293: 782.

Shenhar, R., T. B. Norsten and V. M. Rotello (2004). 'Self-assembly and self-organization'. In Di Ventra, M., S. Evoy and J. R. Heflin (eds.), *Introduction to Nanoscale Science and Technology*. Dordrecht: Kluwer.

Shih, W. M., J. D. Quispe and G. F. Joyce (12 February 2004). 'A 1.7-kilobase single-stranded DNA that folds into a nanoscale octahedron'. *Nature*, 427: 618.

Sirringhaus, H., T. Kawase, R. H. Friend, T. Shimoda, M. Inbasekaran, W. Wu and E. P. Woo (2000). 'High-resolution inkjet printing of all-polymer transistor circuits'. *Science*, 290: 2123.

Smith, B. W. and D. E. Luzzi (2004). 'Carbon nanotubes'. In Di Ventra, M., S. Evoy and J. R. Heflin (eds.), *Introduction to Nanoscale Science and Technology*. Dordrecht: Kluwer.

Solaughter, J. M., M. DeHerrera and H. Durr (2005). 'Magnetoresistive RAM'. In Waser, R. (ed.), *Nano-electronics and Information Technology: Advanced Electronic Materials and Novel Devices*, 2nd edition, p. 589. KGaA: Wiley-VCH Verlag

Storhoff, J. J. and C. A. Mirkin (1999). 'Programmed materials synthesis with DNA'. *Chem. Rev.* 99: 1849.

Stupp, S. I., V. LeBonheur, K. Walker, L. S. Li, K. E. Huggins, M. Keser and A. Amstutz (1997). 'Supramolecular materials: self-organized nanostructures'. *Science*, 276: 384.

Sugano, S. and H. Koizumi (1998). *Microcluster Physics*. Berlin: Springer.

Sun, S. H., C. B. Murray, D. Weller, L. Folks and A. Moser (2000). 'Monodisperse FePt nanoparticles and ferromagnetic FePt nanocrystal superlattices'. *Science*, 287: 1989.

Thursby, M. H., B. G. Grossman, T. Alavie and K.-S. Yoo (1989). 'Smart structures incorporating neural networks, fibre-optic sensors, and solid-state actuators'. In E. Udd (ed.), *Proc. SPIE (Fibre Optic Smart Structures and Skins II)*, 1170: 316.

Tiwari, R. and V. K. Wadhawan (1991). 'Shape-memory effect in the Y-Ba-Cu-O ceramic'. *Phase Transitions*, 35: 47.

Tsallis, C., R. S. Mendes and A. R. Plastino (1998). 'The role of constraints within generalized nonextensive statistics'. *Physica A*, 261: 534.

Turberfield, A. (March 2003). 'DNA as an engineering material'. *Physics World*, 16: 43.

Tzou, H. S. (1993). *Piezoelectric Shells: Distributed Sensing and Control of Continua*. Dordrecht: Kluwer.

Uchida, K. (2005). 'Single-electron devices for logic applications'. In Waser, R. (ed.), *Nanoelectronics and Information Technology: Advanced Electronic Materials and Novel Devices*, 2nd edition, p. 423. KGaA: Wiley-VCH Verlag

Valenzuela, R. (1994). *Magnetic Ceramics*. Cambridge, U. K.: Cambridge University Press.

Wadhawan, V. K. (1988). 'Epitaxy and disorientations in the ferroelastic superconductor $YBa_2Cu_3O_{7-x}$'. *Phys. Rev. B*, 38: 8936.

Wadhawan, V. K. (2000). *Introduction to Ferroic Materials*. Amsterdam: Gordon and Breach.

Wagner, P. (2005). 'Nanobiotechnology'. In R. S. Greco, F. B. Prinz and R. L. Smith (eds.), *Nanoscale Technology in Biological Systems*. New York: CRC Press.

Wallraff, G. M. and W. D. Hinsberg (1999). 'Lithographic imaging techniques for the formation of nanoscopic features'. *Chem. Rev.* 99: 1801.

Waser, R. (2005). 'Logic devices: Introduction'. In Waser, R. (ed.), *Nanoelectronics and Information Technology: Advanced Electronic Materials and Novel Devices*, 2nd edition, p. 319. KGaA: Wiley-VCH Verlag

Weinberg, W. H., C. M. Reaves, B. Z. Nosho, R. I. Pelzel and S. P. DenBaars (2000). 'Strained-layered heteroepitaxy to fabricate self-assembled semiconductor islands'. In Nalwa, H. S. (ed.), *Handbook of Nanostructured Materials and Nanotechnology*. New York: Academic Press, Vol. 1, p.300.

Whitesides, G. M. and M. Boncheva (16 April 2002). 'Beyond molecules: Self-assembly of mesoscopic and macroscopic components'. *PNA, USA*, 99(8): 4769.

Whitesides, G. M. and J. C. Love (September 2001). 'The art of building small'. *Scientific American,* 285: 33.

Whitesides, G. M., J. P. Mathias and C. T. Seto (1991). 'Molecular self-assembly and nanochemistry: A chemical strategy for the synthesis of nanostructures'. *Science*, 254: 1312.

Wilbur, J. L. and G. M. Whitesides (1999). 'Self-assembly and self-assembled monolayers in micro- and nanofabrication', in G. Timp (ed.), *Nanotechnology*. New York: Springer-Verlag.

Xia, Y., J. A. Rogers, K. E. Paul and G. M. Whitesides (1999). 'Unconventional methods for fabricating and patterning nanostructures'. *Chem. Rev.* 99: 1823.

Xia, Y., K. Kamata and Y. Lu (2004). 'Photonic crystals'. In Di Ventra, M., S. Evoy and J. R. Heflin (eds.), *Introduction to Nanoscale Science and Technology*. Dordrecht: Kluwer.

Xu, Y. (1991). *Ferroelectric Materials and Their Applications*. Amsterdam: North-Holland.

Yurke, B., A. J. Turberfield, A. P. Mills, F. C. Simmel and J. L. Neumann (10 August 2000). 'A DNA-fuelled molecular machine made of DNA'. *Nature*, 406: 605.

7

HUMAN INTELLIGENCE

Say not, 'I have found the truth', but rather, 'I have found a truth'.
                                        – Kahlil Gibran, *The Prophet*

Before we can design truly intelligent machines, we must have a good understanding of what human intelligence really is. According to Hawkins and Blakeslee (2004), till recently there was no clearly formulated and formal theory of human intelligence. Hawkins himself has put forward his *memory and prediction theory* of how the human brain functions. We shall briefly describe his work in this chapter. Admittedly, expert opinion on human intelligence ('the last great terrestrial frontier of science') is still divided, and it is debatable whether we really have a good grasp of the intricacies of what goes on inside the human brain. But Hawkins' framework is fairly illustrative of the present thinking, and is sufficient for our simple, very brief, discussion of human intelligence.

The basic idea of Hawkins' theory of intelligence, in his own words, is as follows:

The brain uses vast amounts of memory to create a model of the world. Everything we know and have learnt is stored in this model. The brain uses this memory-based model to make continuous predictions of future events. It is the ability to make predictions about the future that is the crux of intelligence.

7.1 The human brain

The human brain, along with the spinal cord, constitutes the central nervous system, responsible for monitoring and coordinating voluntary and unconscious actions and reactions. Encased in the skull, the top outer portion of the human brain, just under the scalp, comprises the *neocortex*. It has a crumpled appearance, with many ridges and valleys. This soft, thin sheet of neural tissue envelopes most of the rest of the brain.

The rest of the brain, also called the *old brain*, *reptilian-brain*, or *R-brain*, is rather similar in reptiles and humans. Deep inside the R-brain is the *thalamus*. Some other components of the R-brain are: the *fornix*, the *ventricle*, the *hypothalamus*, the *amygdala*, the *pituitary gland*, the *pineal gland*, the *hippocampus*, the *ventricle*, the *corpus callosum*, the *cingulate gyrus*, the *pons*, the *medulla*, and the *cerebellum*. We shall have no occasion or need to go into the details of any of the constituents of the R-brain.

The last mentioned part of the R-brain, namely the cerebellum, is set on the central trunk of the brain; it continues downwards to form the spinal cord. This part of the brain is referred to as the *brain stem*.

The neocortex or the cerebrum is partitioned by the *giant fissure* (or the *sagittal section*) into the *left hemisphere* and the *right hemisphere*. Similarly, the *sulcus* (or the *coronal section*) separates the *frontal lobe* from the *parietal lobe* of the neocortex.

Almost everything we associate with intelligence occurs in the neocortex, with important roles also played by the thalamus and the hippocampus (Sakata, Komatsu and Yamamori 2005).

In old neuroscience, portions of the neocortex had been loosely identified as different *functional regions* or *functional areas*. For example, a portion of the frontal lobe is the *motor area*, responsible for controlling movement and other actuator action. Similarly, there is a *sensory area*, located in a part of the parietal lobe.

The human neocortex (or just cortex, for short), if stretched flat, is the size of a large napkin. Other mammals have smaller cortical sheets. The human neocortex is about 2 mm thick. It has six layers, each of thickness similar to that of a playing card.

The functions of the neocortex are arranged in a branching hierarchy. Out of the six layers, Layer 6 occupies the lowest rung of the hierarchy, and Layer 1 is at the top.

The human brain has $\sim 10^{11}$ nerve cells or neurons, loaded into the neocortex. Most of the neurons have a pyramidal shaped central body (the *nucleus*). In addition, a neuron has an *axon*, and a number of branching structures called *dendrites*. The axon is a signal emitter, and the dendrites are signal receivers (Kaupp and Baumann 2005). A connection called a *synapse* is established when a strand of an axon of one neuron 'touches' a dendrite of another neuron. The axon of a typical neuron makes several thousand synapses. The cell to which the axon belongs is called the *presynaptic neuron*, and the cell the dendrite of which is involved in the synapse is called the *postsynaptic neuron*.

The neocortical tissue is functionally organized into vertical units called *columns*. Neurons within a column respond similarly to external stimuli with a particular attribute.

Just beneath the six-layered cortical sheet there is a layer of white fatty substance called *myelin*. This so-called *subcortical white matter* contains millions of axons and the myelin insulating sheath around them. The path of Layer-6 axons from one cortical column to another is through this white fatty substance, which not only acts as an insulating sheath, but also helps the neural signals travel faster (typically at speeds of 320 km per hour).

When a sensory or other pulse ('spike') involving a particular synapse arrives at the axon, it causes the synaptic vesicles in the presynaptic neuron to release chemicals called *neurotransmitters* into the gap or synaptic cleft between the axon of the first cell and the dendrite of the second. These chemicals bind to the receptors on the dendrite, triggering a brief local depolarization of the membrane of the postsynaptic cell. This is described as a *firing* of the synapse by the presynaptic neuron.

If a synapse is made to fire repeatedly at high frequency, it becomes more sensitive: subsequent signals make it undergo greater voltage swings or spikes. Building up of memories amounts to formation and strengthening of synapses.

The firing of neurons follows two general rules:

1. *Neurons which fire together wire together*. Connections between neurons firing together in response to the same signal get strengthened.
2. *Winner-takes-all inhibition*. When several neighbouring neurons respond to the same input signal, the strongest or the 'winner' neuron will inhibit the neighbours from responding to the same signal in future. This makes these neighbouring neurons free to respond to other types of input signals.

Creation of *short-term memory* in the brain amounts to a stimulation of the relevant synapses, which is enough to temporarily strengthen or sensitize them to subsequent signals.

This strengthening of the synapses becomes permanent in the case of *long-term memory*. This involves the activation of genes in the nuclei of postsynaptic neurons, initiating the production of proteins in them. Thus *learning* requires the synthesis of proteins in the brain within minutes of the training. Otherwise the memory is lost (Fields 2005). The basic processes involved are as follows:

Strong or repeated stimulation temporarily strengthens the synapse, and, in addition, sends a signal to the nucleus of the postsynaptic neuron to make the memory permanent. This signal was once believed to be *carried* by some unknown molecule. It was supposed that this molecule travels and gets into the postsynaptic nucleus, where it activates a protein called CREB. This protein, in turn, activates some genes selectively, causing them to be transcribed into messenger RNA (mRNA), which then comes out of the nucleus. The instructions carried by the mRNA are translated into the production of synapse-strengthening proteins, which then diffuse throughout the cell. Only a synapse already temporarily strengthened by the original stimulus is affected by these proteins, and the memory gets imprinted permanently.

Recent research (Fields 2005) has shown that it is unnecessary to postulate that the synapse-to-nucleus signalling is by some hypothetical molecule. Strong stimulation, either from the repeated firing of a single synapse, or from the simultaneous firing of synapses on a postsynaptic cell, is enough to depolarize the cell membrane. This causes the cell to fire an *action potential* of its own. This, in turn, causes its voltage-sensitive calcium channels to open. The calcium influx into the nucleus of the postsynaptic cell leads to the activation of enzymes which activate CREB. The rest of the mechanism is as described above.

Information meant to become the higher-level or generalized memory, called *declarative memory*, passes through the hippocampus, before reaching the cortex (Fields 2005). The hippocampus is like the principal server on a computer network (Shreeve 2005). It plays a crucial role in consolidating long-term memories and emotions by integrating information coming from sensory inputs with information already stored in the brain.

The main feature of the anatomy of the human brain which distinguishes it from most other living beings is its large neocortex. Reptiles came on the scene at a certain stage in natural evolution. They had sophisticated sensory organs, but no cortex or memory system. Their behaviour, controlled by the old brain, i.e. the reptilian-brain, or R-brain, was (and is) quite sophisticated, but rather rigid because they had no memory of past experiences to go by.

Evolution of a memory system (the cortex), along with a stream of sensory inputs into it, gave the creatures an evolutionary advantage because, when they found themselves in situations they remembered to have faced earlier, a memory recall told them what to expect next, and how to respond effectively and quickly.

With time, the cortex got larger and more sophisticated. Its memory capacity became larger, and it could make predictions based on more complex relationships. In due course, it also started interacting with the motor system of the old brain. As described above, the cortex is situated above the old brain, and covers most of it.

Human beings have the most sophisticated cortex (Mountcastle 1998). In them the cortex has taken over most of the motor behaviour. The human cortex directs behaviour (including motor behaviour) to respond to its own predictions.

The mind and the brain are one and the same thing. The mind is the creation of the neurons in the brain. This is the 'The Astonishing Hypothesis' put forward by Francis Crick (1994). According to Hawkins and Blakeslee (2004), this is a fact, not just a hypothesis.

Functionality of the neocortex is arranged in a branching hierarchy. The primary sensory regions constitute the lowest rung of the hierarchy. This is where raw sensory information first arrives. The sensory region for, say, vision (called V1) is different from that for hearing, etc. V1 feeds information to 'higher' layers called V2, V4 and IT, and to some other regions. The higher they are in the hierarchy, the more abstract they become (cf. Hawkins and Blakeslee 2004).

V2, V4, etc. are concerned with more specialized or abstract aspects of vision. The higher echelons of the functional region responsible for vision have the visual memories of all sorts of objects. Similarly for other sensory perceptions.

In the higher echelons are areas called *association areas*. They receive inputs from several functional regions (Nicolelis and Ribeiro 2006). For example, signals from both vision and audition reach one such association area.

A very important feature of the cortical hierarchy is *feedback*. Information flows both ways (up and down) among the hierarchies.

The functional areas in the frontal lobe of the brain, which generate actuation or motor output, also have a hierarchical structure. *In fact, the hierarchies of sensor areas and actuator areas look remarkably similar*. Although we may be inclined to think of information flowing *up* from the senses to the brain, and flowing *down* from the brain to drive the muscles, the fact is that, because of the very extensive feedback channels, information flows both ways, thus further enhancing the similarity between the sensor and actuator functions. What is feedback in sensory regions is output in actuator regions, and vice versa. This is an important observation from the point of view of designing artificial smart structures.

Although the primary sensor mechanism for, for example, vision is not the same as for hearing, what reaches the brain at a higher level of the hierarchy is qualitatively the same. The axons carry neural signals or spikes which are partly chemical and partly electrical, but their nature is independent of whether the primary input signal was visual or auditory or tactile. They are all action potentials; just patterns.

The input from each of the senses is finally stored in the cortex as *spatio-temporal patterns*. The temporal factor is important, and has been duly emphasized by Hawkins and Blakeslee (2004). As they explain, the cortex stores visual data more like a song than a painting. Similarly, hearing has not only a temporal aspect, but also a spatial aspect. *Patterns from different senses are equivalent inside the brain*.

7.2 Mountcastle's hypothesis

The same types of layers, cell types and connections exist in the entire cortex. Therefore, Mountcastle (1978) put forward the following hypothesis: *There is a common function, a common algorithm, that is performed by all the cortical regions*. What makes the various functional areas different is the way they are *connected*.

In fact, Mountcastle went further to suggest that the reason why the different functional regions look different when imaged is because of these different connections only.

Mountcastle's idea, if validated fully, can have far-reaching consequences in our understanding of how the brain functions. In fact, it forms a basis of Hawkins' theory of human intelligence. What is being said is that the cortex does something universal to all the sensory or motor inputs and outputs. The algorithm of the cortex is expressed in a way that is independent of any particular function or sense.

Our genes and upbringing determine how the cortex is wired. But the algorithm for processing information is the same everywhere.

This has important implications for the future of machine intelligence. If we connect regions of an artificial cortex in a suitable hierarchy, and provide a stream of inputs, it would develop perception and intelligence about the environment, just like humans do (Hawkins and Blakeslee 2004).

The brain is a pattern machine. Our perception of the world is built from these patterns. The cortex does not care whether a pattern originated in vision, hearing or any other sensory organ. The same powerful algorithm is at work. *This means that we can build machines with sensory systems* different *from ours, and still make them intelligent, if the right algorithm can be put in.*

What happens to the various patterns received by the human brain? *Memory* plays a crucial role in the way the human brain functions.

7.3   The cortical memory

Hawkins, whose work we shall outline further in the next section, does not claim complete originality for his theoretical framework of how the neocortex functions. Several features of the formal edifice created by him have been in the air for a long time. Here is a glimpse of the earlier as well as contemporary work:

H. G. Wells (1938) argued that memory is not just an accessory to intelligence, but the substance from which intelligence is formed.

Hebb (1949), in his book *The Organization of Behaviour*, put forward the idea that the brain is constantly making changes to the synaptic connections in the light of new experiences. Certain synapses are strengthened when certain inputs are received, and others get weakened through disuse, and these synaptic patterns are the basis of all learning and memory. Hebb further postulated that the selective strengthening of the synapses causes the brain to get self-organized into *cell assemblies*, i.e. subsets of several thousand neurons in which nerve impulses continue to circulate and reinforce themselves. These cell assemblies were considered the basic building blocks of information. The cell assemblies overlap, in the sense that a given neuron may belong to several of them. Because of this overlap, the blocks are able to organize and reorganize themselves, and develop more complex behaviour and higher-level abstractions.

Holland (1975) provided evidential support (in the early 1950s) for Hebb's *connectionist* theory of the brain by neural-network simulation studies. He employed cellular automata, in which the local rule was a learning rule. The computer simulations revealed emergent behaviour. Coherent, self-reinforcing clusters of neurons evolved, just like the evolution of genetic building blocks. Holland (1998) has developed *a theory of adaptation*, in which a parallel is drawn between long-term genetic adaptation of organisms and short-term neural response inside a brain. Both involve clusters or blocks (genes in one case, neural clusters in the other case), which greatly enhance evolutionary novelty in the first case, and problem-solving capacity in the other.

Stassinopoulos and Bak (1995) modelled the brain as a self-organized criticality (SOC) system (cf. Section 5.5.2 on SOC). As Bak (1996) pointed out, the conventional artificial neural-network (ANN) approach has tended to provide an engineered set of all input and output connections for the artificial neurons to be trained, whereas the fact is that even this must be a self-organized process. One difference between a computer and a brain is that, whereas the computer is designed and made by engineers as per *their* design, the brain does not have an engineer around for making all those synaptic connections. The brain is self-organized. And, as is common for all large self-organized systems, it is self-organized to the *critical* state. In the language of the SOC formalism, a thought in the brain is like an avalanche, triggered by the arrival of some sensory input(s), or even by another thought. However, unlike avalanches in nonthinking systems, the neural avalanches have to be taught (by the conditioning process) to be useful to the organism.

Why should the self-organization of the brain be in a critical configuration? Imagine, say, a visual input. The best response can be possible only if this input is available to all parts of the brain. This is possible only in a critical system, because only such a system can ensure mutual interactions at all 'length' scales. If it is supercritical, the response will be too drastic and meaningless, and if it is subcritical, there will be only a localized response. The critical pathways of firing networks in an otherwise quiet medium enable quick switching from one complicated pattern to another (cf. Chialvo 2006, for a recent update).

Modern cognitive science is well-seized of the fact that the act of perceiving, and the act of recall from memory, both are manifestations of *emergent behaviour*, arising from the interaction and collation of a large number of sensory inputs, spread over time and space. Our brain reconstructs the present, just as it reconstructs the past (memory recall), from a large number of distributed inputs, coming from a large number of interacting sites in the brain (see, e.g. Hasegawa *et al.* 2000; Laughlin and Sejnowski 2003, and references therein). The hive mind (read human mind) is a distributed memory that both perceives and remembers.

7.4   Spin-glass model of the cortical neural network

The efforts that went into understanding the properties of spin glasses (cf. Section 3.5) led to a beautiful theory that is applicable to a large variety of other complex phenomena as well. A neural-network model based on the language of spin-glass mathematics was proposed by Hopfield and Tank (1982, 1984, 1986). We give here an outline of this model. (This section can be skipped in a first reading of the book.)

This model (Hopfield and Tank 1986) for the functioning of the human brain is designed to determine whether a particular neuron in a large network is firing or not. It can be used not only as an aid to understanding associative memory, but also for solving optimization problems (like the travelling-salesman problem).

Hopfield considered circuits consisting of graded-response model neurons. These neurons are taken as organized into networks with effectively *symmetric* synaptic connections. The symmetry features results in a powerful theorem about the behaviour of the system. The theorem shows that a mathematical quantity E, which is interpreted as *computational energy*, decreases during the evolution of the neural pattern with time. This 'energy' is a global quantity, not felt by any individual neuron.

The circuits considered by Hopfield are implementable with electronic components. A large network of amplifiers is considered to simulate a dense network of neurons. The actual demonstration of ANN concepts with real electronic circuits raises hopes that artificial brains with hierarchical cortical intelligence can indeed be built.

The Hopfield model captured the two essential features of biological brains: *analogue processing* and *high interconnectivity*. It therefore marked a definite advance over the McCulloch–Pitts model of neural activity; the latter dealt with only a two-valued variable for the logic gate (cf. Section 2.5.1).

Other essential biological features retained in the Hopfield model are (Khanna 1999):

- The fact that the neuron is a transducer of energy with a smooth 'sigmoid' response up to a maximum level of output.
- The integrative behaviour of the cell membrane.
- Large numbers of excitatory and inhibitory connections.
- The re-entrant or feedback nature of the connections.
- The ability to work with both graded-response neurons and those that produce action potentials.

For the sake of simplicity, the neuron can be taken as being in only two possible discrete states: active and inactive (just like the up and down spin states of a valence electron in a spin glass). The active state of the neuron is that in which it *fires*, i.e. passes on an impulse to another neuron. And the inactive or dormant state is that in which it is not firing.

Whether the two-state neuron remains in a state or switches to the other state depends on the states of all the neurons connected to it via the synapses, so there are *competing interactions* (just like in spin glasses). We thus have a complex assembly of mutually interacting agents (the neurons), and the emergent behaviour (or computational task) is determined by the ever-changing synaptic pattern. There is an element of *randomness*, because there is no information about what the next sensory input is going to be.

If the sum total of action potentials impinging on a neuron exceeds some threshold value, the two-state neuron goes to the other state (cf. Fig. 2.2). There is also the *frustration* feature typical of spin-glasses: Conflicting impulses may arrive at a neuron, wanting it to be in both the states at the same time.

Mydosh (1993) has given a compact description of the spin-glass formalism adopted by Hopfield and Tank, and we outline that here:

Let v_i ($= 1$ or 0) denote the state of a neuron i at time t, T_i the threshold beyond which this neuron will change its state, and C_{ij} the strength of the interaction (the synaptic weight) between neuron i and neuron j. Then

$$v_i = 1 \text{ if } \sum_i C_{ij} v_i > T_i \tag{7.1}$$

$$v_i = 0 \text{ if } \sum_i C_{ij} v_i < T_i \tag{7.2}$$

We introduce new variables:

$$S_i = 2v_i - 1; \; J_{ij} = (1/2)C_{ij}; \; H_i = \sum_i (1/2)C_{ij} - T_i \tag{7.3}$$

The neural network is then described by Ising-like pseudospins:

$$S_i = +1 \text{ if } \sum J_{ij} S_i - H_i > 0 \qquad (7.4)$$

$$S_i = -1 \text{ if } \sum J_{ij} S_i - H_i < 0 \qquad (7.5)$$

These two statements can be combined into a single formula:

$$S_i \left(\sum J_{ij} S_i - H_i \right) > 0 \qquad (7.6)$$

In the formalism of standard spin-glass theory (Section 3.5), the Hamiltonian for the neural network can be written as

$$H = - \sum_{\{ij\}} J_{ij} S_i S_j - \sum H_i S_i \qquad (7.7)$$

In Hopfield's theory, effectively *symmetric* synaptic connections are assumed; i.e. the strength and sign of the synapse from neuron i to neuron j is taken as the same as that of another synapse from neuron j to neuron i. In other words, $J_{ij} = J_{ji}$. Both ferromagnetic and antiferromagnetic pseudo-exchange interactions exist: The synaptic impulses can be both excitatory and inhibitory; J can be both positive and negative. Further, the value of J changes with time: The system not only stores, it also learns.

The set of values of S for all the neurons defines the instantaneous state (the *neural pattern*) of the system. N two-state neurons can form 2^N patterns, and N is typically 10^{11} for the human brain.

In the Hamiltonian defined by eqn 7.7, the frustration feature implies that the solution of this equation is represented by a multivalley phase-space landscape of a very large number of nearly degenerate ground states.

The local energy minima in the solution of the Hamiltonian (one for each ground state) correspond to patterns to be recognized, or memories to be recalled, or some such mental function of the very large collective system.

Each such solution, i.e. each such set of neural connections, labels a valley in the rugged phase space. For the spin-glass problem, the vertical axis for the phase-space plot denotes the energy states of the peaks and valleys. For the neural-network problem, it is the memories to be recalled or the patterns to be recognized.

A sensory or other input determines the initial state of the system (i.e. which neurons are firing and which are not). Depending on the set of neural couplings already ingrained in the brain, it 'relaxes' towards a particular low valley or memorized pattern that corresponds to the class to which the starting impulse belongs. There is a *basin of attraction* around each lowest point of a valley in the phase space (cf. appendices on nonlinear dynamics and on chaos), such that the system moves towards the lowest point if it is anywhere in that valley. Thus, an entire memorized pattern is addressable by any point in the corresponding valley, i.e. by the *association* with a given input pattern.

The synaptic weights J_{ij} in eqn 7.7 are usually taken as given by

$$J_{ij} = (1/p) \sum_{\alpha=1}^{p} \xi_i^{\alpha} \xi_j^{\alpha}, \qquad (7.8)$$

where the ξ_i^{α} (= 1 or −1) are the quenched random variables with prescribed probabilities (corresponding to the correlation lengths in our spin-glass description in Section 3.5). The symbol p corresponds to the p-learned patterns of N-bit words, which are fixed by the p sets of $\{\xi_i^{\alpha}\}$.

It can be shown that the stored patterns will only be stable at $T = 0$ K for negligibly small values of p/N. However, the maximum number of patterns that can be stored in the network can be greatly increased by using an upper limit of *retrieval errors* of learned patterns (Mydosh 1993). One begins by defining a so-called *Hamming distance*, d, between a pure learned pattern S^k and a pattern S' as

$$d(S', S^k) = (1/2)[1 - q(S', S^k)] \qquad (7.9)$$

The Hamming distance represents the *overlap* between the two patterns; it is the number of common bits.

The average number of errors, N_e, corresponds to the number of spins (neurons) which do not align with the learned or embedded pattern:

$$N_e = (N/2)(1 - q) \qquad (7.10)$$

It is found that $p/N \sim 0.14$ for an error limit of $N_e/N \sim 1\%$; for smaller error limits the ratio p/N goes to zero rapidly.

When a perturbation is applied to a system, it recovers from the effect of the perturbation (i.e. it relaxes) according to the dynamics intrinsic to the system. The dynamics assumed in Hopfield's model of the neural assembly is equivalent to the single spin-flip *Glauber dynamics* in a spin glass. (In Glauber dynamics, a spin flip can occur almost independently of other spin variables; see, for example, Wadhawan 2000.) In this scheme, after a spin has been updated, it is the new configuration which is used for updating the next one.

Other relaxation mechanisms have also been investigated. See Mydosh (1993) for more details, as well as for an illustration of the actual emergence of associative memory from this spin-glass formalism.

7.5   Hawkins' theory of human intelligence

Hawkins' theory gives a formal structure and a coherent framework to the existing line of thinking on what the human cortical brain is all about; it also covers fresh ground. In this framework, the neocortical memory differs from that of a conventional computer in four ways:

1. The cortex stores *sequences* of patterns. For example, our memory of the alphabet is a sequence of patterns. It is not something stored or recalled in an instant, or all together. That is why we have difficulty saying it backwards. Similarly our memory of songs is an example of *temporal* sequences in memory.
2. The cortex recalls patterns *auto-associatively*. The patterns are associated with themselves. One can recall complete patterns when given only partial or distorted inputs. During each waking moment, each functional region is essentially waiting for familiar patterns or pattern-fragments to come in. Inputs to the brain link to themselves auto-associatively, filling in the present, and auto-associatively linking to what normally flows next. We call this chain of memories, *thought*.
3. The cortex stores patterns in an *invariant form* (Fig. 7.1). Our brain does not remember *exactly* what it sees, hears, or feels. The brain remembers the important relationships in the world, independent of details. Auto-associative memories are used in ANNs also, but not as invariant representations. Consequently, artificial auto-associative memories fail to recognize patterns if they are moved, rotated, rescaled, etc., whereas our brains handle these variations with ease. We recognize a friend no matter what he/she is wearing, or at what angle is the person visible to us. Even the voice is enough input for recognition.
4. The cortex stores patterns in a *hierarchy*. This we have discussed already.

According to Hawkins and Blakeslee (2004), the problem of understanding how our cortex forms invariant representations (in its higher echelons) remains one of

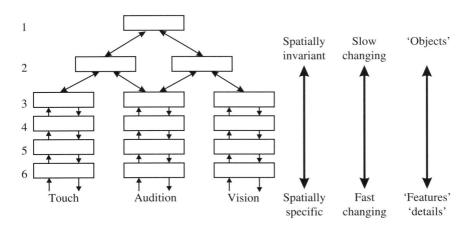

Fig. 7.1 A conceptual depiction of the formation of invariant representations in the human brain for hearing, vision, and touch. The higher cortical regions are the association areas. They receive and integrate inputs from a number of different senses; e.g. hearing plus touch plus vision. In the actual brain, a large number of cortical regions are interconnected in a variety of ways. Association areas are spread all over the cortex. [After Hawkins and Blakeslee (2004).]

the biggest mysteries of science. This will be a great problem to solve, because, as of now, conventional robots and computer programs, like artificial auto-associative memories, have been, by and large, rather terrible at handling variation. There has been some progress recently. For example, the robot *Darwin X* is a brain-based device (see Fox 2005). It is a reasonably intelligent machine because it is actually modelled on the human brain.

Storing sequences, auto-associative recall, and invariant representation are the necessary ingredients for predicting the future based on memories of the past. How this happens is the subject matter of Hawkins' theory of intelligence. Making such predictions is the essence of intelligence.

7.6   Chapter highlights

- Progress in understanding human intelligence will have a direct bearing on our ambition to build truly intelligent artificial smart structures.
- Almost everything we associate with human intelligence occurs in the cortex, with important roles also played by the thalamus and the hippocampus.
- The functions of the cortex are arranged in a branching hierarchy. Out of the six layers, Layer 6 occupies the lowest rung of the hierarchy, and Layer 1 is at the top.
- The cortical tissue is functionally organized into vertical units called columns. Neurons within a column respond similarly to external stimuli with a particular attribute.
- Just beneath the six-layered cortical sheet there is a layer of white fatty substance called myelin, which contains millions of axons and the myelin insulating sheath around them. The path of Layer-6 axons from one cortical column to another is through this white fatty substance, which not only acts as an insulating sheath, but also helps the neural signals travel faster.
- Information meant to become the higher-level or generalized memory, called declarative memory, passes through the hippocampus, before reaching the cortex. The hippocampus is like the principal server on a computer network. It plays a crucial role in consolidating long-term memories and emotions by integrating information coming from sensory inputs with information already stored in the brain.
- In the higher echelons of the cortex are areas called association areas. They receive inputs from several functional regions. For example, signals from both vision and audition reach one such association area.
- A very important feature of the cortical hierarchy is feedback. Information flows both ways (up and down) among the hierarchies.
- Although the primary sensor mechanism for, for example, vision is not the same as for hearing, what reaches the brain at a higher level of the hierarchy is qualitatively the same. The axons carry neural signals or spikes which are partly chemical and partly electrical, but their nature is independent of whether the primary input signal was visual or auditory or tactile. They are all action potentials; just patterns.

- The input from each of the senses is finally stored in the cortex as spatio-temporal patterns. The temporal factor is important. The cortex stores visual data more like a song than a painting. Similarly, hearing has not only a temporal aspect, but also a spatial aspect. Patterns from different senses are equivalent inside the brain.

- The cortex does something universal to all the sensory or motor inputs and outputs. The algorithm of the cortex is expressed in a way that is independent of any particular function or sense. Our genes and upbringing determine how the cortex is wired. But the algorithm for processing information is the same everywhere.

- This has important implications for the future of machine intelligence. If we connect regions of an artificial cortex in a suitable hierarchy, and provide a stream of inputs, it would develop perception and intelligence about the environment, just as humans do.

- The brain is a pattern machine. Our perception of the world is built from these patterns. The human cortex does not care whether a pattern originated in vision, hearing or any other sensory organ. The same algorithm is at work everywhere. This means that we can build machines with sensory systems *different* from ours, and still make them intelligent, if the right algorithm can be devised.

- The act of perceiving, and the act of recall from memory, both are manifestations of emergent behaviour, arising from the interaction and collation of a large number of sensory inputs, spread over time and space. Our brain reconstructs the present, just as it reconstructs the past (memory recall), from a large number of distributed inputs, coming from a large number of interacting sites in the brain. The human mind is a distributed memory that both perceives and remembers.

- The cortex stores patterns in an invariant form, or as irreducible representations, in its higher echelons. Our brain does not remember *exactly* what it sees, hears, or feels. The brain remembers the important relationships in the world, independent of details.

- Storing sequences, auto-associative recall, and invariant representation are the necessary ingredients for predicting the future based on memories of the past. Making such predictions is the essence of intelligence.

References

Bak, P. (1996). *How Nature Works: The Science of Self-Organized Criticality*. New York: Springer.

Chialvo, D. R. (May 2006). 'Are our senses critical?'. *Nature Physics*, 2: 301.

Crick, F. (1994). *The Astonishing Hypothesis: The Scientific Search for the Soul*. London: Simon and Schuster.

Fields, R. D. (Feb. 2005). 'Making memories stick'. *Scientific American,* 292(2): 58.

Fox, D. (5 November 2005). 'Brain Box'. *New Scientist*, p. 28.

Hasegawa, R. P., A. M. Blitz, N. L. Geller and M. E. Goldberg (1 December 2000). 'Neurons in monkey prefrontal cortex that track past or predict future performance'. *Science*, 290: 1786.

Hawkins and Blakeslee, J. (2004). *On Intelligence*. New York: Times Books.

Hebb, D. O. (1949). *The Organization of Behaviour: A Neuropsychological Theory*. New York: Wiley.

Holland, J. H. (1975). *Adaptation in Natural and Artificial Systems*. Ann Arbor: University of Michigan Press.

Holland, J. H. (1998). *Emergence: From Chaos to Order*. Cambridge, Massachusetts: Perseus Books.

Hopfield, J. J. (1982). 'Neural networks and physical systems with emergent collective computational abilities'. *Proc. Nat. Acad. Sci. (USA)*, 79(8): 2554.

Hopfield, J. J. (1984). 'Neurons with graded response have collective computational properties like those of two-state neurons'. *Proc. Nat. Acad. Sci. (USA)*, 81: 3088.

Hopfield, J. J. and D. Tank (1986). 'Computing with neural circuits: A model'. *Science*, 233: 625.

Kaupp, U. B. and A. Baumann (2005). 'Neurons – the molecular basis of their electrical excitability'. In Waser, R. (ed.), *Nanoelectronics and Information Technology: Advanced Electronic Materials and Novel Devices*, 2nd edition, p. 144. KGaA: Wiley-VCH Verlag.

Khanna, T. (1990). *Foundations of Neural Networks*. Reading: Addison-Wesley.

Laughlin, S. B. and T. J. Sejnowski (26 September 2003). 'Communication in neural networks'. *Science*, 301: 1870.

Mountcastle, V. B. (1978). 'An organizing principle for cerebral function: The unit model and the distributed system'. In G. M. Edelman and V. B. Mountcastle (eds.), *The Mindful Brain*. Cambridge, Massachusetts: MIT Press.

Mountcastle, V. B. (1998). *Perceptual Neuroscience: The Cerebral Cortex*. Cambridge, Massachusetts: Harvard University Press.

Mydosh, J. A. (1993). *Spin Glasses: An Experimental Introduction*. London: Taylor and Francis.

Nicolelis, M. A. L. and S. Ribeiro (December 2006). 'Seeking the neural code'. *Scientific American,*, 295(6): 70.

Sakata, S., Y. Komatsu and T. Yamamori (2005). 'Local design principles of mammalian cortical networks'. *Neurosci. Res.*, 51: 309.

Shreeve, J. (March 2005). 'Beyond the brain'. *National Geographic*, 207(3): 2.

Stassinopoulos, D. and P. Bak (1995). 'Democratic reinforcement. A principle for brain function'. *Phys. Rev. E*, 51: 5033.

Wadhawan, V. K. (2000). *Introduction to Ferroic Materials*. Amsterdam: Gordon and Breach.

Wells, H. G. (1938). *World Brain*. New York: Doubleday.

8

SMART SENSOR SYSTEMS

Sensing involves measurement followed by information processing. The sensor may be a system by itself, or it may be a subsystem in a larger system (e.g. a complete smart system).

The evolution of sensing mechanisms like vision and hearing in living systems marked a major milestone in the development of higher-level species. At the top of the hierarchy are, of course, human beings. We have several senses, each of which may not be the best compared to some other animal. But what has put us ahead of all other creatures is the evolution of the large neocortex, which acts as a remarkable memory and information-processing device, and makes us intelligent.

The term *intelligent sensors* has been used by Brignell and White (1994, 1999) for sensors which have at least one digital processor. We prefer to substitute the word 'smart' for 'intelligent'. Only systems with a cortical-like brain (animate or inanimate) should be considered as truly intelligent. A smart-sensor system may well have its local processors (expert systems, or even neural networks) incorporated into it, but it would still not be truly intelligent in the sense described in this book. On the other hand, such a system may be a subsystem of a truly intelligent structure, with the intelligent actions coordinated by the cortical-like brain.

The availability of powerful low-cost digital processors has enormously increased the power and versatility of smart sensors. The near-avoidance of human intervention is a further advantage, as the likelihood of human errors in minimized.

8.1 Measurement and information

Measurement involves assigning numbers to entities or events. Except for measurements involving only counting, one normally maps an entity or event on to a *range* of numbers. There is a range, rather than a single number, because of the inevitability of some imprecision in the measurement process. Errors introduced when the numbers are processed by the processor further add to the total imprecision.

A working definition of information has been given by Brignell and White (1994): *Information is that which decreases uncertainty.*

Information is a quantity which can be identified and measured. It flows in an instrumentation system, and is subject to the rules of continuity.

8.2   Signals and systems

A *signal* is a quantity which can be, in general, represented as a function of time, and which conveys information.

A *system* is a mechanism whereby a causal relationship can be established between the input and output signals. Systems may be classified in various ways: linear/nonlinear; random/deterministic; continuous/discrete; time-varying/unvarying; with memory/without memory; etc.

Signals may also be classified similarly: continuous/discrete; bounded/unbounded; random/deterministic; periodic/aperiodic; stationary/nonstationary; etc.

The most important and easily tractable class of systems are *linear systems*. The full power of linear algebra can be brought to bear on their analysis. These are systems which are additive and homogeneous; i.e. the result of adding two signals together, or multiplying a signal by a constant, is the same whether the operation is performed at the input or the output of the system.

The *response* of a system is the relationship between its input and output signals. Two examples are: the impulse response, $h(t)$, and the frequency response, $H(\omega)$.

8.3   Physical principles of sensing

With reference to a measurement or control system, all transducers are either sensors or actuators: Sensors input information into the system, and actuators output actions into the external world. For electronic processing purposes, the information input by a signal should be in the form of an *electrical* signal. Similarly, actuation normally implies *mechanical* motion.

Materials which can act as efficient media for electromechanical, magnetomechanical or chemomechanical coupling of fields can act as both actuators and sensors. It is just a question of whether the mechanical force is the output signal or the input signal (Busch-Vishniac 1998).

Any transduction of energy by the primary sensor material introduces defects in the signal, and one has to compensate for these. The five major defects are:

1. Frequency-dependence (or time-dependence) of response.
2. Nonlinearity.
3. Noise.
4. Parameter drift.
5. Cross-sensitivity.

The problem of frequency-dependent response is further complicated by the fact that, for digital processing, the input signal (representing the quantity to be measured, i.e. the *measurand*) has to be sampled at a chosen frequency. The compensation strategies for frequency or time response involve postulating a *model* for the response. A merit of smart sensors (cf. Section 8.5 below) is that model-independent compensation can be made.

This model-independent compensation in smart sensors is also available for correcting for nonlinearity and parameter shift. Look-up tables (LUTs) or polynomials provide readily implemented solutions to these problems. Linearization of thermocouple output, as also compensation for zero-point drift, are some familiar examples of this.

Any primary sensor is bound to be sensitive to more than one physical variables, and that leads to the problem of cross-sensitivity. Temperature dependence of all sensor-materials is an example of this.

Various types of compensation techniques are possible: structural compensation; tailored compensation; monitored compensation; and deductive compensation (see Brignell and White 1994, for some details on these).

A variety of transduction mechanisms are possible for sensor systems: mechanical, thermal, chemical, magnetic or radiant forms of energy can be converted to electrical signals. Some common types of sensors are: piezoresistive sensors; capacitive sensors; and piezoelectric sensors. Special mention must be made of interdigital transducers and surface acoustic wave (SAW) sensors in the context of MEMS (Varadan, Jiang and Varadan 2001).

8.4   Optical-fibre sensing and communication

A variety of sensing technologies exist. These are based on silicon, thin films or thick films of ferroics or soft materials, or optical fibres.

Optical fibres are particularly suitable for smart-structure applications, and have been aptly described as *nerves of glass* (Amato 1992; Habel 1999; Lopez-Higuera 2002). There are several reasons for their widespread use:

1. Very wide bandwidth for optical frequencies. A single optical fibre can carry signals corresponding to a million conversations at the same time.
2. Absence of electromagnetic interference.
3. Low dispersion and loss of signal, even for transmission over long distances.
4. The nonmetallic and chemically inert nature of silica from which they are made. This makes it attractive to use them even in corrosive or hazardous environments, as also for medical applications.
5. Thinness, which ensures that they can be embedded in the composite material comprising the smart structure at the fabrication stage itself, without seriously affecting the integrity of the composite.
6. Mechanical flexibility and robustness.
7. High speed of signal processing.

8. Low cost.
9. Light weight.
10. Combination of sensor and telecommunication functions in the same medium.
11. The possibility of their additional use for carrying laser energy for heating the shape-memory-alloy wires used as actuators in smart structures (Chapter 9).

The originally envisaged application of optical fibres was as a telecommunication medium (Kao and Hockham 1966), when it was demonstrated that they could act as waveguides for the transmission of light over long distances. The basic idea is very simple: Light waves undergo total internal reflection from an interface with a lower refractive index medium at suitable angles of incidence, and can therefore propagate in the denser medium (glass or silica) over long distances, free from the vagaries of optical turbulence in the surroundings. The light beam travels as a transverse electric wave along the fibre. The electric field extends weakly outside the higher refractive index medium, decaying exponentially with distance from the axis of the fibre.

It was soon realized that the optical fibre can be also used as a primary sensor medium. The optical waves being guided by an optical fibre can get influenced (*modulated*) in a variety of ways by the environmental parameter to be sensed, so the optical fibre can act as a sensitive and extended sensor also. Intensity, phase, frequency, polarization, and colour of the optical beam are the main quantities analysed for the modulation caused by the measurand (Culshaw 1984).

One makes a distinction between *intrinsic* and *extrinsic* optical-fibre systems. In the former, the fibre itself is the element undergoing modulation of light by the measurand. In the later, the fibre acts only as a waveguide for the light going towards and away from the sensor.

Almost all measurands are accessible to measurement by optical techniques. The accuracy achieved is comparable to, or exceeding, that for other techniques. The use of optical fibres for this purpose requires the incorporation of either a very stable and accurate reference signal, or sophisticated compensation techniques. The latter requirement is well met if the sensor is a smart sensor.

By way of illustration of an extrinsic sensor design, we describe here the development of a smart system to detect and locate impact by foreign bodies on a surface, using neural-network (NN) processing of optical-fibre signals for impact location (Schindler *et al.* 1995). Detection of stress waves emanating from the impact was carried out by fibre-optic strain sensors embedded in the surface. A 1300 nm single-mode silica optical fibre transmitted light from a laser diode to the sensor element through a coupler. At the opposite end of the input fibre, the optical signal was partially reflected and partially transmitted (inside the coupler), exiting the fibre and traversing the air gap separating the ends of the input fibre and an output fibre used solely as a reflector. This signal and the signal reflected from the facing fibre-endface interfered, and propagated back through the input fibre. They finally reached the detector through the coupler.

A commercially available NN simulator, running on a PC, was used to train a network using a back-propagation algorithm (cf. Section 2.5.1). For this a set of hypothetical impact locations were generated randomly and distributed uniformly over the test surface. Data regarding the time-of-flight differences were fed to the NN software for training purposes. The implementation was successful both for isotropic and anisotropic materials.

8.5   Smart sensors

A typical smart sensor is a system by itself, even when it is a subsystem of a larger system. It is not just a primary sensing element, but has additional functionality arising from the integration of microcontrollers, microprocessors, and/or application-specific integrated circuits (ASICs) (Brignell and White 1994; Frank 2000; Swanson 2000; Gardner, Varadan and Awadelkarim 2001; Van Hoof, Baert and Witvrouw 2004).

A typical smart sensor may have the following subsystems (Brignell and White 1994):

1. A primary sensing element.
2. Excitation control.
3. Amplification.
4. Analogue filtering.
5. Data conversion.
6. Monitored compensation.
7. Digital information processing.
8. Digital communication processing.

A thermocouple is an example of a primary sensor that converts energy directly from one form to another, and does not need any further excitation control. In other applications (e.g. in resistivity-based sensors), a probing or excitation signal is given to the measurand, and the response is analysed.

Amplification of the output signal from the sensor is usually needed. Sometimes very high gain may be required.

Analogue filtering is resorted to for effecting a saving on the real-time processing power available. Another reason is the need to overcome the *aliasing effects*: The digital signal-sampling process introduces aliases, not only of the signal, but also of the noise. The aliasing of high-frequency noise into the range of observation can cause serious problems, so pre-filtering of the signals for sampling has to be done.

Digitization of a continuous signal has a distorting and corrupting effect, which should not be lost sight of. An appropriate data conversion subsystem is therefore an important component of the smart-sensor system.

Monitored compensation, of course, is the high point of the idea of introducing a modicum of smartness into sensors. Similarly, *in situ* information processing is a unique feature of smart sensors.

Digital communication processing also is unique to smart sensors in many ways, particularly as it brings in the feature of integrity of communication to smart sensors.

Parity checking and handshake dialogue are some of the forms of redundant coding which can protect the transmission process.

8.6   MEMS

The term MEMS (microelectromechanical systems) has acquired a wider meaning than just what it should literally mean. It is sometimes used as almost a synonym for smart systems. Varadan, Jiang and Varadan (2001) describe MEMS as follows: *MEMS refer to a collection of microsensors and microactuators that can sense their environment and have the ability to react to changes in that environment with the use of a microcircuit control.* They go on to state that MEMS may include, in addition to the conventional microelectronics packaging, an integration of antenna structures for command signals into suitable microelectromechanical structures, for achieving the desired sensing and actuation functions. Micro power supplies, microrelays, and micro signal processing units may also be needed.

Miniaturization makes the sensing and actuation systems cheaper, more reliable, faster, and capable of a higher degree of complexity and sophistication. The very brief description of MEMS given here is based on the works of Varadan, Jiang and Varadan (2001), Varadan, Vinoy and Jose (2003), Santos (2004), and Beeby *et al.* (2004).

MEMS were first developed in the 1970s for use as pressure and temperature sensors, gas chromatographs, microaccelerometers, inkjet printer-heads, micromirrors for projectors, etc. Soon their use was extended to low-frequency applications, essentially as miniature devices using mechanical movement for achieving short-circuit or open-circuit configurations in transmission lines. Modern applications of MEMS include such diverse areas as biomedical engineering, wireless communications, and data storage.

A variety of materials are used for fabricating MEMS. Silicon tops the list because of its favourable mechanical and electrical properties, and also because of the already entrenched IC-chip technology. More recently, there has been a revolution of sorts, brought about by advances in the technology of using multifunctional polymers for fabricating three-dimensional MEMS (Varadan, Jiang and Varadan 2001). Organic-materials-based MEMS are also conceivable now, after the invention of the organic thin-film transistor (Varadan, Jiang and Varadan 2001). In the overall smart systems involving MEMS, there is also use for ceramics, metals, and alloys.

Some of the more integrated and complex applications of MEMS are those in microfluidics, aerospace, biomedical devices, etc. (Fujita 1996, 1998). Mechanical sensors based on the MEMS approach have been discussed comprehensively by Beeby *et al.* (2004).

8.6.1   *Microfabrication techniques for MEMS*

Micromachining of silicon has been a major route for making MEMS. One resorts to either bulk micromachining, or surface micromachining.

Bulk micromachining of silicon

This technique involves selective removal of silicon-wafer material by etching away from the substrate. The thickness range covered is from sub-micron to 500 μm. A variety of shapes of trenches, holes, and other microstructures can be etched, either by employing anisotropically etching agents (which is easy for single-crystal substrates, because of the anisotropic nature of the crystal structure), or by working with selectively doped substrates. Doped portions of silicon etch more slowly.

To assemble complex three-dimensional microstructures of silicon in a monolithic format, bulk micromachining has to be followed by *wafer bonding*. Various approaches are in use: anodic bonding; intermediate-layer-assisted bonding; and direct bonding (Varadan, Jiang and Varadan 2001; Beeby *et al.* 2004).

Surface micromachining of silicon

Unlike bulk micromachining, surface micromachining of silicon *can* build monolithic MEMS directly. One deposits on a silicon substrate (or on the IC chip directly) thin segments of *structural layers* and *sacrificial layers*, and eventually removes the sacrificial layers to obtain the final MEMS configurations. Only thin MEMS can be made by this technique, and this is a major handicap in certain sensing and actuation applications. The main advantage of this approach is the easy integrability of the MEMS with the underlying IC chip.

This technique inherently produces two-dimensional structures. However, three-dimensional MEMS can be assembled by suitable follow-up measures like the use of hinges for connecting polysilicon plates to the substrate, and to one another (Varadan, Jiang and Varadan 2001).

LIGA

LIGA is a German acronym, meaning 'lithography, galvanoforming, moulding'. Deep X-ray lithography is used for mask exposure, followed by galvanoforming to form the metallic parts, and moulding to fabricate parts involving plastics, metals or ceramics (Guckel 1998). Very high aspect ratios can be achieved by this technique, and a wide variety of materials can be incorporated for fabricating highly complex microsensor and microactuator configurations. Moveable microstructures can also be incorporated. But the approach is not suitable for producing curved surfaces.

Micromachining of polymeric MEMS devices

For producing real three-dimensional MEMS with high aspect ratios and even curved surfaces, the best option that has emerged is that of working with polymers. Polymeric devices can be integrated with silicon devices without much difficulty.

Two types of polymers are used: a *structural* one and a *sacrificial* one. The former is usually UV-curable, with urethane acrylate, epoxy acrylate, and acryloxysilane as the main ingredients. The sacrificial polymer is an acrylic resin containing 50% silica, and modified by adding 'Crystal Violet' (Varadan, Jiang and Varadan 2001).

The use of multifunctional polymers adds to the versatility of MEMS. For example, electroactive polymers provide actuator action, usable in micropumps. Biocompatible polymers make it possible to devise MEMS for medical delivery and monitoring systems.

The polymers are so designed that there is a backbone along with functional groups that serve as anchor points for, e.g. metal oxides. Nanoparticles of PZT, PLZT, etc. have active surfaces or functional groups capable of bonding with the polymer chain. Such ferroic materials provide piezoelectric and other functionalities, and the backbone provides the needed mechanical stability and flexibility.

Some of the functional polymers used in MEMS are: PVDF (a piezoelectric); poly(pyrrole) (a conducting polymer); fluorosilicone (an electrostrictor); silicone (an electrostrictor); and polyurethane (an electrostrictor).

Three-dimensional microfabrications

Microscale free-form fabrications offer exceptional flexibility in the design of MEMS. Three-dimensional microstructures are produced in an additive layer-by-layer fashion, with the size and shape of each layer changing in a pre-designed way, resulting in complex three-dimensional creations. The approach also allows for the use of a variety of materials, including functional materials.

In the layer-by-layer approach, one can first generate a complex model of a MEMS by three-dimensional CAD (computer-assisted design), and then produce even complex structures quite rapidly. The term *rapid prototyping* (RP) is used in this context (Varadan, Jiang and Varadan 2001).

From MEMS to NEMS

There is an increasing commercial pressure for making low-cost polymer-based MEMS and NEMS. One would like to fabricate components such as flow channels, reservoirs, and mixing chambers directly in a polymer chip. A variety of modelling studies are already being carried out to make this objective a reality (Pelesko and Bernstein 2003; Santos 2005).

Nanoimprint lithography (NIL) is emerging as a viable, low-cost, high-throughput technology for pushing the limits of micro-device fabrication capabilities to the sub-100 nm regime. In NIL, a material is deformed permanently, with mesoscale reproducibility, with a reusable pre-patterned *stamp*.

The main NIL techniques used for structuring polymers are: hot-embossing lithography (HEL); UV-moulding (UVM); and micro-contact printing (μCP) (Luesebrink *et al.* 2005).

In HEL, patterned structures are transferred into a heat-softened resist, and are then cured by cooling the thermoplastic polymer below its glass-transition temperature, T_g. PMMA (poly(methyl methacrylate)) is the polymer generally used, and the stamp is usually made from silicon or nickel. At the temperature of high-contact-force stamping (above T_g), the polymer is soft enough to enable flow of material for cavity filling. After the pattern has been imprinted into the resist, the usual micro-fabrication steps are employed: de-embossing; de-scumming; etching; deposition and release; template or device formation; and bonding (Luesebrink *et al.* 2005).

UV-moulding is also a stamping process, but can be carried out at room temperature. It is effected in modified mask aligners used traditionally in semiconductor and MEMS fabrication processes. The stamp used should be transparent to UV in the 350–450 nm range. Quartz is used for hard stamps, and certain polymers are used for soft stamps. A monomer-coated carrier substrate (e.g. Si wafer), as well as the transparent stamp, are loaded into the aligner, and fixed by vacuum on their respective chucks. The usability of an optical aligner is behind the high level of resolution achievable by this technique. After alignment, the substrate and the stamp are brought into contact. The imprinting force required is not as high as in HEL. The UV-curing results in cross-linking (polymerization) of the monomers.

In a micro-contact-printing process, molecules like the thiols are transferred from an 'inked stamp' to another surface, typically gold or silver. The stamp is inked by putting a drop on the stamp surface, or by applying the material in an inked pad.

8.6.2   *RF MEMS*

There are several advantages of employing MEMS-based fabrication technologies for integrated circuits working in the radio-frequency (RF) and microwave regime. As the frequency increases, the sizes of the microwave components go down. For mm-wave systems, it is necessary that the dimensions of most of the components be in the sub-mm range. The high-precision fabrication offered by micromachining provides a viable solution. System integration is an added advantage. Reduction of insertion losses and increase of bandwidth becomes possible for surface-micromachined devices such as RF switches, tuneable capacitors, and microinductors (Varadan, Vinoy and Jose 2003; Santos 2004).

The cantilever beam was the first MEMS device developed for use in a switch (Petersen 1979). The cantilever and the doubly supported beam are the two basic electrostatically actuated, surface-micromachined MEMS, and their key applications in RF MEMS are as switches and resonators (Larson *et al.* 1991). Santos (2004) has discussed a variety of applications of MEMS in electrostatically actuated microwave switches, transmission lines, and passive lumped fixed-value and tuneable circuit elements.

Two basic requirements have to be met for the efficient functioning of portable, low-cost and light-weight wireless communication systems: very-low-loss *passive* components; and very efficient *active* components. (Transmission lines are examples of passive components; they do not require a standby battery backup. Amplifiers and digital circuits are active components; they do require a backup power supply.) The current technology for RF MEMS has the potential for meeting both these requirements. Here are a few examples (Santos 2004):

(i) Micromachining of the substrate beneath inductors and transmission lines improves their quality factor and extends their frequency of operation.

(ii) MEM techniques can produce on-chip variable capacitors with potentially higher quality factor than semiconductor-based varactors.

(iii) MEM switches have the potential for very low insertion losses, high bandwidth, and zero standby dc power consumption.

8.7 Perceptive networks

Vivisystems like beehives (cf. Section 2.3.1) continue to inspire more and more inventions in various branches of human activity and research. Perceptive networks are another such example.

Smart sensor systems, small and cheap enough to be produced in large numbers and distributed over the area to be monitored, have been developed. Very simple computers and operating systems are linked by radio transceivers and sensors to form small autonomous nodes called *motes*. Running on a specially designed operating system called TinyOS (cf. Ananthaswamy 2003; Culler and Mulder 2004), each mote links up with its neighbours. Although each unit by itself has only limited capabilities, a system comprising hundreds of them can spontaneously emerge as a *perceptive* network. Such networks, like other vivisystems, work on the principle that *people learn to make sense of the world by talking with other people about it* (Kennedy 2006).

Perceptive networks are already being manufactured, and have been used for helping biologists in studying seabird nests and redwood groves. Their near-future applications include monitoring of vibrations in manufacturing equipment, strain in bridges, and people in retirement homes. Defence applications are an obvious next step, as also mapping and prediction of global weather patterns.

A network of motes is *not* just a network of miniaturized PCs, although each member does have some simple computing capability. The whole approach is based on the following considerations:

- Cut costs.
- Conserve power.
- Conserve space (miniaturize).
- Have wireless communication (networking) among the nodes or agents, like in a beehive.
- Let collective intelligence emerge, like in a beehive, even though each agent is a very simple 'dumb' member of the network.
- Ensure robust, efficient, and reliable programming of a large and distributed network of motes.
- Incorporate regrouping and reprogramming, again like in a beehive.
- Include redundancy of sensor action to increase the reliability of the motes, keeping in view the fact that they may have to operate in hostile environmental conditions.

At present, a typical agent of the swarm is a very simple thumb-sized gadget incorporating sensors, actuators, microprocessors, a small memory, a radio transceiver, and an operating system (namely TinyOS). It can do some pre-processing of sensor data, so that only the 'conclusions' are beamed to other agents or to the next higher-level data processor.

Motes come in various shapes, sizes, and capabilities. Three of them have been described by Culler and Mulder (2004):

1. *Mica*, a second-generation mote (Hill and Culler 2002), is in use in some 500 research projects. It is the size of a matchbox. It incorporates an interface for different sensors like magnetometers, barometers, temperature sensors, etc. It is powered by two AA batteries, and can support up to eight sensors.
2. *Mica2Dot* is one-fourth the size of Mica, and incorporates four kilobytes for data, 128 kilobytes for programs, and a 900-megahertz radio transceiver. There are layers of sensors connected to the processing board.
3. *Smart Dust*, a prototype mote, is only 5 mm^2 in area. It has an ultra-efficient radio, and an analogue-to-digital converter. It would not need batteries, and would run on the energy harvested from ambient light or vibrations (as done by self-winding watches).

Let us now see how motes are able to meet the objectives set out above. The main approach is to imitate bees in a beehive. Each mote has its own TinyOS (like the tiny brain of a bee). This software runs on microprocessors that require very little memory (as little as eight kilobytes). It manages efficiently and economically the various sensors, the radio links to other motes, and the power supply.

The software has to be very stingy about power consumption. This is achieved in various ways, one of which is to make the mote 'sleep' 99% of the time. In the sleep state, all functions except sensing are in a standby mode. The system wakes up, say once very second, and spends ~50 μs collecting data from the sensors, and another 10 ms exchanging data with other motes.

Another strategy for saving power (and for achieving something else with far-reaching consequences) is that of *multihop networking*. It makes each mote have extremely short-range radio transmitters, which saves power. A multitiered network is established, with motes in a particular tier or layer communicating only with those in the next lower and the next upper layer. Information hops from a mote to another mote only by one layer-step. This hierarchy is reminiscent of the neo-cortex of the human brain, although it is much simpler in concept and connectivity. It also makes room for parallelism, like in the human brain, and like in a beehive. If, for example, a particular mote stops functioning, there is enough redundancy and parallelism in the network that other motes reconfigure the connectivity to bypass that mote.

In a simple experiment conducted in 2001 (cf. Ananthaswamy 2003), six motes were fitted with magnetometers, wrapped in Styrofoam, and then dropped by a drone plane, one at a time, from a height of 50 m. Their assigned job was to monitor the movement of tanks and other vehicles on the road.

On hitting ground, the motes woke up from sleep, established a wireless network among themselves, and synchronized their clocks. The presence and movement of vehicles was sensed via the distortion of Earth's magnetic field. The direction and speed of vehicles was calculated, using information about the location of the motes, and the results were stored in the memory. The magnetometers were interrogated at regular intervals, and the sensor data were processed, after which the motes went to sleep for pre-assigned periods. The whole sequence of operations was repeated periodically. After an hour, the drone plane flew by again, receiving the transmitted data from the motes onto a laptop.

To update or replace the software for a network of motes, the method used was similar to how viruses or worms spread in PCs via the Internet. The new software is placed only in the 'root' mote, which then 'infects' the neighbouring motes with it, and so on, up the line.

A variety of applications are being planned and implemented for such perceptive distributed networks of smart sensors (cf. Culler and Mulder 2004; also see Grabowski, Navarro-Serment and Khosla 2003, for a somewhat different approach to building up and deploying armies of small robots).

8.8   The human sensing system

The human sensing system is the dream and the envy of the smart-structures aspirant. However, the conventional notion that we have only five senses is changing. The actual number of senses is now agreed to be much larger.

In terms of the nature of the sensor input, we have just three types: chemical (taste and smell); mechanical (touch and hearing); and radiant (vision).

However, the *sensations* experienced by the brain are many more than just three or five. Touch, for example, involves pressure, temperature, pain, and vibration.

'Sense' could be defined as a system consisting of a specialized cell type, responding to a specific type of signal, and reporting to a particular part of the cortex. Vision, for example, could be taken as one sense (light), or two (light and colour), or four (light, red, green, blue). Similarly, the sense of taste could be split into sweet, salty, sour, bitter, etc.

The human body has an entire system of sensors (called *the proprioceptive system*) that inputs the brain about our joint angles and bodily position. Similarly, our sense of balance is governed by the vestibular system in the inner ear.

The total number of human senses, according to a section of expert opinion, is at least 21 (see, e.g. Durie 2005).

It has been realized for quite some time that, more than the sensations, it is the feelings or perceptions generated by the human sensory system which are really important or relevant. We do not just see light, shade and colour; we form *invariant representations* of objects, spaces, people, and their mutual relationships (Hawkins and Blakeslee 2004). This occurs in the higher echelons of the cortex, which get input, not only from the lower echelons, but also from the collateral layers. This value-addition to the raw sensory data, done by collation of data from the various senses, including memory recall, finally gets stored as generalizations or invariant representations in the brain, and constitutes perception or feeling.

8.9   Humanistic intelligence

A variety of techniques are being employed for *intelligent signal processing* (ISP) (Haykin and Kosko 2001). ISP employs learning to extract information from incoming signal and noise data. The approach is model-independent. Artificial neural networks are the most important tool used in ISP. Some other tools used are fuzzy logic, genetic algorithms, and expert systems.

The human brain is the finest available neural network. Mann (2001) has described the development and testing of a new type of intelligent camera based on the 'humanistic intelligence' (HI) approach. HI is a signal-processing framework in which the processing apparatus is inextricably intertwined with the capabilities of our brain.

The HI approach recognizes the fact that it is not going to be easy to construct a really intelligent artificial brain modelled on the human brain. Therefore, it aims at using the human brain directly by making it a part of the overall signal-processing and learning strategy. The user turns into a part of an intelligent control system and is an integral part of the feedback loop.

The HI equipment constitutes a symbiotic relationship with the human user, in which the high-level intelligence arises because of the existence of the human host, and the lower-level computational workload is borne by the signal-processing hardware.

The HI equipment is already so compact that it can be easily concealed within ordinary-looking eyeglasses and clothing. It is always 'on', and may be worn continuously during all facets of day-to-day living. Through long-term adaptation, it begins to function as an extension of the human mind and body.

8.10   Bioelectronic sensors

Bioelectronics is the science of integrating biomolecules and electronic circuits into functional devices (Willner 2002; Fromherz 2005; Willner and Katz 2005; Woo, Jenkins and Greco 2005). Several types of such devices have been fabricated, or are being developed: bioelectronic sensors; biofuel cells; templates for nanofabrication (Rothemund 2006); and DNA-based computers (Adleman 1998). Out of these, biosensors have been the most successful so far.

A bioelectronic device may operate in one of two possible modes: There can be a biological input, giving an electronic output in the form of current, potential, impedance or electrical power. Alternatively, an electronic input may transduce biosensor activities like biocatalysis and biorecognition, as also ion transport and information storage or computing. The strong motivation for using biomolecules in such applications comes from the fact that they have been *evolution-optimized* by Nature for tasks like catalysis, ion pumping and self-assembly.

Biosensing devices electronically transduce recognition events between biomolecules. Their possible applications include clinical diagnostics, environmental analysis, detection of chemical and biological warfare, as also forensic applications.

8.11 Chapter highlights

- Sensing involves measurement followed by information processing. Smart sensors are those that incorporate at least one digital processor. They are often an important component of sophisticated smart structures.
- Any transduction of energy by the primary sensor material introduces defects in the signal, and one has to compensate for these. The compensation strategies for frequency or time response normally involve postulating a model for the response. A merit of smart sensors is that model-independent compensation can be made.
- Optical fibres, the 'nerves of glass', are very versatile extended sensors, which can be embedded in composites at the fabrication stage itself, and can be put to a variety of uses in smart structures.
- MEMS comprise of a collection of microsensors and microactuators that can sense their environment and have the ability to react to changes in that environment with the use of a microcircuit control. MEMS may include, in addition to the conventional microelectronics packaging, an integration of antenna structures into suitable microelectromechanical structures, for achieving the desired sensing and actuation functions. Micro power supplies, microrelays, and micro signal processing units may also be incorporated.
- Silicon tops the list of materials used for making MEMS because of its favourable mechanical and electrical properties, and also because of the already entrenched IC-chip technology. More recently, there has been a revolution of sorts, brought about by advances in the technology of using multifunctional polymers for fabricating three-dimensional MEMS.
- There are several advantages of employing MEMS-based fabrication technologies for integrated circuits working in the RF and microwave regime. As the frequency increases, the sizes of the microwave components go down. For mm-wave systems, it is necessary that the dimensions of most of the components be in the sub-mm range. The high-precision fabrication offered by micromachining provides a viable solution. System integration is an added advantage. Reduction of insertion losses and increase of bandwidth becomes possible for surface-micromachined devices such as RF switches, tuneable capacitors, and microinductors.
- Perceptive networks are modelled on vivisystems like beehives and ant colonies. Their applications, which will have far-reaching consequences, have just begun. They work on the principle that people learn to make sense of the world by talking to other people about it.
- The human sensing system is the dream and the envy of the smart-structures aspirant. More than sensations, it is the feelings or perceptions generated by the human sensory system which are really important or relevant. We do not just see light, shade and colour; we form invariant representations of objects, spaces, people, and their mutual relationships. This occurs in the higher echelons of the cortex, which get input, not only from the lower echelons, but also from the collateral layers. This value-addition to the raw sensory data, done by collation of

data from the various senses, including memory recall, finally gets stored as generalizations or invariant representations in the brain, and constitutes perception or feeling.

- The field of humanistic intelligence (HI) attempts to get the best of both the worlds: It incorporates the decision-making capabilities of the human brain into the overall design of the smart sensing system, leaving the lower-level computational workload to the signal-processing hardware. The HI approach is based on the realization that it is not going to be easy to construct a really intelligent artificial brain modelled on the human brain. Therefore, it makes the human brain itself a part of the overall signal-processing and learning strategy. The user turns into a part of an intelligent control system, and is an integral part of the feedback loop. The HI equipment is always 'on', and may be worn continuously by the human user during all facets of day-to-day living. Through long-term adaptation, it begins to function as an extension of the human mind and body.

- Bioelectronics is the science of integrating biomolecules and electronic circuits into functional devices. Several types of devices have been fabricated, or are being developed: bioelectronic sensors; biofuel cells; templates for nanofabrication; and DNA-based computers. A bioelectronic device may operate in one of two possible modes: There can be a biological input, giving an electronic output in the form of current, potential, impedance or electrical power. Alternatively, an electronic input may transduce biosensor activities like biocatalysis and biorecognition, as also ion transport and information storage or computing. The very strong motivation for using biomolecules in such applications comes from the fact that they have been *evolution-optimized* by Nature for tasks like catalysis, ion pumping and self-assembly.

References

Adleman, L. M. (Aug. 1998). 'Computing with DNA'. *Scientific American,* 279: 34.

Amato, I. (17 March 1992). 'Animating the material world'. *Science* 255: 284.

Ananthaswamy, A. (23 August 2003). 'March of the motes'. *New Scientist,* p. 26.

Beeby, S., G. Ensell, M. Kraft and N. White (2004). *MEMS Mechanical Sensors.* Boston: Artech House.

Brignell, J. and N. White (1994). *Intelligent Sensor Systems.* Bristol: Instituted of Physics Publishing.

Brignell, J. and N. White (1999). 'Advances in intelligent sensors', in H. Janocha (ed.), *Adaptronics and Smart Structures: Basics, Materials, Design, and Applications.* Berlin: Springer-Verlag.

Busch-Vishniac, I. J. (July 1998). 'Trends in electromechanical transduction'. *Physics Today,* p. 28.

Culler, D. E. and H. Mulder (June 2004). 'Smart sensors to network the world'. *Scientific American,*, p. 85.

Culshaw, B. (1984). *Optical Fibre Sensing and Signal Processing.* London: Peter Peregrinus Ltd.

Durie, B. (2005). 'Doors of perception'. *New Scientist,* 185(2484): 34.

Fujita, H. (1996). 'Future of actuators and microsystems'. *Sensors and Actuators,* A56: 105.

Fujita, H. (1998). 'Microactuators and micromachines'. *Proc. IEEE,* 86(8): 1721.

Frank, R. (2000). *Understanding Smart Sensors,* 2nd edition. Norwood, MA: Artech House.

Fromherz, P. (2005). 'Neuroelectronic interfacing: Semiconductor chips with ion channels, nerve cells, and brain'. In Waser, R. (ed.), *Nanoelectronics and Information Technology: Advanced Electronic Materials and Novel Devices,* 2nd edition, p. 777. KGaA: Wiley-VCH Verlag

Gardner, J. W., V. K. Varadan and O. O. Awadelkarim (2001). *Microsensors, MEMS and Smart Devices*. Chichester: Wiley.

Grabowski, R., L. E. Navarro-Serment and P. K. Khosla (November 2003). 'An army of small robots'. *Scientific American,*, p. 63.

Guckel, H. (1998). 'High-aspect-ratio micromachining via deep X-ray lithography'. *Proc. IEEE*, 86(8): 1552.

Habel, W. (1999). 'Fibre optic sensors', in H. Janocha (ed.), *Adaptronics and Smart Structures: Basics, Materials, Design, and Applications*. Berlin: Springer-Verlag.

Hawkins, J. and S. Blakeslee (2004). *On Intelligence*. New York: Times Books (Henry Holt).

Haykin, S. and B. Kosko (eds.) (2001). *Intelligent Signal Processing*. New York: IEEE Press.

Hill, J. and D. Culler (2002). 'Mica: A wireless platform for deeply embedded networks'. *IEEE Micro*, 22(6): 12.

Kao, C. K. and G. A. Hockham (1966). 'Dielectric fibre surface waveguides for optical frequencies'. *Proc. IEE*, 113: 1151.

Kennedy, J. (2006). 'Swarm intelligence'. In Zomaya, A. Y. (ed.), *Handbook of Nature-Inspired and Innovative Computing: Integrating Classical Models with Emerging Technologies*. New York: Springer, p. 187.

Larson, L. E., R. H. Hackett, M. A. Meledes and R. F. Lohr (1991). 'Micromachined microwave actuator (MIMAC) technology – a new tuning approach for microwave integrated circuits'. In *Microwave and Millimetre-Wave Monolithic Circuits Sym. Digest*. Boston, MA, June 1991.

Lopez-Higuera, J. M. (ed.) (2002). *Handbook of Optical Fibre Sensing Technology*. West Sussex, England: Wiley.

Luesebrink, H., T. Glinsner, S. C. Jakeway, H. J. Crabtree, N. S. Cameron, H. Roberge and T. Veres (2005). 'Transition of MEMS technology to nanofabrication'. *J. Nanosci. and Nanotech.* 5: 864.

Mann, S. (2001). 'Humanistic intelligence: "Wear Comp" as a new framework and application for intelligent signal processing'. In S. Haykin and B. Kosko (eds.), *Intelligent Signal Processing*. New York: IEEE Press.

Pelesko, J. A. and D. H. Bernstein (2003). *Modeling MEMS and NEMS*. London: Chapman and Hall/CRC.

Petersen, K. E. (1979). 'Microelectromechanical membrane switches on silicon'. *IBM J. Res. Develop.* 23(4): 376.

Rothemund, P. W. K. (16 March 2006). 'Folding DNA to create nanoscale shapes and patterns'. *Nature*, 440: 297.

Santos, H. J. de Los (2004). *Introduction to Microelectromechanical Microwave Systems* (2nd edition). Boston: Artech House.

Santos, H. J. de Los (2005). *Principles and Applications of NanoMEMS Physics*. Dordrecht: Springer.

Schindler, P. M., J. K. Shaw, R. G. May and R. O. Claus (1995). 'Neural network processing of optical fibre sensor signals for impact location'. In George, E. P., S. Takahashi, S. Trolier-McKinstry, K. Uchino and M. Wun-Fogle (eds.), *Materials for Smart Systems*. MRS Symposium Proceedings, Vol. 360. Pittsburgh, Pennsylvania: Materials Research Society.

Swanson, D. C. (2000). *Signal Processing for Intelligent Sensor Systems*. New York: Marcel Dekker.

Van Hoof, C., K. Baert and A. Witvrouw (5 November 2004). 'The best materials for tiny, clever sensors'. *Science*, 306: 986.

Varadan, V. K., X. Jiang and V. V. Varadan (2001). *Microstreolithogrphy and Other Fabrication Techniques for 3D MEMS*. New York: Wiley.

Varadan, V. K., K. J. Vinoy and K. A. Jose (2003). *RF MEMS and Their Applications*. London: Wiley.

Willner, I. (20 December 2002). 'Biomaterials for sensors, fuel cells, and circuitry'. *Science*, 298: 2407.

Willner, I and E. Katz (eds.) (2005). *Bioelectronics: From Theory to Applications*. KGaA: Wiley-VCH Verlag GmbH and Co.

Woo, R. K., D. D. Jenkins and R. S. Greco (2005). 'Biomaterials: Historical overview and current directions'. In R. S. Greco, F. B. Prinz and R. L. Smith (eds.), *Nanoscale Technology in Biological Systems*. New York: CRC Press.

9

SENSORS AND ACTUATORS FOR SMART STRUCTURES

Suppose a cause X produces an effect Y in a material, and Y is a mechanical force. Then the material can function as an actuator. For example, X may be an electric field. Then Y can arise through the inverse piezoelectric effect, and also through other effects by a coupling of properties.

If the roles of X and Y are reversed, the same material can also function as a sensor of mechanical force (through the direct piezoelectric effect). Thus, if either X or Y is a mechanical force, the same material can function as both a sensor and an actuator. There is thus a *sensor–actuator duality* with respect to certain properties.

That is not all. The same X (electric field in our example) may produce not only an effect Y (mechanical force in our example), but also an effect Z. For instance, Z can be an induced magnetic moment (through the inverse magnetoelectric effect in the same material). Measurement of the magnetic moment is then one more way of sensing the electric field. The same material, through the direct magnetoelectric effect, can act as a sensor of magnetic field (by sensing the electric dipole moment induced by the magnetic field). Thus the sensor action of a material can arise in a variety of ways.

In condensed-matter physics, there are three types of forces to deal with: electric, magnetic, and mechanical, each giving a distinct type of signal. Two more types of signals to consider are optical and thermal. If we extend our operations to soft materials, chemical signals become another significant possibility. This is because soft materials are generally 'open' systems: they exchange matter with the medium they are surrounded by. In certain situations, chemical signals can have a significant interaction with 'closed' systems also; e.g. by modifying the surface. Such closed systems can then be used as sensors.

Coming to actuators, two types can be distinguished. The first is the conventional type, in which the mechanical thrust is produced by the conversion of other forms of energy by hydraulic, pneumatic or electrical means (see, e.g. Funakubo 1991). Examples include linear motors, hydraulic and pneumatic actuators, servo motors, AC and DC control motors, and stepper motors. It has been difficult to miniaturize such actuators, although the advent of MEMS and NEMS has changed the scenario (Amato 1998).

The other class of actuators use *functional materials*. In them, the actuator material itself possesses the energy-conversion mechanism (Lindberg 1997). Very often, the same material can function as both actuator and sensor. An example is that of a superelastic alloy like Ni-Ti. Its (smart) actuation action comes from its drastic deviation from Hooke's law. And its stress–strain relationship is temperature dependent, so if the material is constrained, e.g. by embedding in a matrix, the temperature-dependent force exerted by it can serve as a sensor of temperature.

For all types of actuators and sensors, there has been a steady trend towards more and more miniaturization. There are several reasons:

1. There is a saving in the consumption of materials for making a device.
2. There is a saving in power consumption.
3. The speed of devices can go up.
4. Assembly-line production of micro- and nanodevices improves reproducibility, and can drastically lower the rejection rate. One also gets better selectivity, sensitivity, and dynamic range, as also a higher accuracy and reliability.
5. Because of the above reasons, one gets better performance per unit cost.
6. A whole array of devices can be put in a small volume. This not only effects a saving of space, one can even provide for redundancy.
7. Miniaturization is also conducive to integration, which helps give a structure (smart or not) an identity (Judy 2001).
8. Integration with electronics becomes easier and more efficient. This is particularly true for MEMS and NEMS.

For the sake of a structured discussion, we shall divide sensors and actuators into three classes: those based on ferroic materials; those exploiting size-related effects at small sizes; and those based on soft matter.

9.1   Ferroic materials for sensors and actuators

9.1.1   *Single-phase materials*

Quartz is the archetypal high-frequency transducer material. Its piezoelectric response is not very large, but it is linear, unhysteretic, and stable.

Since quartz is a ferroic crystal (it is a ferrobielastic, as well as a ferroelastoelectric), it is possible to detwin it (provided the twinning involved is a result of the β-to-α ferroic phase transition (cf. Wadhawan 2000, for details). Detwinned crystals are needed for most applications of quartz. A major application is in crystal resonators for the stabilization of oscillators. Its use as a transducer also occurs in telephone speakers, sonar arrays, a variety of sensors and actuators, and more recently in MEMS.

Apart from quartz, the four most widely used and investigated crystalline materials for sensor and actuator applications in smart structures are also all ferroics:

1. Lead zirconate titanate (PZT) and related piezoelectric materials.
2. Lead magnesium niobate (PMN), lead zinc niobate (PZN), and related relaxors.
3. Terfenol-D.
4. Nitinol.

Each of these has its advantages and disadvantages from the point of view of device applications, and therefore has a specific use and niche. There is also a surprising degree of commonality of certain features of these materials (cf. Section 9.1.3).

PZT

PZT, like quartz, is a piezoelectric material; the strain produced in it depends linearly on the applied electric field, and is therefore sign-dependent: Reversing the polarity of the field changes compression to extension, or vice versa.

PZT cannot be grown in the form of good large single crystals. Therefore it is used as a (poled) polycrystalline ceramic for sensor and actuator applications. The best piezoelectric response is exhibited by compositions close to the morphotropic phase boundary (MPB), i.e. around PZT(52/48). At the MPB, the crystal structure changes, apparently from tetragonal (PT-rich) to rhombohedral (PZ-rich), via an intermediate monoclinic phase (Glazer *et al.* 2004; Damjanovic 2005).

The occurrence of a monoclinic phase in PZT in the vicinity of the MPB was first reported by Noheda *et al.* (1999). Soon after that, Vanderbilt and Cohen (2001) established that it is not enough to retain, in the Landau expansion, terms only up to the sixth degree in the order parameter, for explaining the occurrence of a monoclinic phase in PZT and other perovskite oxides, and that terms up to the eighth order must be included.

Glazer *et al.* (2004) have presented electron diffuse scattering data to argue that, on a small enough length-scale, PZT is monoclinic (at a local level) throughout the phase diagram (except for the end members PZ and PT), and that the succession of apparent symmetries from rhombohedral to monoclinic to tetragonal is simply a manifestation of short-range order (locally monoclinic, but averaging to rhombohedral) changing over to long-range order (monoclinic), and then back to short-range order (locally monoclinic, but averaging to tetragonal).

Over the years, several reasons have been put forward for explaining the superior performance of compositions of PZT close to the MPB. One reason is the availability of several competing phases (tetragonal, rhombohedral, monoclinic), each with its own set of directions along which the spontaneous polarization vector can be reoriented during the poling process. The large number of available ferroelectric switching modes makes the poling process very efficient: The tetragonal phase has six possible orientation states, the rhombohedral phase has eight, and the monoclinic phase has 24. Another conceivable reason is that the MPB signifies a *composition-induced phase transition*, with the attendant high values of dielectric and electromechanical response functions associated with the vicinity of a phase transition.

The discoveries related to the occurrence of one or more monoclinic phases in the phase diagram have led to the well-founded suggestion that the high dielectric and electromechanical response functions of PZT in the vicinity of the MPB are due to the fact that the monoclinic phase has *Cm* symmetry, and therefore the spontaneous polarization vector can lie along any direction in the plane *m* of this phase, thus enabling this vector to *rotate* (while remaining in the plane *m*) under the action of the probing or poling electric field (Fu and Cohen 2000; Noheda *et al.* 2002; Glazer *et al.* 2004; Noheda and Cox 2006). The mirror plane *m* in *Cm* is the only symmetry element common to the space group *R3m* of the *R*-phase (*R* for rhombohedral) and the space group *P4mm* of the *T*-phase (*T* for tetragonal). The monoclinic phase thus plays the role of a *bridging phase* between the *R*-phase and the *T*-phase.

Suitable dopants can modify the performance of PZT in a variety of ways. *Soft* PZT can be obtained by doping with donor cations, e.g. putting Nb^{5+} ions at some of the Ti/Zr^{4+} sites. Such a material has a higher d_{33} coefficient, but is less stable. *Hard* PZT has the opposite characteristics, and can be produced by doping with an acceptor cation like Fe^{3+}.

A comparison of the various transducer options can be made by defining η, the *field-limited energy density*, of a material (Lindberg 1997):

$$\eta = Ye^2/2 \qquad (9.1)$$

Here Y is the Young's modulus, and e the maximum strain obtainable at the practical limit of the electric driving field that can be applied. This parameter permits comparison of materials on a per-unit-volume basis. For PZT this energy density is $\sim$500 J m^{-3}. This serves as a benchmark for checking the performance of other types of materials for actuator applications.

PZT as an actuator material has the advantage of being familiar; its long usage has led to the development of some auxiliary technologies and practices. Any alternative new material and technology has to be worthwhile enough for replacing the existing technology. PZT is low-cost, and has a rather good electromechanical coupling coefficient k_{33} of $\sim$ 0.7. Being a ferroelectric, its disadvantage is that its response is hysteretic.

The piezoelectric response of a material has an *intrinsic* component and an *extrinsic* component. The former refers to the effect of the crystal structure and the atomic mechanism of the piezoelectric response (Bellaiche and Vanderbilt 1999). The latter can have a number of extraneous contributing factors like the domain structure, or the defect structure, or even a field-induced phase transition.

Superpiezoelectric response (a term coined in this chapter of the book by analogy with superelasticity) is extrinsic piezoelectric response (by definition): The electric field may cause a phase transition and/or cause a movement of domain boundaries and phase boundaries, all leading to a net deformation which is in addition to (and often much larger than) the intrinsic deformation attributable directly to the crystal structure.

Whereas the intrinsic piezoelectric response of PZT is a linear effect, the extrinsic or superpiezoelectric response has a nonlinear and hysteretic character (Li, Cao and Cross 1991; Damjanovic 2005). The d_{33} coefficient for PZT can be in the range 200–700 pC N p^{-1}, depending on factors like composition, doping, and the proximity or otherwise of the temperature of the ferroelectric phase transition. Application of a field of the order of 10 kV cm^{-1} causes domain-wall movement, as well as a phase transition, resulting in a net piezoelectric strain of the order of 0.1% (Park and Shrout 1997).

Apart from domain-wall movement, another contributor to the extrinsic piezoelectric response of PZT ceramic is the pinning of domain walls by defects (Damjanovic 1997). This leads to a field-dependence of the piezoelectric coefficients. This feature may be a nuisance for certain purposes, but it also provides *field-tuneability* of piezoelectric response in an otherwise linear-response material. Field-tuneability raises the possibility of *adaptability*, a central feature of smart materials.

PMN-PT and PZN-PT

From the point of view of device applications, PMN and related relaxors are best known for their large electrostriction response and large dielectric permittivity. Naturally, their main application areas are in microactuators (Uchino 1986; Wheeler and Pazol 1991), and in miniaturized multilayer capacitors (Shrout and Halliyal 1987; Husson and Morell 1992; Kim and Yang 1996). Since electrostrictive strain depends on the square of the applied electric field, it is invariant to a change of sign of the field. Moreover, the field *vs.* strain curve shows little or no hysteresis. Applications of PMN-based relaxors include deformable mirrors and bistable optical devices (Uchino *et al.* 1981; Nomura and Uchino 1983).

The energy density η (cf. eqn 9.1) is typically ~5000 J m^{-3} for PMN, an improvement over PZT by a factor of 10. But the k_{33} for the polycrystalline ceramic is rather small (~0.37). Its use as an actuator also requires the application of a dc bias for linear operation (Lindberg 1997). Ageing effects are another problem needing further investigation and tackling. Its mechanical properties also need to be improved.

PMN is seldom used in its undoped form. It readily forms a solid solution (PMN-PT) with PT (lead titanate) (Noheda *et al.* 2002; Zekria and Glazer 2004). Unlike PZT, it is possible to grow PMN-PT in the form of fairly large single crystals.

Another solid solution involving PT and a relaxor, which has been grown in the form of large single crystals and investigated extensively, is PZN-PT (Park and Shrout 1997); here PZN stands for lead zinc niobate, $Pb(Zn_{1/3}Nb_{2/3})O_3$.

Investigations on single crystals of PMN-PT and PZN-PT, grown for compositions near their respective MPBs, have revealed much better performance characteristics compared to the polycrystalline forms of these materials (Park and Shrout 1997). The MPB for PMN-PT is near 65/35 (Noblanc, Gaucher and Calvarin 1996; Guo *et al.* 2003), and that for PZN-PT is near 92/08 (La-Orauttapong *et al.* 2002).

Single crystals of PZN-PT(91/09) are found to have the largest known electromechanical coupling factor (k_{33}) of 95% (Uchino 1997; Park and Shrout 1997). Important factors determining this and other anomalously large response functions near the MPB include the direction of poling and the prevailing domain structure.

Before the successful growth of large single crystals of PMN-PT and PZN-PT, the main available single crystals for electromechanical applications were quartz (SiO_2), lithium niobate ($LiNbO_3$), and lithium tantalate ($LiTaO_3$). None of them has a d_{33} coefficient greater than ~ 50 pC N^{-1}. By contrast, the value reported for the charge produced per unit force for a single crystal of PZN-PT(92/08) is as high as ~ 2500 pC N^{-1}. This is loosely referred to as the d_{33} coefficient (Park and Shrout 1997), although the effect involved is electrostriction (a tensor property of rank 4), rather than piezoelectricity (a tensor property of rank 3).

The electrostrictive strain observed for PZN-PT(92/08) is as large as $\sim 0.6\%$, with the additional advantage that, unlike the case for PZT (which is a piezoelectric), there is little or no hysteresis to contend with.

Like PZN-PT, single crystals of the near-morphotropic composition PMN-PT(65/35) also exhibit a large 'd_{33}' value (~ 1500 pC N^{-1}).

In the phase diagrams of PZN-PT and PMN-PT, the MPB 'nominally' separates the rhombohedral phase from the tetragonal phase. The high d_{33} values mentioned above for PZN-PT and PMN-PT are for the rhombohedral phase, and along the $\langle 001 \rangle$ directions, even though the polar directions are $\langle 111 \rangle$. In fact, compared to the value 2500 pC N^{-1} for the d_{33} coefficient of PZN-PT(92/08) along $\langle 001 \rangle$, the value along $\langle 111 \rangle$ is just 84 pC N^{-1}.

As in the case of PZT, monoclinic phases have also been discovered near the MPB for PZN-xPT (Noheda et al. 2001) and PMN-xPT (Singh and Pandey 2001). For PMN-xPT, the symmetry at room temperature is $R3m$ for $x < 0.26$; Cm for $0.27 < x < 0.30$; Pm for $0.31 < x < 0.34$; and $P4mm$ for $x > 0.35$ (Singh and Pandey 2003). Vanderbilt and Cohen (2001) had predicted the occurrence of three monoclinic phases (M_A, M_B, M_C) in the vicinity of the MPB. M_A and M_B have the space-group symmetry Cm, and M_C has the symmetry Pm. If the spontaneous-polarization vector, **P**, is written in terms of its components along the axes of the pseudo-cubic unit cell, then $P_z > P_x$ for the M_A phase, and $P_z < P_x$ for the M_B phase.

An orthorhombic phase (O-phase) of symmetry $Bmm2$ is also expected to occur between the M_B and the M_C phases of PMN-xPT. It occurs around $x \sim 0.30$–0.31, when the angle β of the monoclinic phase touches the value $90°$ (Singh and Pandey 2003).

A similar sequence of phases also occurs for PZN-xPT in the vicinity of the MPB (Noheda et al. 2001). Viehland (2000) has discussed the occurrence of an orthorhombic phase in single crystals of PZN-0.08PT, and interpreted it in terms of symmetry-adaptive ferroelectric mesostates. This analysis draws analogies from the situation in martensite formation, wherein there is no uniqueness of lattice parameters; instead, the lattice parameters of the adaptive phase depend on those of the parent phase and of the ferroelastic phase undergoing repeated twinning to create an adaptive 'phase' and an invariant-plane strain (Khachaturyan, Shapiro and Semenovskaya 1991; Jin et al. 2003). It is argued that the large piezoresponse of the material in the vicinity of the MPB indicates that the crystalline anisotropy of the polarization direction is vanishingly low, thus making it easy to create domain walls with very little energy cost. This, in turn, makes easy the occurrence of highly twinned adaptive phases. The existence of special intrinsic relationships among the lattice parameters of the monoclinic phases and the tetragonal phase lends further credence to this model (Wang 2006). In any case, the Landau theory put forward by Vanderbilt and Cohen (2001) for a homogeneous ferroelectric state cannot explain this relationship, not to mention the fact that, by going to the eighth-order term in the order parameter, the theory fits as many as eight parameters to the data.

X-ray diffuse scattering studies, combined with the application of high pressure, have provided important insights into the nature of order and disorder in the structure of PMN and NBT ($Na_{1/2}Bi_{1/2}TiO_3$) (Chaabane *et al.* 2003; Kreisel *et al.* 2003). It is found that the diffuse scattering, associated with the disorder in the structure, disappears at a high pressure of ∼4 GPa, and a new crystal structure appears. In a general way, high pressure or reduced volume is expected to cause a crossover from a normal ferroelectric state to a relaxor state (Samara and Venturini 2006).

So far we have considered only the occurrence of *intrinsic* electrostriction in PMN-PT and PZN-PT, i.e. the strain arising from the effect of a small probing electric field ($e \sim E^2$), or the inverse effect manifesting as charge separation caused by mechanical force (measured in units of pC N^{-1}). But if the applied field is large enough, it can also cause a movement of domain walls, or even a phase transition. This *extrinsic* effect, which we can call *superelectrostriction* (by analogy with superelasticity), has been found to produce dramatically large effects in single crystals of PZN-PT. A superelectrostrictive strain as high as 1.7% has been reported by Park and Shrout (1997) for a single crystal of PZN-PT(95.5/4.5) oriented along a ⟨001⟩ direction. Accompanying this performance are features like very high dielectric permittivity, large k_{33} (>90%), and low dielectric loss (<1%).

The large and field-tuneable electrostriction exhibited by relaxors enables them to function as smart materials which can act as *tuneable transducers* (Newnham and Amin 1999). One monitors the change in the capacitance of the material, and this signal generates two feedbacks, one a dc field and the other an ac field. The dc field is used as a bias for tuning the magnitude and linearity or nonlinearity of the electrostrictive response. And the ac field drives the material as an actuator (around the applied bias value). Thus, three coupled effects arise: (i) change in the dielectric permittivity with stress; (ii) field-dependence of the piezoelectric voltage; and (iii) electrically driven mechanical strain.

An interesting effect for creating an 'actuator with a memory' has been reported for PMN-PT(65/35) SME ceramic (Wadhawan, Pandit and Gupta 2005). This composition is close to the MPB (Zekria and Glazer 2004). After just one shape-memory thermal cycle, a thin bar of the material remembers its shape (degree of bending) at every temperature in a range of temperatures in the martensitic or relaxor–ferroelectric phase (Figs. 9.1 and 9.2). This is a single material and not, say, a bimetallic strip. A variety of applications can be possible such a 'trained' very smart material.

There could be several contributing factors responsible for the *temperature-imprinted reversible behaviour* depicted in Fig. 9.1. This includes the possibility

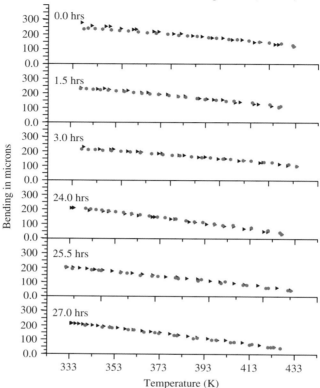

Fig. 9.1 A thin bar of the PMN-35PT ceramic remembers its shape over a whole range of temperatures (for both heating and cooling runs), after going through just one shape-memory cycle. The different pairs of curves show the absence of ageing effects for the time period investigated (Wadhawan, Pandit and Gupta 2005). Figure reproduced with permission of the publisher (Elsevier).

of canonical-glass-type memory associated with the nonergodic nature of the relaxor phase.

Terfenol-D

The term Terfenol-D stands for $Tb_xDy_{1-x}Fe_2$ (typically $Tb_{0.3}Dy_{0.7}Fe_2$): 'Ter' for the element Tb, 'fe' for Fe, 'nol' for Naval Ordnance Laboratory (where this material was developed in the USA), and D for the element Dy. Certain rare-earth elements like terbium and dysprosium had been known for long to exhibit large magnetostriction strains ($\sim$10,000 microstrain, or 10,000 ppm, or 1%), but only at low temperatures (see du Tremolet de Lacheisserie 1993 and Clark 1995 for updates). Clark and Belson (1972) discovered that room-temperature strains as large as 2000 microstrain

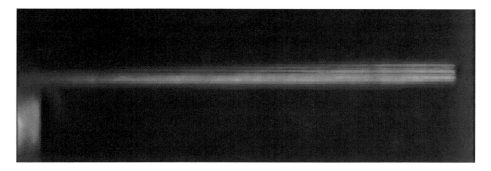

Fig. 9.2   Direct visualization of the experiment described in Fig. 9.1. A video movie was made of the bending and unbending behaviour of the PMN-35PT ceramic bar as a function of temperature, and a superposition of three of the snapshots (taken at 333, 383, and 433 K) is shown in this picture. The length of the bar is 48 mm, and the spontaneous movement at its tip is $\sim$1.8 mm. This is probably the first visual demonstration of the temperature-reversible shape-memory effect in a bar-shaped specimen of a ceramic. [Picture courtesy P. Pandit and A. Saxena, RRCAT, Indore.]

can occur in TbFe$_2$ ('Terfenol') and in DyFe$_2$. Making a solid solution of the two materials offered the great advantage of choosing the best composition for improving the performance characteristics, and the result was Terfenol-D. The favoured composition is near $x = 0.7$. The phase boundary separating the rhombohedral and the tetragonal phases is somewhat like an MPB, although not strongly vertical. Commercially available Terfenol-D ceramic gives a magnetostrictive strain of $\sim$1500 ppm or 0.15%.

Like PZT, Terfenol-D has cubic symmetry at high temperatures. This paramagnetic phase changes to the ferromagnetic rhombohedral phase on cooling, with spins pointing spontaneously along $\langle 111 \rangle$. The symmetry below room temperature is tetragonal. Near room temperature, the structure is poised on an instability: The structure is rhombohedral, but rather close to adopting tetragonal symmetry at the 'MPB'. This fact makes its magnetostrictive response function large.

The availability of Terfenol-D has led to a renewal of interest in designing magnetostrictive actuators cum sensors. The sensor configurations based on this material can measure motion, stress, torque, and magnetic fields. Similarly, the actuators can produce displacement, force, and acoustic waves (Flatau, Dapino and Calkins 2000).

The actuators using the magnetostrictive effect in Terfenol-D have the advantage of high force per unit area (>50 MPa), high-precision, direct control of the movement, and practically no fatigue (Gerver *et al.* 1997). Moreover, they work at rather low fields, and, compared to piezoelectric actuators, can be used at much lower temperatures.

The energy density, η, is $\sim$20,000 J m^{-3} for Terfenol-D. The highest known value of $\eta \sim 50,000$ J m^{-3} is obtained for the alloy Tb$_{0.6}$Dy$_{0.4}$ for operation at 77 K, assuming $Y = 25$ GPa, and $e = 1980$ ppm for an applied field of 22 kA m^{-1} (rms). The e.m. coupling coefficient k_{33} for this material is 0.70. The high energy density and ruggedness make it very suitable for use in vibration-suppression and in high-power sonar.

Use of Terfenol-D in single-crystal form, rather than in polycrystalline form, offers some distinct advantages. Single crystals of $Tb_{0.3}Dy_{0.7}Fe_2$, oriented along the [111] direction, give a strain as high as 3600 ppm (Clark 1995).

Actuators using Terfenol-D generate very large strains compared to those based on piezoelectricity or electrostriction. However, they are of comparatively massive proportions, and also present problems related to nonlinearity and hysteresis (Banks, Smith and Wang 1996).

Nitinol

SMAs can generate large deflection and forces. They have the highest known force/weight ratio among actuators. They also have good damping properties. Their applications have been particularly striking in clamps, switches, and springs; and all these applications involve going across the martensite–austenite phase transition. Other areas in which deployment of SMAs is under active research and development are (Boller 2000): adaptive wings of aircraft and missiles; active control of rotorblades in rotorcraft; in-orbit control of the structure of antennas in satellites; hinge-free solar panels; robot arms; and the general area of SMA-embedded composites.

Bellouard *et al.* (2000) have described the development of a monolithic SMA microactuator. It is made from one single piece of material, and is a complete mechanical system by itself, rather than being part of a larger mechanism. In this sense, it qualifies to be called a smart *material*, rather that a smart structure, because it meets Culshaw's (1996) information-reduction criterion. The two-way SME (TWSME) is one of the features it exploits for achieving reversible motion without the intervention of an external force. It has all the elements of an active mechanical system: a force generator, a coupling mechanism, a transmission mechanism, a guiding mechanism, and an output function. All these are integrated into a single monolithic block. A laser beam is used for varying the temperature of the material locally. The material can function as a micro-gripper and a micro-switch.

One of the most extensive and successful applications of Ni-Ti SMAs and the superelasticity exhibited by them has been in the medical industry (Duerig 1995). The biocompatibility of the material is one reason for this. Some of the prominent medical applications are: guidewires; laparoscopic surgical instruments; implants; stents; retrieval baskets; and bone anchors. Most of these applications exploit the superelastic behaviour of the material at body temperature. For example, a spring made of the superelastic material exerts a *constant* stress for deformations as large as a few percent. Orthodontic archwires exploit this property (see Duerig 1995 for a review).

Many of these binary intermetallics exhibit a phase transition from a high-temperature partially ordered bcc phase (space group $Im\bar{3}m$) to a partially ordered austenitic phase of symmetry $Pm\bar{3}m$ (CsCl structure) (Newnham and Amin 1999). On further cooling, a second phase transition (the martensitic phase transition) occurs to a distorted phase of symmetry $P4/mmm$. For this phase, the spontaneous strain is usually large, and therefore there is heavy twinning, occurring in a self-accommodative manner.

Magnetic shape-memory alloys

The best magnetic shape-memory materials, used in magnetically controlled actuators, are the Ni_2MnGa-based *Heusler alloys*, which show reversible strains as large as 6%, under a magnetic field of 1 T. The properties of these alloys are a sensitive function of composition. Nearly stoichiometric Ni_2MnGa undergoes a ferromagnetic (nonstructural) phase transition at 376 K, which is not far above room temperature. The chemically cubic ferromagnetic phase changes to a martensitic tetragonal phase at 202 K.

For $Ni_{2+x}Mn_{1-x}Ga$ the temperature of the ferromagnetic transition decreases with increasing x, while the temperature of the low-temperature martensitic transition increases. For the composition range $x = 0.18$–0.20 the temperatures for the nonstructural and the structural phase transitions merge (cf. Vassiliev 2002). This common temperature is still above room temperature, which is very suitable for device applications.

Shape-memory ceramics

Shape-memory ceramics (like PMN-PT(70/30)), being insulators, offer the possibility of control of their performance by electric fields, apart from the use of mechanical loads (Pandit, Gupta and Wadhawan 2006).

Such materials offer some unique advantages, not fully exploited yet (Uchino 1998; Melton 1998):

1. Since they can be influenced by electric fields, there is great potential for their integration into electronically controlled devices.
2. Compared to mechanical loads, response times to electric fields are quicker ($\sim$ms).
3. They entail a lower consumption of energy ($\sim 1/100$) compared to SMAs.
4. Although their main limitation is the rather low recoverable shape-strain ($\sim 0.2\%$), the use of multilayer configurations should overcome this problem substantially.
5. The use of multilayer devices, which are connected in series mechanically, and in parallel electrically, can result in a huge lowering of the driving voltages required.

Uchino (1998) has described some applications of shape-memory ceramics ('latching relays', 'mechanical clampers') based on the field-induced phase transition in certain antiferroelectrics.

Piezoelectric polymers

For sensors and actuators, a proper interfacing with the environment is an important requirement. *Impedance matching* is a familiar concept in electronics. For underwater applications, polymer transducers have an advantage over ceramics because of their better *acoustic*-impedance matching with water (Howarth and Rittenmyer 1995). The most familiar polymer in this regard is PVDF (polyvinylidene fluoride, or PVF_2). Another polymer traditionally used in hydrophone applications is the copolymer of PVF_2 with trifluoroethylene (PVF_2EF_3, or P(VDF-TrFE)). The piezoelectric response of these polymers is not as large as that of the conventionally used ceramics. Moreover, the dielectric loss is rather high, and the e.m. coupling factor is low.

A detailed comparative analysis of various types of single-phase transducer materials, from the point of view of actual field applications, has been given by Lindberg (1997).

9.1.2 *Composites*

Embedded piezoelectrics

PZT has a high piezoelectric coefficient d_{33}, and is therefore an excellent transducer material. But, as mentioned above, the use of this and other ceramics in underwater applications presents some interfacing problems. Its density (7.9 g cm^{-3}) is too high compared to that of water, so the acoustic matching is poor for hydrophone applications, or for applications involving the human tissue. Moreover, it is brittle. Use of composites based on PZT and polymers can provide an answer to these problems (Newnham *et al.* 1995).

In hydrophone applications, the typical range of frequencies is of the order of a few kHz or less. At these frequencies, the wavelength of acoustic waves is much larger than the size of the transducer. Therefore, the stress impinging on the transducer is essentially hydrostatic stress. The hydrostatic piezoelectric charge coefficient, d_{3h}, is therefore given by $(d_{31} + d_{32} + d_{33})$, which can be often approximated as

$$d_{3h} = d_{33} + 2d_{31} \qquad (9.2)$$

For PZT, d_{3h} has a low value (because d_{31} has a negative sign, and $d_{33} \sim -2d_{31}$). Moreover, the relative dielectric permittivity of PZT is rather high (~ 1800), so the piezoelectric voltage coefficients g_{33} and g_{3h} are low (cf. the appendix on tensor properties for some relevant definitions). For all these problems, making a carefully designed composite of PZT with a polymer, with an appropriate connectivity of the two phases, helps. The 1–3 connectivity (cf. the appendix on composites), wherein (poled) parallel PZT fibres are embedded in a polymer matrix, has been found to be particularly useful. For hydrophone applications, the product $d_{3h}g_{3h}$ serves as a good figure of merit:

$$d_h g_h = (d_{33} + 2d_{31})^2 / \varepsilon_{33} \qquad (9.3)$$

In the 1–3 connectivity pattern, there is a reduction of the d_{31} coefficient, as well as of the average permittivity ε_{33}, while the large value of d_{33} is retained. This leads to an effective increase in d_{3h} and g_{3h}.

The 1–3 composites based on PZT have also found applications in nondestructive testing and medical diagnosis. Biomedical transducers require low acoustic impedance, a large bandwidth, and high electromechanical coupling coefficients. The operating resonance frequencies are typically in the 1–10 MHz range. These requirements have been met in the commercially available transducers based on PZT composites (Preu *et al.* 1991).

Embedded magnetostrictives

Problems related to acoustic impedance matching, as also the need for flexibility of design, have spurred research into the possible use of Terfenol-D as an active embedded phase in a composite in which a resin is the matrix phase. Duenas, Hsu and Carman (1997) have reported work on a 1–3 composite of this type. Particles of Terfenol-D are blended with a binder resin, and cured in the presence of a magnetic field to obtain 1–3 connectivity. It has been found possible to do *in situ* preloading of the active material, which results in domain orientations giving a performance comparable to that of monolithic Terfenol-D.

Embedded shape-memory alloys

When a pseudoplastically deformed SMA is heated to the austenitic phase, there is shape recovery. If the material is embedded in, say, an epoxy matrix, there is a clamping effect from the matrix, and the SMA exerts a recovery stress, which can be several hundred MPa. This has a stiffening effect on the composite, changing its natural frequency of vibration, among other things. The system thus offers the possibility of temperature-controlled, active, and reversible change of the vibrational behaviour of the composite. Bidaux, Manson and Gotthardt (1997) embedded prestrained NiTi fibres in an epoxy matrix. It was found that the prestraining could be used for fine-tuning the fraction of the R-phase forming along with the ordinary martensitic phase. The R-phase is favoured for this application, not only because of the large recovery stress generated by it, but also because of the low transformation strain, which is conducive in maintaining the integrity of the bonding between the two phases in the composite. The thermomechanical modelling of such composite systems has been critically reviewed by Stalmans, Delaey and Humbeeck (1997).

Realistic modelling of the *in situ* behaviour of SMAs continues to be a challenge. Although sensing and actuation are readily achieved with these materials, bringing the third element of smart behaviour (i.e. control) into the material itself (rather than in the structure built by using the material) is still an illusive proposition. The main reason is the inherent complexity of their constitutive behaviour, brought about by the fact that temperature is an additional parameter (apart from stress and strain).

Nevertheless, efforts are already underway for integrating SMA wires into composite structures (Balta *et al.* 2000), and investigating their behaviour. Interfacial phenomena (e.g. rough interface *vs.* smooth interface), and differences in thermal expansion of the constituent phases of the composite are some of the problems which need to be systematically studied.

9.1.3 *Exploiting instability*

There are some underlying themes responsible for the extraordinary performance of many of the actuator materials described in this chapter. Usually, such materials are *poised on an instability*, resulting in very large response functions.

In the phase diagram of a material, a phase transition implies instability. The star performers described in this chapter are materials with usually two or more phase transitions, fine tuned to exploit the consequences of these phase transitions, as also their mutual interactions and coupling.

All the single-phase materials described above in some detail are primary ferroics: PZT and PMN/PZN are ferroelectrics; Terfenol-D is ferromagnetic; and Nitinol has features similar to those of a ferroelastic. A material in a ferroic phase has domain structure. The domain boundaries, as also the phase boundaries (if present), can move under the influence of a driving field, and this gives nonlinear response, and therefore field-tuneability of properties. The domain boundaries are particularly mobile (unstable) in the vicinity of the ferroic phase transition. The external field can also *cause* a phase transition, with the attendant consequences (e.g. anomalously large responses).

For PZT and Terfenol-D, the T_c is high, and the actuator is operated near the MPB.

PMN-xPT and PZN-xPT also operate near the MPB, and, in addition, have a partially ordered atomic structure. The charge imbalance experienced by every unit cell of PMN/PZN results in frustration (like in a spin glass), with the consequential competing ground states. The partial ordering in these compositionally disordered perovskites results in a diffuse phase transition and coexisting phases.

Nitinol also has a partially ordered structure, and a heavily twinned martensitic phase, which responds to stress in a reversible way, as if it has memory of what it has been through. Coexisting phases is another feature it shares with PMN/PZN.

It is not surprising that polycrystalline PMN-xPT exhibits the shape-memory effect (SME), like Nitinol. PMN-35PT also exhibits *reversible* shape change over a whole range of temperatures below T_c, after just one SME cycle of 'training' (cf. Fig. 9.1) (Wadhawan, Pandit and Gupta 2005). In this temperature range, the specimen has a specific shape at every temperature, irrespective of whether that temperature is reached by cooling or by heating.

Not all SME systems may have the same kind of shape-recovery mechanisms. The composition PMN-35PT is particularly interesting. It is close to the MPB. It also marks some kind of a dividing line between relaxor and normal ferroelectric behaviour. Compositions with $x > 35\%$ exhibit domain structure and hysteresis loops typical of *normal* ferroelectrics. However, the fact remains that nothing short of 100% PT can completely solve the local charge-imbalance problem faced by the atomic structure of PMN-xPT. The mesoscopic clusters in PMN-35PT therefore have to coexist with the macroscopic domain structure, and there is evidence that they coexist throughout the bulk of the specimen (Shvartsman and Kholkin 2004), rather than concentrating near the domain walls. The survival of these mesoscopic clusters in an otherwise normal ferroelectric has been postulated as providing one of the mechanisms for the SME, as also for the temperature-reversible shape recovery mentioned above (Wadhawan, Pandit and 2005): Mechanical fields (just like electrical fields) have an ordering effect on the clusters, tending to increase their average size, and this effect is not only reversible, it is also largely temperature-independent, thus explaining why there is a whole range of temperatures over which the clusters have a reversible and unique configuration at every temperature. This reversibility provides a restoring force for the movement of the macroscopic domain walls for creating a unique macroscopic shape of the specimen at every temperature.

There are other factors also for the SME in PMN-35PT. Relaxors like PMN-xPT have a glassy character, and glassiness has been linked to long-term memory ('remanence'), arising from nonergodicity (Stillinger 1995). The rough and rugged energy landscape of a nonergodic system results in some retention of the memory of a field-cooling (or training) episode. The memory may be in the form of nucleation sites favouring the formation of a particular variant of the ferroelectric–ferroelastic phase, thus contributing to the 'two-way' reversibility of shape memory.

PMN-35PT is a good example of a system with *persistent disequilibrium* over a wide range of temperatures (because of the remanent relaxor character, and because of the vicinity of the MPB). This staying on *the hairy edge between equilibrium and disequilibrium*, or between order and chaos (Kelly 1994), as also between normal and relaxor character, leads to rapid, large, and reversible response. This material has a learning and memory feature one associates in a general way with all strongly interacting, quasi-autonomous, large, distributed systems (Kelly's *vivisystems*). The members ('agents') of the system in this case are the mesoscopic positively and negatively charged clusters.

A fairly common theoretical framework has been built up for describing spin glasses and their electrical and mechanical analogues, as also neural networks, protein folding, and many other vivisystems (Anderson 1988, 1989; Angell 1995). Relaxor ferroelectrics are *very* smart materials because they are vivisystems of the same genre.

Multifunctional composites offer additional degrees of freedom, and further enhancement or fine-tuning of properties. For example, a polymer can be combined with a polycrystalline ferroelectric, each with its own phase transition (Newnham 1997). The phase transition in the polymer entails large changes in the elastic properties, while that in the ferroelectric results in a large dielectric response. The coupling between the two kinds of instability in the same composite material enables one to build especially efficient sensors and actuators (Newnham 1997).

The properties of a composite depend in a crucial way on its connectivity pattern. *Connectivity transitions* can result in dramatic new effects (Newnham 1986).

9.2   Sensors and actuators based on nanostructured materials

The concept of *integrative systems* has gained ground steadily. Such systems combine information-processing functionality with sensing and actuation, preferably on a single IC chip. The terms MEMS (micro-electro-mechanical systems) and MST (microsystems technology) are used in this context (Gad-el-Hak 2006a, b). MEMS refers mainly to systems employing electrostatically actuated mechanical components. MSTs incorporate thermal, fluidic, chemical, biological and optical components also.

MEMS are used as both sensors and actuators (Amato 1998; Judy 2001; Datskos, Lavrik and Sepaniak 2004; Allen 2005; Bao 2005). An example of the former is the *smart accelerometer*, which senses sudden changes in acceleration in an automobile, and activates the very quick inflation and release of airbags to minimize injury from a collision (Poole and Owens 2003). An example of the latter is the Digital Light Projector (DLP), in which each pixel is a tiny mirror with a movement-controlling IC chip underneath (Gaitan 2004).

Integration (i.e. monolithic integration of mechanical systems with the microelectronics) offers several advantages: batch fabrication; micro-scale devices; large arrays that can work independently or together; and high reliability.

Gaitan (2004) has listed the following MEMS and MST products for which the market will grow: Read-write heads; ink-jet heads; heart pacemakers; biomedical diagnostics; hearing aids; pressure sensors; accelerometer sensors; gyroscopes; IR sensors, thermopiles, bolometers; flow sensors; microdisplays; drug delivery systems; MST devices for chemical analysis; optical mouse; inclinometers; microspectrometers; optical MEMS; RF MEMS; fingerprint biomimetic sensors; micromotors; micro-optical scanners; electronic paper.

Single-walled carbon nanotubes have all their atoms on the surface. Therefore, their electronic properties are very sensitive to surface charge transfers and to changes in the environment. Even simple adsorption of several kinds of molecules has a strong effect on their properties. Therefore they have been found to be very suitable as sensors, particularly biochemical sensors (Kong *et al.* 2000).

Till date there are not many commercial sensors and/or actuators based on deliberately designed NEMS, although the trend is changing (cf. Luesebrink *et al.* 2005; Lyshevski 2005). An important example of where it has indeed happened, and which has already found industrial applications, is based on GMR (cf. Chapter 3). The useably large GMR effect was discovered in Fe-Cr multilayers (i.e. alternating layers of a ferromagnetic and a nonferromagnetic metal), with each layer having a thickness of ~ 1 nm. Commercial devices now use Co-Fe ferromagnetic layers and Cu metal spacers. The applications have become possible because the effect is several orders of magnitude larger in the composite multilayers than in bulk metals (cf. Heinonen 2004).

The better sensor performance for disc drives has allowed more bits to be packed on the surface of each disc. Introduced in 1997, this has been an enabling technology for the multibillion-dollar storage industry. GMR and TMR are also finding uses in MRAMs (magnetoresistive random access memories). Apart from compactness, they have the advantage of being *nonvolatile*; i.e. the memory is not erased when the power is switched off.

9.3 Sensors and actuators based on soft matter

9.3.1 *Gels*

Biological materials are wet and soft, whereas metals and ceramics are dry and hard. Anticipating that smart structures of the future are likely to be more and more bio-inspired and bio-like, research on soft-matter actuators is bound to pay rich dividends. Another factor to take note of is that biological systems tend to be 'open systems' in relation to their environment (e.g. the matrix fluid in which they operate), in the sense that they can exchange chemicals with it and change their molecular state. Which artificial materials fit this description?

One answer is: *synthetic polymer gels* (Hoffman 2004). They have been described as *chemomechanical systems* because they can undergo large and reversible deformations in response to chemical stimuli. They are able to transform the chemical part of the free energy directly to mechanical work, and may therefore function as *artificial muscles* in smart structures (cf. Section 9.3.5 below). The isothermal conversion of chemical energy to mechanical work forms the basis of motility in practically all living organisms.

Osada *et al.* (Osada, Okuzaki and Hori 1992; Osada and Ross-Murphy 1993; Osada, Gong and Narita 2000) have reported some interesting work on such materials. For example, shape-memory effect was observed in a water-swollen hydrogel comprising of a layered-aggregate structure made by copolymerizing acrylic acid and stearyl acrylate (Osada, Gong and Narita 2000). The occurrence of the effect was explained in terms of the order–disorder transition of the layered structure of the alkyl side-chains.

These authors also reported the occurrence of electric-field-driven quick responses in a polymer gel, with worm-like motility (Osada, Okuzaki and Hori 1992). The material is a hydrogel made of weakly cross-linked poly(2-acrylamido-2-methyl propane) sulphonic acid (PAMPS). The material, when dipped in a dilute solution of a surfactant, namely n-dodecyl pyridinium chloride ($C_{12}PyCl$) containing 3×10^{-2} M sodium sulphate, swells by a factor of 45 compared to its dry weight. The gel network is anionic. Therefore the positively charged surfactant molecules can bind to its surface. This induces a local shrinkage by decreasing the difference in osmotic pressure between the gel interior and the solution outside. Electric field can be used to direct the surfactant binding selectively to one side of the gel, thus inducing contraction and curvature in a strip of the gel. If the field polarity is reversed, the attached surfactant molecules go back into the solution, and contraction and bending occurs in the opposite direction as the surfactant molecules attach to the other side of the gel strip. If the gel strip is suspended in the solution from a ratchet (sawtooth) mechanism, the alternating field can make it move with a worm-like motion. Both electrostatic and hydrophobic interactions play a role in determining the net contraction of the gel.

9.3.2 *Magnetorheological fluids*

Magnetorheological (MR) fluids change their rheological behaviour on the application of a magnetic field. The main manifestation of this property is the field-dependent change of viscosity (by a factor as large as 10^5–10^6). They offer a simple, quiet and rapid (~ms) interface between electronic control and mechanical response (Ginder 1996; Jolly 2000).

A typical MR fluid consists of Fe or other 'soft' magnetic particles suspended in some carrier fluid like silicone oil, or even water. A surfactant is also added. The magnetically polarizable Fe or Fe_3O_4 particles are typically in the size range 100–10000 nm. And their volume fraction can be from 10 to 50%. The magnetization induced by the magnetic field makes the suspended particles line up as columnar structures parallel to the applied field. This configuration impedes the flow of the carrier fluid. The mechanical pressure needed to make these chain-like structures yield increases as the magnetic field is increased, and can be as high as ~100 kPa. The possible yield stress scales quadratically with the saturation magnetization of the magnetic particles (Phule, Ginder and Jatkar 1997).

The composition of MR fluids is similar to that of *ferrofluids*, but the dimensions of the particles in MR fluids are one to three orders of magnitude larger than the colloidal ferrofluid particles. The micron-sized particles in MR fluids sustain hundreds of ferromagnetic domains, and domain rotation by the field contributes to interparticle attraction.

Commercial applications of MR fluids include: exercise equipment; vehicle seat vibration control; primary automotive suspension applications (shock absorbers); flow-control valves; and industrial motion control (Phule, Ginder and Jatkar 1997).

The technology of controllable fluids holds great promise for smart-structure applications.

9.3.3 *Electrorheological fluids*

ER fluids exhibit the ER effect: They change their viscosity (even solidify to a jelly-like object) when an electric field is applied to them (Winslow 1949; Gandhi, Thompson and Choi 1989; Aldersey-Williams 1990). The response time is very quick (~ms), and the fields required are ~3 MV m^{-1}. The stiffening is proportional to the field applied, and is reversible.

Their typical composition is that of particles of polarizable molecules, dispersed in a nonconducting liquid. Silicone oil is used frequently. A variety of particles have been investigated: starch; zeolites; oxide- or polymer-coated metal particles; aluminium silicate; and polythene quinine (semiconducting polymer).

The applied field polarizes the particles, which then line up in a chain-like fashion, parallel to the lines of force of the electric field. The lined up chains resist shearing deformation, just as a solid does.

Vibration control in smart structures (e.g. automobiles) is one of the applications of ER fluids. Their fast response time makes them suitable for replacing electromagnets in high-speed machines; the fluid can be switched on and off at least ten times faster than an electromagnet.

Certain other applications of ER fluids require a high strength, which may be typically 5–10 kilopascals. Increasing the particle fraction helps in this. To prevent high-concentration fluids from segregation and settling down, a dispersant liquid is sometimes introduced, in addition to the main carrier liquid. The total fluid fraction can be made as low as 5%. Such high-concentration ER fluids are candidates for use as the interfacing medium in car clutches.

Less strong ER fluids can be used in artificial limbs for humans or robots, and in clutches in photocopiers.

ER fluids are likely to be essential components in smart structures of the future. Their main use at present is as actuators in large load-bearing structures. Rather large forces can be generated with relatively small electric fields.

Temperature dependence of the ER response is a drawback, as also the lack of long-term stability of the response.

9.3.4 *Shape-memory polymers*

Shape-memory polymers (SMPs) (cf. Section 4.3.4) as actuators offer some advantages over shape-memory alloys (SMAs) (cf. Ashley 2001). SMPs like the polyurethanes are low-cost and biodegradable. They can undergo shape recovery for deformations as high as 400%. Their phase-transition temperature can be tailored quite easily by introducing some small compositional cum structural changes, and this temperature does not have to be high: shape-recovery temperatures ranging from −30°C to 70°C have been realized. The shaping procedures are easy, and the deformed shapes are stable. Even bulk compression and recovery can be achieved.

Their main drawback compared to SMAs is the relatively low recovery force.

9.3.5 *Artificial muscle*

Natural muscle is an actuator par excellence (Nachtigall 1999). It can undergo billions of work cycles, with strains exceeding 20%, at a rate of 50% per second. It can generate stresses of ∼0.35 MPa, and can change its strength and stiffness as needed. The efficiency of energy conversion (with adenosine triphosphate (ATP) as fuel) can be as high as 40%. Natural muscle also exhibits *scale invariance*: the underlying mechanism works equally efficiently in a whole range of length-scales.

Various options are available for making an artificial equivalent of natural muscles, but none of them gives all the desired properties in one material.

Shape-memory alloys (SMAs) can generate strains up to 8%, as well as massive actuation stresses, but are poor in speed and efficiency because they require conversion of electrical energy to thermal energy, and then conversion of the latter to mechanical energy.

Piezoelectric, electrostrictive and magnetostrictive materials do provide large stresses, high speeds, and fairly good efficiency of energy conversion, but only small strains.

Certain dielectric elastomers (silicones and acrylics) have emerged as the nearest artificial thing we have to natural muscle (Pelrine *et al*. 2000; Ashley 2003). They are called *electroactive polymers* (EAPs). Application of a high electric field can produce strains as large as 30–40% in them. The mechanism involved is based on *Maxwell stress*: Typically, a film of the dielectric elastomer is coated on both sides with a compliant electrode material. This assembly behaves like a rubbery capacitor. When a voltage is applied, the oppositely charged electrodes attract each other, and squeeze out the polymer insulator substantially, which responds by expanding in area. The typical film thickness used is 30–60 microns. 'Prestraining' of the film is found to further improve the performance (Pelrine *et al*. 2000). By this approach, actuator strains as large as 117% were achieved for silicone, and 215% for acrylic elastomer. Enormous progress has been reported recently (cf. http://eap.jpl.nasa.gov): 120% strain, 3.2 MPa stress, and a peak strain rate of 34,000% per second for a strain of 12%.

Actuation devices using EAPs have the advantage that there is no need for motors, drive shafts, gears, etc. The reaction time is quick, and the electromotive force generated by the electric field is large. The main drawback at present is the need for applying rather high electric fields, although alternatives exist for getting over this problem to some extent (cf. Zhang, Bharti and Zhao 1998).

Whereas the EAPs described above function essentially as *capacitors*, actuation devices based on conducting polymers like polyaniline or polypyrrole function as *batteries*. Typically, the conducting polymer constitutes an electrode (in contact with an electrolyte) that changes its dimensions by the electrochemical incorporation of solvated dopant ions (Smela, Inganas and Lundstrom 1995; Lu *et al*. 2002; Baughman 2005). This happens because, when electrons move in or out of the conducting polymer chain, compensating ionic species get extracted or inserted. Currently used artificial muscles based on conducting polymers can operate at only a few volts, and yet generate strains of the order of 26%, strain rates as high as 11% per second, and stresses ranging from 7 to 34 MPa (cf. Baughman 2005). Their limitation at present is that the performance degrades after only a few thousand cycles if the actuator is operated at high strains and high applied loads; moreover, their energy-conversion efficiency is only ∼1%.

Cho (2006) has reviewed the progress made in improving the strength, speed, flexibility, biocompatibility (for use in prosthetic limbs), and efficiency of 'machine muscles'. One would like to do away entirely with the use of battery-powered artificial muscles. The performance of such muscles is severely restricted because of the limitations on the duration of their usage before battery recharge becomes necessary. Like in so many other things in the field of smart structures, the best approach is to emulate Nature and opt for *fuel-powered* artificial muscles (Ebron *et al*. 2006). One would like to develop artificial muscles which can 'perform such tasks as leaping, climbing or travelling for long stretches with only enough fuel to fill a lunch box' (Cho 2006). Like in natural muscle, the best approach (at least in the long run) is going to be that which converts the chemical energy of high-energy-density fuels to mechanical energy.

Ebron *et al.* (2006) have described two such approaches, based on fuels like hydrogen, methanol, and formic acid. The first uses changes in stored electrical charge for actuation. A catalyst-containing carbon-nanotube electrode simultaneously acts as a muscle, a fuel-cell electrode, and a supercapacitor electrode. Such a muscle converts chemical energy in the fuel to electrical energy. This electrical energy can be used either for immediate actuation, or for storage for future actuation. It can also meet other energy requirements of the system.

The second type of artificial muscle developed by Ebron *et al.* (2006) operates as a 'continuously shorted fuel cell'. It converts the chemical energy of the fuel to thermal energy that can do useful work, namely actuation. The redox reactions occur on a catalyst-coated, mechanically loaded shape-memory alloy (Nitinol), which gets heated and passes to the austenitic phase, resulting in mechanical movement through its contraction. In this arrangement, both fuel and oxidant (oxygen or air) are given simultaneously to a single electrode, namely the Pt-coated SMA wire, which functions as a shorted electrode pair. When the fuel supply is interrupted, the SMA wire cools back to the martensitic phase and expands in the process. Such a fuel-powered muscle could support a stress of ~ 150 MPa, while undergoing a contraction of $\sim 5\%$. The fuel used was a mixture of oxygen and either methanol vapour or formic-acid vapour, or a mixture of hydrogen in inert gas.

9.4   Chapter highlights

• Sensors and actuators have a mutually complementary relationship.
• Ferroic materials are the natural choice as sensors and actuators in a variety of smart-structure applications. They are inherently nonlinear in their response, thus offering scope for field-tuneability of their properties.
• The terms superpiezoelectricity and superelectrostriction have been introduced in this chapter, by analogy with the well-known term superelasticity (cf. Chapter 3).
• An analysis of the star performers among ferroics (PZT, PMN-PT, PZN-PT, Terfenol-D, Nitinol) shows that their superlative performance derives from staying on the hairy edge between equilibrium and disequilibrium, or between order and chaos.
• PMN-35PT is a particularly good example of such a system with persistent disequilibrium over a wide range of temperatures (because of the remnant relaxor character, and because of the vicinity of the MPB). This staying on the hairy edge between normal and relaxor character leads to rapid, large, and temperature-reversible response. This material has learning and memory features one associates in a general way with all strongly interacting, quasi-autonomous, large, distributed systems (i.e. vivisystems).

- Shape-memory ceramics like PMN-PT offer exciting possibilities in smart-structure applications. This is because most devices are electronically operated and controlled. Integration of shape-memory ceramics (dielectrics) into them is a more attractive proposition, compared to the integration of highly conducting shape-memory alloys.
- The above statement is also valid for the integration of piezoelectric ceramics into electronic devices.
- The availability of Terfenol-D has led to a renewal of interest in designing magnetostrictive actuators cum sensors. The sensor configurations based on this material can measure motion, stress, torque, and magnetic fields. Similarly, the actuators can produce displacement, force, and acoustic waves. Actuators based on Terfenol-D offer the advantages of high force per unit area, high-precision, direct control of the movement, and practically no fatigue. Moreover, they work at rather low fields, and, compared to piezoelectric actuators, can be used at much lower temperatures.
- SMAs like Ni-Ti can generate large deflection and forces. They have the highest known force/weight ratio among actuators. They also have good damping properties. Their applications have been particularly striking in clamps, switches, and springs; and all these applications involve going across the martensite–austenite phase transition. Other areas in which deployment of SMAs is under active research and development are: adaptive wings of aircraft and missiles; active control of rotorblades in rotorcraft; in-orbit control of the structure of antennas in satellites; hinge-free solar panels; robot arms; and the general area of SMA-embedded composites. One of the most extensive and successful applications of Ni-Ti SMAs and the superelasticity exhibited by them has been in the medical industry. The biocompatibility of the material is one reason for this. Some of the prominent medical applications are: guidewires; laparoscopic surgical instruments; implants; stents; retrieval baskets; and bone anchors.
- For underwater applications, polymer transducers offer an advantage over ceramics because of their better acoustic-impedance matching with water. The most familiar polymer in this regard is PVDF. Another polymer traditionally used in hydrophone applications is P(VDF-TrFE). However, the piezoelectric response of these polymers is not as large as that of the conventionally used ceramics. Their dielectric loss is also rather high, and the e.m. coupling factor is low.
- Embedding of ferroic materials into cleverly designed composites offers great scope for fine-tuning their usefulness, and for improving properties like impedance matching. PZT-polymer composites, Terfenol-D-resin composites, and Nitinol-epoxy composites are some examples of this approach.
- *Multifunctional* composites offer additional degrees of freedom, and further enhancement or fine-tuning of properties. For example, a polymer can be combined with a polycrystalline ferroelectric, each with its own phase transition. The phase transition in the polymer entails large changes in the elastic properties, while that in the ferroelectric results in a large dielectric response. The coupling

between these two instabilities enables one to build especially efficient sensors and actuators.

- The concept of integrative systems has been gaining ground steadily. Integration offers the advantages of batch fabrication, micro-scale devices, large arrays that can work independently or together, and high reliability.

- MEMS, an important example of integrated devices, are mainly systems employing electrostatically actuated mechanical components. MSTs incorporate thermal, fluidic, chemical, biological and optical components also. Nanoferroics are very often an important part of the overall design of MEMS and MSTs.

- Single-walled carbon nanotubes have all their atoms on the surface. Therefore, their electronic properties are very sensitive to surface charge transfers and to changes in the environment. Therefore they have been found to be very suitable as sensors, particularly biochemical sensors.

- Biological materials are wet and soft, whereas metals and ceramics are dry and hard. Anticipating that smart structures of the future are going to be more and more bio-like, research on soft-matter actuators assumes great significance. Another factor to take note of is that biological systems tend to be 'open systems' in relation to their environment, in the sense that they can exchange chemicals with it and change their molecular state. Synthetic polymer gels fit this description. Several actuators based on them have been developed.

- MR fluids change their rheological behaviour on the application of a magnetic field. The main manifestation of this property is the field-dependent change of viscosity (by a factor as large as 10^5-10^6). They offer a simple, quiet and rapid ($\sim$ms) interface between electronic control and mechanical response.

- ER fluids are likely to be essential components in several smart structures of the future. Their main use at present is as actuators in large load-bearing structures. Rather large forces can be generated with relatively small electric fields.

- Development of fuel-powered artificial muscles holds great promise for rapid progress in the field of smart structures.

References

Aldersey-Williams, H. (17 March 1990). 'A solid future for smart fluids'. *New Scientist* 125(1708): 37.

Allen, J. A. (2005). *Micro Electro Mechanical System Design*. London: Taylor and Francis.

Amato, I. (16 October 1998). 'Fomenting a revolution, in miniature'. *Science*, 282: 402.

Anderson, P. W. (1988, 1989). *Physics Today*. A series of six articles: January 1988; March 1988; June 1988; September 1988; July 1989; September 1989.

Angell, C. A. (31 March 1995). 'Formation of glasses from liquids and biopolymers'. *Science*, 267: 1924.

Ashley, S. (May 2001). 'Shape-shifters'. *Scientific American,* 284(5): 15.

Ashley, S. (October 2003). 'Artificial muscles'. *Scientific American,* p. 53.

Balta, J. A., M. Parlinska, V. Michaud, R. Gotthardt and J.-A. E. Manson (2000). 'Adaptive composites with embedded shape memory alloy wires'. In Wun-Fogle, M., K. Uchino, Y. Ito and R. Gotthardt (eds.), *Materials for Smart Systems III*. MRS Symposium Proceedings, Vol. 604. Warrendale, Pennsylvania: Materials Research Society.

Banks, H. T., R. C. Smith and Y. Wang (1996). *Smart Material Structures: Modeling, Estimation, and Control*. New York: Wiley.

Bao, M. (2005). *Analysis and Design Principles of MEMS Devices*. Amsterdam: Elsevier.

Baughman, R. H. (1 April 2005). 'Playing Nature's game with artificial muscles'. *Science*, 308: 63.

Bellaiche, L. and D. Vanderbilt (1999). 'Intrinsic piezoelectric response in perovskite alloys: PMN-PT versus PZT'. *Phys. Rev. Lett.* 83(7): 1347.

Bellouard, Y., T. Lehnert, T. Sidler, R. Gotthardt and R. Clavel (2000). 'Monolithic shape memory alloy actuators: A new concept for developing smart micro-devices'. In Wun-Fogle, M., K. Uchino, Y. Ito and R. Gotthardt (eds.), *Materials for Smart Systems III*. MRS Symposium Proceedings, Vol. 604. Warrendale, Pennsylvania: Materials Research Society.

Bidaux, J.-E., J.-A. E. Manson and R. Gotthardt (1997). 'Active stiffening of composite materials by embedded shape-memory-alloy fibres'. In George, E. P., R. Gotthardt, K. Otsuka, S. Trolier-McKinstry and M. Wun-Fogle (eds.), *Materials for Smart Systems II*. MRS Symposium Proceedings, Vol. 459. Pittsburgh, Pennsylvania: Materials Research Society.

Boller, C. (2000). 'Shape memory alloys – their challenge to contribute to smart structures'. In Wun-Fogle, M., K. Uchino, Y. Ito and R. Gotthardt (eds.), *Materials for Smart Systems III*. MRS Symposium Proceedings, Vol. 604. Warrendale, Pennsylvania: Materials Research Society.

Chaabane, B, J. Kreisel, B. Dkhil, P. Bouvier and M. Mezouar (2003). 'Pressure-induced suppression of the diffuse scattering in the model relaxor ferroelectric $PbMg_{1/3}Nb_{2/3}O_3$'. *Phys. Rev. Lett.* 90: 257601–1.

Cho, D. (12 August 2006). 'Wrestle a robot'. *New Scientist*, 191: 39.

Clark, A. E. (1995). 'High power magnetostrictive materials from cryogenic temperatures to 250 C'. In George, E. P., S. Takahashi, S. Trolier-McKinstry, K. Uchino and M. Wun-Fogle (eds.), *Materials for Smart Systems*. MRS Symposium Proceedings, Vol. 360. Pittsburgh, Pennsylvania: Materials Research Society.

Clark, A. E. and Belson (1972). 'Giant room-temperature magnetostrictions in $TbFe_2$ and $DyFe_2$'. *Phys. Rev. B*, 5: 3642.

Culshaw, B. (1996). *Smart Structures and Materials*. Boston: Artech House.

Damjanovic, D. (1997). 'Logarithmic frequency dependence of the piezoelectric effect due to pinning of ferroelectric–ferroelastic domain walls'. *Phys. Rev. B*, 55: R649.

Damjanovic, D. (2005). 'Piezoelectricity'. In F. Bassini, G. L. Liedl and P. Weyder (eds.), *Encyclopedia of Condensed Matter Physics*, Vol. 4. Amsterdam: Elsevier.

Datskos, P. G., N. V. Lavrik and M. J. Sepaniak (2004). 'Micromechanical sensors'. In Di Ventra, M., S. Evoy and J. R. Heflin (eds.), *Introduction to Nanoscale Science and Technology*. Dordrecht: Kluwer.

Duenas, T. A., L. Hsu and G. P. Carman (1997). 'Magnetostrictive composite material systems: Analytical/experimental'. In George, E. P., R. Gotthardt, K. Otsuka, S. Trolier-McKinstry and M. Wun-Fogle (eds.), *Materials for Smart Systems II*. MRS Symposium Proceedings, Vol. 459. Pittsburgh, Pennsylvania: Materials Research Society.

Duerig, T. W. (1995). 'Present and future applications of shape memory and superelastic materials', in E. P. George, S. Takahashi, S. Trolier-McKinstry, K. Uchino and M. Wun-Fogle (eds.), *Materials for Smart Systems*. MRS Symposium Proceedings, Vol. 360. Pittsburgh, Pennsylvania: Materials Research Society.

du Tremolet de Lacheisserie, E. (1993). *Magnetostriction: Theory and Applications of Magnetoelasticity*. London: CRC Press.

Ebron, von Howard, Z. Yang, D. J. Seyer, M. E. Kozlov, J. Oh, H. Xie, J. Razal, L. J. Hall, J. P. Ferraris, A. G. MacDiarmid and R. H. Baughman (17 March 2006). 'Fuel-powered artificial muscles'. *Science*, 311: 1580.

Flatau, A. B., M. J. Dapino and F. T. Calkins (2000). 'On magnetostrictive transducer applications', in M. Wun-Fogle, K. Uchino, Y. Ito and R. Gotthardt (eds.), *Materials for Smart Systems III*. MRS Symposium Proceedings, Vol. 604. Warrendale, Pennsylvania: Materials Research Society.

Fu, H. and R. E. Cohen (20 January 2000). 'Polarization rotation mechanism for ultrahigh electromechanical response in single-crystal piezoelectrics'. *Nature*, 403: 281.

Funakubo, H. (ed.) (1991). *Actuators for Control*. London: Gordon and Breach.

Gad-el-Hak, M. (2006a) (ed.), *MEMS: Design and Fabrication* (2nd edition). London: CRC Press, Taylor and Francis.

Gad-el-Hak, M. (2006b) (ed.), *MEMS: Applications* (2nd edition). London: CRC Press, Taylor and Francis.

Gaitan, M. (2004). 'Introduction to integrative systems'. In Di Ventra, M., S. Evoy and J. R. Heflin (eds.), *Introduction to Nanoscale Science and Technology*. Dordrecht: Kluwer.

Gandhi, M. V., B. S. Thompson and S. B. Choi (1989). 'A new generation of innovative ultra-advanced intelligent composite materials featuring electro-rheological fluids: An experimental investigation'. *J. Composite Materials*, 23: 1232.

Gerver, M. J., J. R. Berry, D. M. Dozor, J. H. Goldie, R. T. Ilmonen, K. Leary, F. E. Nimblett, R. Roderick and J. R. Swenbeck (1997). 'Theory and applications of large stroke Terfenol-D actuators'. In George, E. P., R. Gotthardt, K. Otsuka, S. Trolier-McKinstry and M. Wun-Fogle (eds.), *Materials for Smart Systems II*. MRS Symposium Proceedings, Vol. 459. Pittsburgh, Pennsylvania: Materials Research Society.

Ginder, J. (1996). 'Rheology controlled by magnetic fields'. In Immergut, E. (ed.) (1996), *Encyclopaedia of Applied Physics*, Vol. 16, pp. 487–503. New York: VCH.

Glazer, A. M., P. A. Thomas, K. Z. Baba-Kishi, G. K. H. Pang and C. W. Tai (2004). 'Influence of short-range and long-range order on the evolution of the morphotropic phase boundary in $Pb(Zr_{1-x}Ti_x)O_3$'. *Phys. Rev. B*, 70: 184123.

Guo, Y., H. Luo, D. Ling, H. Xu, T. He and Z. Yin (2003). 'The phase transition sequence and the location of the morphotropic phase boundary region in $(1-x)[Pb(Mg_{1/3}Nb_{2/3})O_3]-x PbTiO_3$ single crystal'. *J. Phys.: Condens. Matter*, 15: L77.

Heinonen, O. (2004). 'Magnetoresistive materials and devices'. In Di Ventra, M., S. Evoy and J. R. Heflin (eds.), *Introduction to Nanoscale Science and Technology*. Dordrecht: Kluwer.

Hoffman, A. S. (2004). 'Applications of "smart polymers" as biomaterials'. In B. D. Ratner, A. S. Hoffman, F. J. Schoen and J. E. Lemons (eds.), *Biomaterials Science: An Introduction to Materials in Medicine*, 2nd edn. Amsterdam: Elsevier.

Howarth, T. R. and K. M. Rittenmyer (1995). 'Transduction applications'. In Nalwa, H. S. (ed.), *Ferroelectric Polymers: Chemistry, Physics, and Applications*. New York: Marcel Dekker.

Husson, E. and A. Morell (1992). 'Ferroelectric materials with ferroelectric diffuse phase transitions'. *Key Engg. Materials*, 68: 217.

Jin, Y. M., Y. U. Wang, A. G. Khachaturyan, J. F. Li and D. Viehland (2003). 'Conformal miniaturization of domains with low domain-wall energy: Monoclinic ferroelectric states near the morphotropic phase boundaries'. *Phys. Rev. Lett.* 91: 197601-1.

Jolly, M. R. (2000). 'Properties and applications of magnetorheological fluids'. In Wun-Fogle, M., K. Uchino, Y. Ito and R. Gotthardt (eds.), *Materials for Smart Systems III*. MRS Symposium Proceedings, Vol. 604. Warrendale, Pennsylvania: Materials Research Society.

Judy, J. W. (2001). 'Microelectromechanical systems (MEMS): fabrication, design and applications'. *Smart Mater. Struct.* 10: 1115.

Kelly, K. (1994). *Out of Control: The New Biology of Machines, Social Systems, and the Economic World*. Cambridge: Perseus Books.

Khachaturyan, A. G., S. M. Shapiro and S. Semenovskaya (1991). 'Adaptive phase formation in martensitic transformation'. *Phys. Rev. B*, 43: 10832.

Kim, T.-Y. and H. M. Jang (1996). 'Improvement on the thermostability of electrostriction in $Pb(Mg_{1/3}Nb_{2/3})O_3$-$PbTiO_3$ relaxor system by La-doping'. *Ferroelectrics*, 175: 219.

Kong, J., N. R. Franklin, C. Zhou, M. Chapline, S. Peng, K. Cho and H. Dai (2000). 'Nanotube molecular wires as chemical sensors'. *Science*, 287: 622.

Kreisel, J., P. Bouvier, B. Dkhil, P. A. Thomas, A. M. Glazer, T. R. Welberry, B. Chaabane and M. Mezouar (2003). 'High-pressure x-ray scattering of oxides with a nanoscale local structure: Application to $Na_{1/2}Bi_{1/2}TiO_3$'. *Phys. Rev. B*, 68: 014113.

La-Orauttapong, D., B. Noheda, Z.-G. Ye, P. M. Gehring, J. Toulouse, D. E. Cox and G. Shirane (2002). 'Phase diagram of the relaxor ferroelectric $(1 - x)$Pb$($Zn$_{1/3}$Nb$_{2/3})$O$_3$-xPbTiO$_3$'. *Phys. Rev. B*, 65: 144101.

Li, S., W. Cao and L. E. Cross (1991). 'The extrinsic nature of nonlinear behaviour observed in lead zirconate titanate ferroelectric ceramic'. *J. Appl. Phys.* 69(10): 7219.

Lindberg, J. F. (1997). 'The application of high energy density transducer materials to smart systems'. In George, E. P., R. Gotthardt, K. Otsuka, S. Trolier-McKinstry and M. Wun-Fogle (eds.), *Materials for Smart Systems II*. MRS Symposium Proceedings, Vol. 459. Pittsburgh, Pennsylvania: Materials Research Society.

Lu, W., A. G. Fadeev, B. Qi, E. Smela, B. R. Mattes, J. Ding, G. M. Spinks, J. Mazurkiewicz, D. Zhou, G. G. Wallace, D. R. MacFarlane, S. A. Forsyth and M. Forsyth (9 August 2002). 'Use of ionic liquids for π-conjugated polymer electrochemical devices'. *Science*, 297: 983.

Luesebrink, H., T. Glinsner, S. C. Jakeway, H. J. Crabtree, N. S. Cameron, H. Roberge and T. Veres (2005). 'Transition of MEMS technology to nanofabrication'. *J. Nanosci. and Nanotech.* 5: 864.

Lyshevski, S.E. (2005). *Nano- and Micro-Electromechanical Systems: Fundamentals of Nano- and Microengineering*, 2nd edn. London: CRC Press.

Melton, K. N. (1998). 'General applications of SMA's and smart materials'. In Otsuka, K. and C. M. Wayman (eds.), *Shape Memory Materials*. Cambridge, U. K.: Cambridge University Press.

Nachtigall, W. (1999). 'The muscle as a biological universal actuator in the animal kingdom', in H. Janocha (ed.), *Adaptronics and Smart Structures: Basics, Materials, Design, and Applications*. Berlin: Springer-Verlag.

Newnham, R. E. (1986). 'Composite electroceramics'. *Ann. Rev. Mater. Sci.* 16: 47.

Newnham, R. E. (1997). 'Molecular mechanisms in smart materials'. *MRS Bulletin*, May 1997.

Newnham, R. E. and A. Amin (1999). 'Smart systems: Microphones, fish farming, and beyond'. *Chemtech*, 29(12): 38.

Newnham, R. E., J. F. Fernandez, K. A. Murkowski, J. T. Fielding, A. Dogan and J. Wallis (1995). 'Composite piezoelectric sensors and actuators'. In George, E. P., S. Takahashi, S. Trolier-McKinstry, K. Uchino and M. Wun-Fogle (eds.), *Materials for Smart Systems*. MRS Symposium Proceedings, Vol. 360. Pittsburgh, Pennsylvania: Materials Research Society.

Noblanc, O., P. Gaucher and G. Calvarin (1996). 'Structural and dielectric studies of Pb$($Mg$_{1/3}$Nb$_{2/3})$O$_3$-PbTiO$_3$ ferroelectric solid solutions around the morphotropic boundary'. *J. Appl. Phys.* 79(8): 4291.

Noheda, B. and D. E. Cox (2006). 'Bridging phases at the morphotropic boundaries of lead oxide solid solutions'. *Phase Transitions*, 79: 5.

Noheda, B., D. E. Cox, G. Shirane, J. A. Gonzalo, L. E. Cross and S.-E. Park (1999). 'A monoclinic ferroelectric phase in Pb$($Zr$_{1-x}$Ti$_x)$O$_3$ solid solution'. *Appl. Phys. Lett.* 74: 2059.

Noheda, B., D. E. Cox, G. Shirane, S.-E. Park, L. E. Cross and Z. Zhong (2001). 'Polarization rotation via a monoclinic phase in the piezoelectric 92% PbZn$_{1/3}$Nb$_{2/3}$O$_3$ - 8% PbTiO$_3$'. *Phys. Rev. Lett.* 86: 3891.

Noheda, B., D. E. Cox, G. Shirane, J. Gao and Z.-G. Ye (2002). 'Phase diagram of the ferroelectric relaxor $(1 - x)$PbMg$_{1/3}$Nb$_{2/3}$O$_3$-xPbTiO$_3$'. *Phys. Rev. B*, 66: 054104-1.

Nomura, S. and K. Uchino (1983). 'Recent applications of PMN-based electrostrictors'. *Ferroelectrics*, 50: 197.

Osada, Y., H. Okuzaki and H. Hori (1992). 'A polymer gel with electrically driven motility'. *Nature* 355: 242.

Osada, Y. and S. B. Ross-Murphy (1993). 'Intelligent gels'. *Scientific American,* 268(5): 42.

Osada, Y., J. P. Gong and T. Narita (2000). 'Intelligent gels'. In Wun-Fogle, M., K. Uchino, Y. Ito and R. Gotthardt (eds.), *Materials for Smart Systems III*. MRS Symposium Proceedings, Vol. 604. Warrendale, Pennsylvania: Materials Research Society.

Pandit, P., S. M. Gupta and V. K. Wadhawan (2006). 'Effect of electric field on the shape-memory effect in Pb$[($Mg$_{1/3}$Nb$_{2/3})_{0.770}$Ti$_{0.30}]$O$_3$ ceramic'. *Smart Materials and Structures* (2006), 15: 653.

Park, S.-E. and R. Shrout (1997). 'Ultrahigh strain and piezoelectric behaviour in relaxor based ferroelectric single crystals'. *J. Appl. Phys.* 82(4): 1804.

Pelrine, R., R. Kornbluh, Q. Pei and J. Joseph (4 February 2000). 'High-speed electrically actuated elastomers with over 100% strain'. *Science*, 287: 836.

Phule, P. P., J. M. Ginder and A. D. Jatkar (1997). 'Synthesis and properties of magnetorheological (MR) fluids for active vibration control'. In George, E. P., R. Gotthardt, K. Otsuka, S. Trolier-McKinstry and M. Wun-Fogle (eds.), *Materials for Smart Systems II*. MRS Symposium Proceedings, Vol. 459. Pittsburgh, Pennsylvania: Materials Research Society.

Poole, C. P. and F. J. Owens (2003). *Introduction to Nanotechnology.* New Jersey: Wiley.

Preu, G., A. Wolff, D. Crames and U. Bast (1991). 'Microstructuring of piezoelectric ceramics'. *Euro-Ceramics II*, 3: 2005.

Samara, G. A. and E. L. Venturini (2006). 'Ferroelectric/relaxor crossover in compositionally disordered perovskites'. *Phase Transitions*, 79: 21.

Shrout, T. R. and A. Halliyal (1987). 'Preparation of lead-based ferroelectric relaxors for capacitors'. *Am. Ceram. Soc. Bull.* 66(4): 704.

Shvartsman, V. V. and A. L. Kholkin (2004). 'Domain structure of $0.8Pb(Mg_{1/3}Nb_{2/3})O_3$-$0.2PbTiO_3$ studied by piezoresponse force microscopy'. *Phys. Rev. B*, 69: 014102-1.

Singh, A. K. and D. Pandey (2001), "Structure and the location of the morphotropic phase boundary region in $(1 - x)[Pb(Mg_{1/3}Nb_{2/3})O_3]$-$xPbTiO_3$". *J. Phys.: Condens. Matter*, **13**, L931.

Singh, A. K. and D. Pandey (2003). 'Evidence for M_B and M_C phases in the morphotropic phase boundary region of $(1 - x)[Pb(Mg_{1/3}Nb_{2/3})O_3]$-$xPbTiO_3$: A Rietveld study'. *Phys. Rev. B*, 67: 064102–1.

Smela, E., O. Inganas and I. Lundstrom (23 June 1995). 'Controlled folding of micrometer-size structures'. *Science*, 268: 1735.

Stalmans, R., L. Delaey and J. van Humbeeck (1997). 'Modelling of adaptive composite materials with embedded shape memory wires'. In George, E. P., R. Gotthardt, K. Otsuka, S. Trolier-McKinstry and M. Wun-Fogle (eds.), *Materials for Smart Systems II*. MRS Symposium Proceedings, Vol. 459. Pittsburgh, Pennsylvania: Materials Research Society.

Stillinger, F. H. (31 March 1995). 'A topographic view of supercooled liquids and glass formation'. *Science*, 267: 1935.

Uchino, K. (1986). 'Electrostrictive actuators: Materials and applications'. *Am. Ceram. Soc. Bull.* 65(4): 647.

Uchino, K. (1997). 'High electromechanical coupling piezoelectrics – How high energy conversion rate is possible'. In George, E. P., R. Gotthardt, K. Otsuka, S. Trolier-McKinstry and M. Wun-Fogle (eds.), *Materials for Smart Systems II*. MRS Symposium Proceedings, Vol. 459. Pittsburgh, Pennsylvania: The Materials Research Society.

Uchino. K. (1998). 'Shape memory ceramics'. In Otsuka, K. and C. M. Wayman (eds.), *Shape Memory Materials*. Cambridge, U. K.: Cambridge University Press.

Uchino, K., Y. Tsuchiya, S. Nomura, T. Sato, H. Ishikawa and O. Ikeda (1981). 'Deformable mirror using the PMN electrostrictor'. *Appl. Optics*, 20: 3077.

Vanderbilt, D. and M. H. Cohen (2001). 'Monoclinic and triclinic phases in higher-order Devonshire theory'. *Phys. Rev. B*, 63: 094108-1.

Vassiliev, A. (2002). 'Magnetically driven shape-memory alloys'. *J. Magn. Magn. Mater.* 242–245: 66.

Viehland, D. (2000). 'Symmetry-adaptive ferroelectric mesostates in oriented $Pb(BI_{1/3}BII_{2/3})O_3$-$PbTiO_3$ crystals'. *J. Appl. Phys.* 88: 4794.

Wadhawan, V. K. (2000). *Introduction to Ferroic Materials.* Amsterdam: Gordon and Breach.

Wadhawan, V. K., P. Pandit and S. M. Gupta (2005). 'PMN-PT based relaxor ferroelectrics as very smart materials'. *Mater. Sci. and Engg. B*, 120: 199.

Wang, Yu. U. (2006). 'Three intrinsic relationships of lattice parameters between intermediate monoclinic M_C and tetragonal phases in ferroelectric $Pb[(Mg_{1/3}Nb_{2/3})_{1-x}Ti_x]O_3$ and $Pb[(Zn_{1/3}Nb_{2/3})_{1-x}Ti_x]O_3$ near morphotropic phase boundaries". *Phys. Rev. B*, 73: 014113-1.

Wheeler, C. E. and B. G. Pazol (1991). 'Multilayer electrodisplacive actuators'. *Am. Ceram. Soc. Bull.* 70(1): 117.

Winslow, W. M. (1949). 'Induced fibration of suspensions'. *J. Appl. Phys.* 20: 1137.

Zekria, D. and A. M. Glazer (2004). 'Automatic determination of the morphotropic phase boundary in lead magnesium niobate titanate $Pb(Mg_{1/3}Nb_{2/3})_{(1-x)}Ti_xO_3$ within a single crystal using birefringence imaging'. *J. Appl. Cryst.* 37: 143.

Zhang, Q. M., V. Bharti and X. Zhao (26 June 1998). 'Giant electrostriction and relaxor ferroelectric behaviour in electron-irradiated poly(vinylidene fluoride-trifluoroethylene) copolymer'. *Science*, 280: 2101.

10

MACHINE INTELLIGENCE

We now see that no logic except bio-*logic*
can assemble a thinking device,
or even a workable system of any magnitude.
– Kevin Kelly, *Out of Control*

10.1 Introduction

The ultimate 'artificial' smart structures will completely blur the present distinction between the living and the nonliving. They will be truly intelligent machines; i.e. they will have sensors, actuators, a body, a communication system, and a truly intelligent artificial brain.

The question of machine intelligence (MI) is linked to the question of artificial life (AL). But what is life, 'real' or 'artificial'? A still lingering viewpoint is that life begins with the presence of a self-replication mechanism. The fact is that self-replication is not a necessary condition for the origin or propagation of life, although its presence can speed up evolution through inheritance and natural selection (Dyson 1985).

We need not adopt the slow-paced trial-and-error Darwinian evolution model for evolving our machines. An appeal to Lamarckian mechanisms can yield quicker results.

It is tempting to speculate that, like so much else that has fallen under the spell of self-organized criticality (SOC) (cf. Section 5.5.2), progress in machine intelligence is also a similar process, governed by an inverse power law for the average magnitude of progress, but unpredictable about the magnitude and timing of the next great leap forward. There are great expectations from nanotechnology, so far as development of faster and larger-capacity computers is concerned. Perhaps the next breakthrough in this field will give us computing systems closer to the human brain in terms of capacity and connectivity.

We shall discuss the present situation in this chapter. The general feeling is that the task is indeed formidable (Horgan 2004), unless we settle for some easier but still very worthwhile options. Then there are staunch optimists like Moravec (1988, 1998, 1999a, b), whose work will also be discussed in this chapter.

Hawkins and Blakeslee (2004), whose model for human intelligence we outlined in Chapter 7, have made some specific projections about the shape of things to come in the field of truly intelligent machines. So have others like Minsky (1986), Moravec (1988, 1998, 1999a, b), Kelly (1994), Chislenko (1996), Dyson (1997), and Kurzweil (1998).

In this chapter we take a look at the present status of machine intelligence, and assess the problems that need to be tackled. We then discuss autonomous robots, artificial life, and distributed supersensory systems. There is also the question of why at all we should be chasing the dream of developing truly intelligent machines; some reasons are discussed.

The first point to note is that, whereas an artificial cortex can perhaps be developed inside a large computing system, making the artificial equivalent of the *old brain* (also called the reptilian brain, or the R-brain) is a different matter altogether. However, there has been some progress in that direction also (Krichmar *et al.* 2005).

Hawkins and Blakeslee (2004) take the view that there is really no need for trying to implement the difficult, costly, and even undesirable proposition of developing the R-brain, and we should focus mainly on the cortical equivalent of the human brain. If we accept this, there would be no scope for emotions and feelings and other similar human characteristics and failings in the artificial brains we build, which is perhaps just as well.

Unlike what the Turing test stipulated (from 1950 onwards), *predictive ability, and not humanlike behaviour*, of a computer or a robot is the hallmark of intelligence. Our artificial-life machines need not be humanoids (or androids), and, if we plan our research carefully, they will not have the capability of interacting with us, and with one another, in human-like emotion-laden ways (unless there are accidental developments, beyond our control).

As an aside, there can be a very positive fallout of this scenario: Machine-brains without emotions and feelings (and with no self-replicating capabilities, unless we want this for certain purposes) are not likely to pose a threat to our existence. They cannot possibly be scheming creatures with ulterior motives, or other human failings like fragile and brittle egos, vindictiveness, or the urge for self-aggrandizement, who may one day get the idea of running the world better than we have been able to.

There is also the other viewpoint about the *inevitability* of evolution of machine intelligence to such an extent that we humans will be left far behind (in all respects), whether we want it or not (Moravec 1988, 1999a; Kurzweil 1998, 2005).

The development of artificial machine-brains will involve three basic steps (we present Hawkins' scenario first):

(i) Provision for a set of sensors and actuators – the input–output peripherals for the brain. These sensors and actuators need not be similar to those possessed by human beings.

(ii) A hierarchical memory, attached to the sensors and actuators with a high degree of interconnectivity and feedback, and modelled on the human cortex.

(iii) Real-world training of the artificial memory, the way one trains a child (Scassellati *et al.* 2006). The term *autonomous mental development* (AMD) is employed in this context (Weng *et al.* 2001). After the basic hardware has been put in place, the machine-brain will develop its own neural synapses on a continual and autonomous basis (Weng and Hwang 2006; Thompson 2006), depending on the nature of its sensor capabilities and the inputs made available to it. Unlike the classical artificial-intelligence (AI) approach, no programming rules, no databases, no expert systems will be incorporated. The machine-brain will evolve its own memory, predictive capabilities, and world-view (quite different from ours because its sensor inputs will be different), and provide solutions to new problems (cf. Fox 2005). In due course, not only the software, but also the hardware will become evolvable (cf. Sipper 2002).

There will be some kind of a 'body' to house all this, as also a communication system. But it need not look or behave like a humanoid robot. In fact, the 'body' need not even be in one piece (for certain applications); it can have a distributed existence if necessary. Hawkins and Blakeslee (2004) give the example of a truly intelligent security system of the future, which may have sensors distributed throughout a factory or town, emitting signals to a central hierarchical memory system (the 'brain'), situated securely in some building. We shall discuss other examples when we consider robots, as well as distributed intelligent systems, in this chapter.

10.2   Truly intelligent artificial brains: A SWOT analysis

Let us carry out what is called, in the parlance of management science, a SWOT analysis of the question of building truly intelligent machine brains (MBs).

Strengths

1. We can make the MBs *bigger* than the human brain. The size of the human brain is restricted by certain biological and anatomical constraints. In any case, the blind and slow evolutionary processes of Nature, in which there is no place for intelligent design or purpose (Dawkins 1986), have resulted in the size the human brain has, and there is no likelihood of any drastic changes in the near future (unless human beings intervene). But there are no serious limits on how large a machine-brain can be, although, beyond a certain level of complexity even size can start being counterproductive.

2. We can make the MBs operate much *faster* than the human brain. Hawkins and Blakeslee (2004) advocate the use of silicon-based IC chips (also see Smith 2006). We shall discuss this and other possibilities in more detail below, but let us consider here the question of speed of operations if we agree to work with silicon. At present, a typical operation in an IC chip takes a nanosecond, whereas neurons in the human brain work on a millisecond time-scale. The synapse-strengthening mechanisms in the human cortex are mediated chemically by molecules like neurotransmitters, and, the way human beings have happened to evolve, Nature had no pressing need for making things happen any faster than they do at present. So silicon-based brains will be a million-fold faster than the organic brains Nature has evolved. We shall discuss this some more when we consider computers of the future.

3. We can give the MBs sensors with ranges and types of sensitivity *different* from ours, so that we end up extending and supplanting our own sensing capabilities. A different set of sensors would mean that the world-view the machine-brain develops through experience and through extracting patterns from the environment would be different from ours.

4. We can build *a whole variety* of MBs, each type having its own set of sensing and actuation capabilities.

5. Machine-brains, by being bigger, faster, and different from our brains, will be incredibly quick and intelligent at *recognizing complex patterns* in the phenomena around them, something that we humans cannot even imagine.

6. Because the MB will be a distributed memory (a vivisystem), it will be *fault-tolerant* and *robust*. This also will be discussed below in more detail.

Weaknesses

1. The present *high cost* of building an artificial cortical brain, using a set of currently available computers, is a weakness. In fact, as Moravec (1999a) has argued, lack of adequate financial support and incentive has been the major reason why low-cost general-purpose computing power has not increased over the years as it could have. But he also believes that things are going to change rapidly in the near future.

2. The present *low level of understanding* of how the human brain forms invariant representations in its hierarchical memory is a weakness.

3. The necessity of building a massively parallel computing machine like the human brain is a weakness. This problem of *connectivity* is going to be a tough nut to crack (see below).

Opportunities

1. Machine-brains served by a set of sensors different and better than the human sensing apparatus will have *a world-view very different from ours*, and they will be able to tackle a variety of problems which are considered tough at present. Dependable weather forecasting is an example of this. We discussed perceptive networks (motes) in Section 8.7. The motes could span the globe, and be put

under the overall control of a truly intelligent weather-brain, resulting in a qualitative improvement in their performance.

2. Apart from the macroscopic opportunities described above, there will also be an availability of opportunities at the *microscopic* and *nanoscopic* level. Nanosensors could be let loose inside the biological cells of living beings. Such truly intelligent sensor systems will see patterns in physical and biological phenomena which the human brain has not been able to perceive. The result will be new discoveries and understanding, leading to the development of more effective drugs for diseases.

3. Our machines with a cortex larger than ours, with more than just six layers which our cortex has, will have a huge memory and a far better capacity at absorbing information and solving problems.

4. We could unite a whole bunch of such superintelligent brains in a grand hierarchy, which may end up having *supernatural powers*, as also a deep understanding of the universe. In due course, deep understanding will lead to *wisdom*.

Threats

1. If we cannot resist the temptation to imbue our intelligent machines with emotions and feelings (or if there are accidental developments, beyond our control), there is a danger that, at some stage, things may really get out of our control (Joy 2000; Fukuyama 2002).

2. If we are going to play gods, shall we be worthy gods? Shall we give ourselves the right to 'kill' the life-like but possibly truant creatures we shall make?

3. Moravec (1999a) has given strong reasons for the inevitability of a scenario in which robots will excel our physical and mental abilities in about 50 years from now. What will happen after that? The intelligent robots, of course, will continue to evolve rapidly, and soon a stage will come when humans will perforce be reduced to live a life of 'idle retirement'. The intelligent robots will control everything.

There are two main challenges to be overcome for fabricating intelligent machines. One is to gain a good understanding of how the human brain functions? For example, how does it form what Hawkins and Blakeslee (2004) call invariant representations (Chapter 7)? The other challenge is to have adequate low-cost computing power to imitate the massively parallel connectivity of the human neocortex, as also the memory requirements. We go into some details of the second problem here. The first problem is really a big question mark at present, although some experts are certainly quite optimistic about it.

10.3   Building of machine brains

> *Ours is the age proud of machines that think*
> *and suspicious of men who try to.*
> — H. Mumford Jones

Hawkins and Blakeslee (2004) have given an estimate of what it would take to build a machine brain having a hierarchical-predictive memory modelled on the human cortex. The human cortex has ~32 trillion synapses. Suppose we represent each synapse by using only two bits. Taking each byte as having 8 bits, we would get four possible synapses per byte, with four possible configurations per synapse.

That means that ~8 trillion bytes of memory will be needed. This can be managed with 80 hard drives of the type used in modern PCs. Not a formidable task at all, although one would like to make things more compact.

In fact, things are even easier than that if we remember that there is no need to build the equivalent of the *entire* human cortex. A smaller cortex, using a fewer number of types of sensors, will be still truly intelligent (provided there are no other serious problems like connectivity to be tackled).

Thus capacity is not an issue, if we are not too ambitious about what we want to build. But the present high cost is. According to Hawkins and Blakeslee, we may start by using hard drives and optical discs, but we should aim at making the truly intelligent memories from silicon. Silicon chips are expected to be small, low-power, and rugged (Smith 2006). There is another substantial advantage that Hawkins and Blakeslee have pointed out: IC chips used in a conventional computer are made from single-crystal wafers of silicon. Large crystals of silicon are grown from its melt, and then thin wafers are cut for use in IC chips. It is impossible to grow a completely perfect crystal, and only a very small number of faults in a given chip are acceptable. So there is a rather high rejection rate of silicon chips meant for use in conventional computers, adding to the overall cost. But a truly intelligent memory is not at all going to be like the memory of a conventional computer. It is going to be a distributed memory, a fault-tolerant vivisystem, a robust memory. Recall the archetypal vivisystem, a beehive. If a fraction of the bees are made to just disappear, the hive will adjust quickly and carry on merrily, as if nothing significant had happened. Similarly, the human memory is a distributed memory: Practically no single component of it holds indispensable or exclusive information or data. A substantial number of neurons die every day, and yet a person's brain carries on merrily as if nothing of much significance has happened. In a like manner, the intelligent machine-brain we shall build will be *fault-tolerant*. This means several things. For one, it would be possible to use larger but slightly faulty wafers, which would have been unacceptable for use in a conventional computer memory. Secondly, it would be possible to pack the circuit elements a little more densely than in a conventional chip. Cost reduction can be expected, apart from the fact that it would also be a *robust* system.

The robustness of the artificial brain will not be affected seriously by ageing also. Older persons continuing to be mentally fit, in spite of the fact that the brain loses a few thousand neurons each day.

Apart from capacity, which seems feasible, the other big issue is *connectivity*, and that is a really big challenge at present.

The human cortical brain is a massively parallel system. There are millions of axons, submerged in subcortical white matter (Chapter 7), which comprise the parallel memory and computing architecture of the real brain. Each neuron connects to some five to ten thousand other neurons. By contrast, what we have at present in the semiconductor industry is the essentially planar technology outlined in Chapters 6 and 8: The IC chips are made as a succession of layers, some insulating and some conducting. Such technology cannot allow for the massively parallel wiring needed for making something like the cortical brain, in silicon.

A possible way out has been discussed by Hawkins for tackling the connectivity problem, which would exploit the very high speed with which electrical signals can be transmitted in an IC chip. The analogy given is that of how a telephone service works. In a telephone system, a not-too-large number of high-capacity lines share a huge number of telephone connections. Such a system will also be very compact and low-power, as well as resistant to heat, cold and magnetism.

Nevertheless, some real breakthroughs in the computing power, memory capacity, and architecture of computers are needed. Before we discuss the possibilities, let us first get conversant with another genre of smart structures, namely smart robots.

10.4   Smart robots

Advancing computer performance is like water slowly flooding the landscape. A half century ago it began to drown the lowlands, driving out human calculators and record clerks, but leaving most of us dry. Now the flood has reached the foothills, and our outposts there are contemplating retreat. We feel safe on our peaks, but, at the present rate, those too will be submerged within another half century. I propose that we build Arks as the day nears, and adapt a seafaring life! For now, though, we must rely on our representatives in the lowlands to tell us what water is really like.

– Hans Moravec (1999a)

The word 'robot' comes from a Czech word meaning 'compulsory labour'. It first appeared in Karl Capek's play *R.U.R.*, which stands for 'Rossum's Universal Robots'. The play was first staged in Czechoslovakia in 1921.

There are two main types of robots: *industrial* robots, and *autonomous* robots. Industrial robots do useful work in a structured or predetermined environment. They do repetitive jobs like fabricating cars, stitching shirts, or making computer chips, all according to a set of instructions programmed into them. The first industrial robot was employed by General Motors in 1961.

Autonomous or smart robots, by contrast, are expected to work in an *unstructured* environment. They *move around* in an environment that has not been specifically engineered for them (although there are serious limitations at present), and do useful and 'intelligent' work. The work may involve avoiding obstacles, sensing the environment continuously (even doing image analysis competently, and doing it quickly), and solving problems. They have to interact with a dynamically changing and complex environment, with the help of sensors and actuators and a brain centre (Grabowski, Navarro-Serment and Khosla 2003; Katic and Vukobratovic 2003). Some supervision by humans is also required, at least for some decades to come (Weng *et al.* 2001; Marks 2006).

In view of the fact that only limited success has been achieved by techniques of classical AI (cf. the appendix on AI), there has been a trend to move towards *evolutionary or adaptive robotics* (Nolfi and Floreano 2000; Krichmar *et al.* 2005; Fox 2005; Thurn, Burgard and Fox 2005). Adaptive robotics is about imbuing robots with creative problem-solving capabilities. This means that they must *learn* from real-life experiences, so that, as time passes, they get better and better at problem-solving. The analogy with how a small child learns and improves as it grows is an apt one. The child comes into the world equipped with certain sensors and actuators (eyes, ears, hands, etc.), as well as that marvel of a learning and memory apparatus, the brain. Through a continuous process of experimentation (unsupervised learning), as well as supervised learning (from parents, teachers, etc.), and reinforced learning (the hard knocks of life, and rewards for certain kinds of action), the child's brain performs *evolutionary computation.*

Can we build robots which can do all that? That is a somewhat distant dream at present, but there has been progress. The robot *Darwin X*, developed by Krichmar *et al.* (2005), is a brain-based device (BBD) (also see Fox 2005). It is an intelligent machine, modelled on the human brain. Experience with an unstructured environment makes it learn and remember, just as a child does. But there is still a long way to go.

Another such developmental robot is SAIL (Weng 1998). Its goal is to autonomously generate representations and architectures which can be scaled up to more complex capabilities in unconstrained environments. Like babysitters for human children, it needs human *robot-sitters* for supervised learning. Such robots can 'live' in our company, and become smarter with time, in an autonomous manner (albeit under some human supervision).

In the traditional AI approach to robotics, the processing for robot control is decomposed into a chain of information-processing modules, proceeding from sensing to action. In a radically different approach adopted by Brooks (1991), a parallel is drawn from how coherent intelligence (swarm intelligence) emerges in a beehive (cf. Section 2.3.1) or an ant colony, wherein each agent is a simple device interacting with the world with real sensors, actuators, and a very simple brain. In Brooks' *subsumption architecture*, the decomposition of the robot-control process is in terms of *behaviour-generating modules*, each of which connects sensing to action. Like an individual bee in a beehive, each behaviour-generating module directly generates some part of the behaviour of the robot. The tight (proximity) coupling of sensing to action produces an intelligent network of simple computational elements that are broad rather than deep in perception and action.

There are two further concepts in this approach: *situatedness*, and *embodiment*. Situatedness means the incorporation of the fact that the robot is situated in the real world, which directly influences its sensing, actuation, and learning processes. Embodiment means that the robot is not some abstraction inside a computer, but has a body which must respond dynamically to the signals impinging on it, using immediate feedback.

Although adaptive robots need not be *humanoid robots*, research and development work is progressing in that direction also (cf. Schaub 2006). Three important characteristics of such a 'creature' are walking (Cuesta and Ollero 2005; Ball 2006; Hollis 2006), talking (Boyd 2006), and manipulation (Huang 2006). Gait, voice, and optimal gripping are tough problems to tackle in robotics, although they are, quite literally, child's play in the case of humans.

There is an evolutionary underpinning to this situation (Fig. 10.1). The evolutionary natural-selection processes, spread over hundreds of millions of years, gave our body and large brain a highly developed proficiency for locomotion, navigation, talking, manipulation, recognition, and commonsense reasoning. These were essential for the battle for survival against competitors and predators. Only those of our early ancestors survived and evolved who were good enough at doing certain things repeatedly and very well: procure food, outsmart predators, procreate, and protect the offspring. For achieving this, the human brain evolved into a *special* kind of

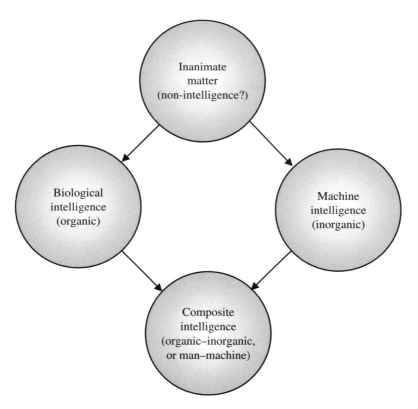

Fig. 10.1 Evolution of intelligence. Biological or organic intelligence, and machine or inorganic intelligence, each has its strengths and limitations. Perhaps the future course of evolution of intelligence will be towards a composite (organic–inorganic, or man–machine) kind of intelligence.

computer, rather than a general-purpose or *universal* computer. By contrast, skills like 'number crunching' are of recent origin, and therefore our brains are no match for present-day universal computers. Mathematical skills were not necessary for the survival of our early ancestors. For our machines, on the other hand, calculating is easier than reasoning, and reasoning is easier than perceiving and acting (Moravec 1988, 1999a, b).

Darwinian-evolution principles are now being used for evolving robot brains for specific tasks like obstacle avoidance. The robot adapts to the surroundings under the control of an algorithm which is not provided beforehand, but must evolve on an ongoing basis. The important thing about adaptive robots is that, although the various candidate algorithms (e.g. for obstacle avoidance, combined with the criteria for fastest movement and farthest foray from a reference point) are still created inside a computer, actual tests for evolutionary fitness are carried out in the real world, *outside the computer*. In one of the approaches, this is done by mapping each of the candidate algorithms onto the robot brain and checking for fitness.

There have been several distinct or parallel approaches to the development of machine intelligence (Fig. 10.2). The classical AI approach attempted to imitate some aspects of rational thought. Cybernetics, on the other hand, tended to adopt the human nervous-system approach more directly. Evolutionary robotics embodies a convergence of these two approaches.

Moravec (1999a) has discussed in substantial detail the progress in the development of autonomous robots. Evolutionary robotics involves a good deal of simulation and generalization work, rather like the creation of invariant representations in the human brain. Now, suppose the robot is able to continuously update, using simulation algorithms, knowledge of its own configuration, as well as that of the environment. Suppose further that the robot carries out the updation simulation a little bit faster than the real rate of change in the physical world. The smart robot can then see the consequences of its intended action *before* taking the action! If the simulated consequences are not desirable, the robot would change its 'mind' about what would be a more appropriate course of action under the circumstances (Moravec 1999a). Shall we describe such a robot as having an *inner life* or consciousness?

How do robot brains *coevolve* when a number of robots are confined in a small arena? *Cooperation*, as also other types of collective behaviour, are seen to emerge (Axelrod 1984; Nolfi and Floreano 1998; Uchibe and Asada 2006). Such research is throwing important light on the dynamics of human societies as well (Balch *et al.* 2006). Webb (1996) developed a female cricket robot, and studied 'her' behaviour when there were mating calls from two male cricket robots. Knowledge gained from investigations on such *robosects* can find important applications. For example, one can send an army of them to the site of a building crash, looking for survivors.

Rapid development of impressively intelligent robots and other smart machines will require the low-cost availability of more computing power than what we have at present. Let us examine this issue in some detail.

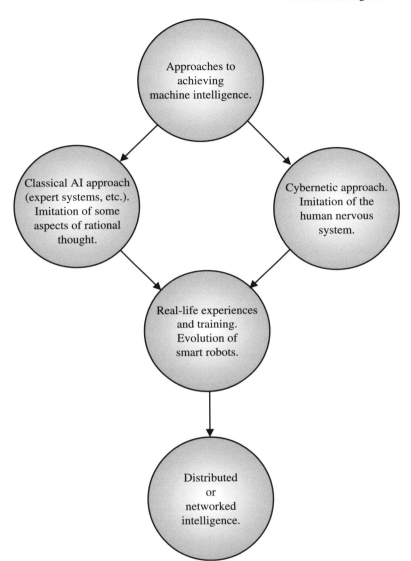

Fig. 10.2 Approaches to machine intelligence. The classical AI approach and the cybernetics approach get merged when we expose our autonomous robots to real-life experiences, and make them learn and evolve as intelligent machines. When there are several such interacting intelligences, distributed over space, and communicating and sharing information and inferences, a superintelligence can evolve in due course.

10.5   MIPS, megabytes, and cost

This book projects humanlike competence in thousand-dollar computers in forty years.

– Hans Moravec (1999a)

Moravec (1999a) has argued quite convincingly that, although progress in the development of smart robots has been slow because of a lack of adequate economic incentive for the robotics industry, there has been a major upswing in the recent past. There are three parameters to consider: The processing power (speed) of a computer; its memory size; and the price for a given combination of processing power and memory size.

The processing power or computing power can be quantified in terms of 'million instructions per second' or MIPS. The processing needed for adding two eight-digit numbers is a typical example of an instruction. The size of the available memory is specified in megabytes.

It turns out that, by and large, the MIPS and the megabytes for a computer cannot be chosen independently: Barring some special applications, they should have a certain degree of compatibility (per unit cost), for reasons of *optimal performance in a variety of applications.* An analysis by Moravec (1999a) reveals that, for general-purpose or *universal computers,* the ratio of memory (the megabytes) to speed (the MIPS) has remained remarkably constant during the entire history of computers. A 'time constant' can be defined as roughly the time it takes for a computer to scan its own memory once. One megabyte per MIPS gives one second as the value of this time constant. This ratio has remained amazingly constant as universal computing machines have been developed over the decades. Machines having too much memory for their speed are too slow (for their price), even though they can handle large programs. Similarly, lower-memory higher-speed computers are not able to handle large programs, in spite of being fast.

Special jobs require specially designed computers, entailing a departure from the above universal time constant, and a higher cost. For example, IBM's Deep Blue computer, developed for competing with the chess legend Garry Kasparov (in 1996–97) had more speed than memory (~3 million MIPS and ~1000 megabytes, instead of the 'universally' optimal combination of, say, 1000 MIPS and 1000 megabytes). Similarly, the MIPS-to-megabytes ratio for running certain aircraft is also skewed in favour of MIPS.

Examples of the other kind, namely slow machines (less MIPS than megabytes), include time-lapse security cameras and automatic data libraries.

Moravec estimated, in 1999, that the most advanced supercomputers available at that time were within a factor of 100 of having the power to mimic the human brain. But then such supercomputers come at a prohibitive cost. Costs must fall if machine intelligence is to make much headway. This has indeed been happening, as we shall see in some detail in the next section.

Although low-cost universal robots will be run by universal computers, their proliferation will have more profound consequences than those engendered by low-cost universal computers alone. Computers only manipulate symbols; i.e. they basically do 'paperwork' only, although the end results of such paperwork can indeed be used for, say, automation. A sophisticated universal robot goes far beyond such paperwork. It goes into perception and action in real-life situations. Since there is a far greater diversity of situations in the real world than just paperwork, and in far greater numbers, there would be a much larger number of universal robots in action than universal computers. This, of course, will happen only when the cost per unit capability falls to low levels. 'Capability, numbers sold, engineering and manufacturing quality, and cost effectiveness will increase in a mutually reinforcing spiral' (Moravec 1999a).

10.6   Moore's law and beyond

In April 1965, Gordon Moore, cofounder of Intel, gave his now-famous axiom that the number of transistors on a chip of silicon would double every year, an estimate changed later by him to 'every two years' (cf. Ulaby 2006). Moore's law has held its ground for 40 years, and may do so for another 10–20 years. In fact, the pace has quickened. In the 1980s the doubling time for the number of transistors on a chip fell to 18 months, contracting further to ~12 months in the late 1990s.

It is becoming increasingly difficult to pack more and more transistors in smaller and smaller silicon chips. In 2004, silicon chips measured 90 nm^2. In 2005, Intel could reduce this to 65 nm^2, and it was projected that the ultimate chip would be ~5 nm on a side. Transistors will approach atomic levels below that size, raising problems such as those concerning heat dissipation, nanolithography, and quantum tunnelling.

The need for miniaturization is driven by factors such as lower inertia, higher speed at a lower energy cost, and denser packing of components. Cost (per unit computing power and memory) also goes down correspondingly. The cost of computation has fallen rapidly for a whole century.

What about the future? How long can this go on? For quite a while, provided technological breakthroughs, or new ideas for exploiting existing technologies, keep coming. An example of the latter is the use of *multicore processors*. Multicore chips, even for PCs, are already in the market. Recently, Nvidia has introduced a chip, GeForce 8800, capable of a million MIPS speed, and low-cost enough for use in commonplace applications like displaying high-resolution video. It has 128 processors (on a single chip) for specific functions, including high-resolution video display. In a multicore processor, two or more processors on a chip process data in tandem. For example, one core may handle a calculation, a second one may input data, while a third one sends instructions to an operating system. Such load-sharing and parallel functioning improves speed and performance, and reduces energy consumption and heat generation.

Nanotechnology holds great promise as the next-generation solution to faster and cheaper computation. We discussed QDs, RTDs, and SETs in Section 6.9 (Goldhaber-Gordon *et al.* 1997). There are other possibilities (Lieber 2001; Stix 2005; Ekert 2006). Nanotubes, for example, have extraordinary tensile strength and resilience. Application of an electric field makes them change their length a little bit. Experiments are underway for using the sagging and straightening of these tubes to represent the 1 and 0 states of binary logic (Appenzeller and Joselevich 2005). If successful and commercially viable, this approach could provide a memory which combines the speed of a static random-access memory (SRAM), the low cost of a dynamic random-access memory (DRAM), and the nonvolatility of a flash memory.

What is very relevant in the present context is the fact that, unlike the top-down approach currently adopted in conventional microelectronic fabrication (starting from a single crystal of silicon and working downwards to the final IC chip, passing through steps like wafering, deposition, lithography, and etching), the techniques for the fabrication of nanoelectronic structures are headed for the bottom-up approach (Lieber 2001). Methods of synthetic chemistry are used to produce building blocks by the mole, and to assemble (or rather induce them to self-assemble) portions of them into progressively larger structures. In such a freshly developing technology, researchers will consciously attack the connectivity problem innovatively. Connectivity is a major problem for putting together an artificial brain modelled on the human neocortex. Building of a massively fanned out and delicately intermeshed cortical structure looks more tractable by a bottom-up strategy.

DNA computing is another approach being investigated (Adleman 1994, 1998; Lipton 1995; Shapiro and Benenson 2006). The technique has the potential for massive parallelism.

10.7   Distributed supersensory systems

The notions of situatedness and embodiment were introduced in the context of robots in Section 10.4. But robots, as also motes (Section 8.7), need to communicate with one another, and with a centralized 'brain'. Distribution of sensing, actuation, and control over space, along with a provision for adequate (wireless) communication among the various units or 'agents', can lead to the emergence of additional intelligence (swarm intelligence), not present in individual agents. We got a feel for this when we discussed motes in Section 8.7.

In any case, it is not mandatory that the brain and the sensors of a smart or intelligent structure be housed in one monolithic edifice. It is also not necessary for the sensors in our artificial smart structures to be the same as the 'natural' sensors possessed by humans. Let us see what can be the implications of these two features of *pervasive computing* (Kumar and Das 2006) for applications concerning terrestrial phenomena like weather forecasting, animal and human migrations, etc.

Suppose we let loose weather sensors all over the globe, all communicating with local brain centres, and eventually with a centralized brain. Since these are truly intelligent artificial brains, they will form a world-view with the passage of time, as more and more sensory data are received, processed, and generalized (i.e. irreducible representations are formed). Since these sensors need not be like those of humans, the pattern formation in the artificial superbrain will be different from that in our brains.

The machine-brain will recognize local patterns and global patterns about winds, etc., and there will be a perspective about weather patterns at different time-scales: hours, days, months, years, decades. The crux of the matter is that the intelligent machine-brain will see patterns we humans have not and cannot.

It follows from the memory-prediction model of intelligence that these predictions about weather would be more accurate, as it would become second nature for the artificial organism to anticipate weather changes.

Why restrict to weather only? Such global coverage by supersensory distributed systems could also be applied to the study of animal migrations, spread of disease, and demographic changes of various types.

Distributed supersensory systems will not only have swarm intelligence (like the bees in a beehive have it), they will also undergo *evolution* with the passage of time. Like in the rapid evolution of the human brain, both the gene pool and the meme pool will be instrumental in this evolution of *distributed intelligence*. This ever-evolving superintelligence and knowledge sharing will be available to each agent of the network, leading to a snowballing effect (Fig. 10.2).

10.8 What will the truly intelligent machines do?

The answer is: They will complement and surpass our own very limited capabilities. And we shall also make them do things we find boring, or difficult.

10.8.1 *Near-term applications*

Three near-term applications of truly intelligent MBs are: speech recognition; vision; and smart cars.

Speech recognition is an area in which the conventional AI approach has not had spectacular successes (however, see Mullin 2005, for a description of a modern approach to AI.). This is because it attempts at matching auditory patterns to the word-templates stored in the memory of the computer, without knowing much about the *meaning* of words, and the *context* in which they are spoken. By contrast, a truly intelligent MB would understand not only words, but also sentences, and the context of a statement (Jelinek 1997). This would become possible because the artificial brain would already have a model of the world, and it would be able to *anticipate* ideas, phrases, and individual words.

Machine vision is another area in which the classical AI approach has not been able to go very far. At present there is no machine that, when presented with a picture, can describe its contents with a high degree of sophistication. Compare this with what even a child can do. An intelligent machine will be able to do a good job here because it would have formed and stored in its memory a world-view, just as an intelligent child or adult does.

Smart cars are already there, but are they truly smart? Can they, like human drivers, understand and anticipate the behaviour of other drivers on the road? Can they have contingency plans for sudden, totally unanticipated, situations? Such capabilities would require the inclusion of a cortical algorithm, trained over a period of time by the inputs received from an incorporated set of sophisticated sensors, which can, in principle, be even superior to the human senses in terms of range, sensitivity and capability. Moreover, the training of the artificial cortex could be either a one-time training in the factory, or a continuous, evolving process. In the latter case the car could become a true (but unemotional!) buddy of the driver, even sensing his/her moods and suggesting corrective or preventive actions.

10.8.2 *Long-term applications*

The possible long-term applications of artificial cortical brains can be discussed in terms of some fundamental characteristics.

Speed makes a difference

The electronic brains we build and train will be able to absorb information and analyse it a million times faster than us. That would be a case of 'more is different'. They would be able to read all the e-books in a library in a matter of minutes. Such machines would be able to solve mental problems so quickly that their capacity for comprehending and solving problems will become *qualitatively* superior to ours. Such lightning-speed minds could solve several types of scientific and mathematical problems in a jiffy.

More is really different

If we add large memory capacity to high speed of operations, we can expect truly mind-boggling consequences. Why limit ourselves to an area of the artificial brain similar to the area of the human cortex (the size of a large napkin)? A larger area would make the artificial brain to remember more details, and develop a more refined set of sensory perceptions.

Similarly, adding a few more *layers* of cortical hierarchy (beyond the present six in a real brain) would result in unimaginably high levels of pattern perception and depth of abstraction.

Standing on the shoulders of giants

> *You live and learn . . . then you die and forget.*
> – Noel Coward

Each human being, after he or she comes to this world, has to learn all the basics like the mother tongue, the rules of society, arithmetic, arts, science, whatever. This learning process takes several years or decades. When the person dies, practically all the acquired knowledge and experience is destroyed, and there is no way we can transplant it from an old, mature person to a toddler, so that the toddler does not have to start the learning process from scratch. MBs will not have that limitation. They will really be able to 'stand on the shoulders of giants' (namely the already developed machine-brains), and thence be 'taller' for all practical purposes.

Intelligent MBs could replicate like software. Modules of specialized learning, memory and intelligence could be swapped or implanted, with incredible consequences (Kennedy 2006). When this happens, it would be the artificial equivalent of the evolution of the 'meme pool' in Nature, as propounded by Dawkins (1989, 1998) in his extension of the classical theory of Darwinian evolution. Such a process would also give centre-stage to Lamarckism, with its attendant higher speed of evolution.

10.9   Brave new world of intelligent structures

This human/robot 'conflict' looks like a typical generation gap problem. The machines, our 'mind children', are growing up and developing features that we find increasingly difficult to understand and control. Like all conservative parents, we are puzzled and frightened by processes that appear completely alien to us; we are intermittently nostalgic about the good old times, aggressive in our attempts to contain the 'children', and at the same time proud of their glorious advance. Eventually, we may retire under their care, while blaming them for destroying our old-fashioned world. And only the bravest and youngest at heart will join the next generation of life.

– A. Chislenko (1996)

The power of artificial intelligent structures to discern patterns not conceivable by humans could be extended to the micro- and nanoworld also (Hawkins and Blakeslee 2004). Why do proteins fold the way they do? If our intelligent machines could tell this by simply looking at the sequence of amino acids in the primary structure of a protein, that would be a major step forward for drug design. Suitable nanosensor assemblies will have to be first designed for this purpose.

We live in a three-dimensional world; or do we? Our senses and brains do not allow us to see or visualize more than three dimensions. But scientific theories abound which require the existence of more than three dimensions. String theory is an example. Our intelligent machines will have no difficulty in forming a worldview involving more than three dimensions, if we let them train themselves with relevant information.

The superintelligent systems could be combined into a hierarchy, just like the one existing in the neocortex. Such an assembly could see incredibly complex patterns and analogies which escape our comprehension. The end result will be a dramatic increase in our knowledge and understanding of the universe.

It is difficult to imagine what such a brave new world of intelligent structures would be like. In any case, opinions tend to differ strongly when we try to look too far into the future. We refer here to the writings of Moravec in this regard. Moravec (1988, 1999a, b) takes the robotics route, and ends up in a scenario in which robotic intelligence has advanced to a level where it is more mind than matter, suffusing the entire universe. We humans, of course, are left far behind, and perhaps disappear altogether from the cosmic scene.

Moravec (1998) does not appear to be unduly worried about the fact that human intelligence is still an enigma. He takes the line that easy availability of mass-produced computing power, which is presently available only in ultra-expensive supercomputers, will be instrumental in overcoming the presently perceived barriers to an understanding of how the human brain really functions. He estimates that $\sim$100 million MIPS of computing power, along with a commensurate memory size in universal computers, should suffice to do something comparable to the performance of the human brain. Or perhaps it is not even necessary that the intelligent robots of the future will have brains modelled on the human brain. A moot point indeed.

Moravec also makes a case that the human body and brain (made of protein) are fragile stuff, not at all suitable for surviving in harsh conditions, like those in outer space. By contrast, robots can indeed be built of materials which do not depend on carbon chemistry for 'life'. He expects that, 'before long, conventional technologies, miniaturized down to the atomic scale, and biotechnology, its molecular interactions understood in detailed mechanical terms, will have merged into a seamless array of techniques encompassing all materials, sizes, and complexities. Robots will then be made of a mix of fabulous substances, including, where appropriate, living biological molecules' (Moravec 1988).

We humans will assist this artificial evolution, and it would be millions of times faster than what biological evolution has been. Beyond a certain crossover point (perhaps in the present century itself), machine intelligence will be able to evolve without our help, because it would have attained superiority over us in every aspect. Beyond this point, evolution of machine intelligence will become exponentially rapid because of what Kurzweil (1998, 2005) calls the *law of accelerating returns*. Since computer technology is an evolutionary process that feeds on its own progress, its speed of evolution is exponential.

Kurzweil (1998, 2005) predicts that, before the end of the present century, humans will be able to *coevolve* with their intelligent machines via neural implants that will enable them to 'upload' their carbon-based neural circuitry into the prevailing hardware of the intelligent machines. The distinction between the 'living' and the 'nonliving' will be blurred. Humans will simply merge with the intelligent machines.

Habitation of outer space by intelligent robots will be the inevitable next step (Moravec 1999b). All along, there will be a parallel evolution of *distributed intelligence*. Widely separated intelligences will communicate with one another, leading to the evolution of an omnipresent superintelligence.

10.10 Chapter highlights

- Machine intelligence modelled on the neocortex can have unprecedented consequences for the progress and well-being of humankind.

- The human brain is a massively parallel computer, with an extremely complex and ever-changing connectivity pattern. Building of (or the self-assembly of) an artificial equivalent of such a structure is a highly nontrivial task. But even minor successes in that direction can lead to the creation of smart structures with far-reaching consequences.

- It may be neither necessary nor desirable (nor possible?) to develop the complete equivalent of the mammalian brain. In particular, a large part of the old brain or the R-brain can perhaps be left out, and one can focus on putting together an artificial neocortex.

- Since the sensors installed in an MB need not be the same as those possessed by humans, the worldview evolved by it will be different from that of humans.

- Such machine brains (MBs) will have the potential of being not only different, but also bigger and faster than the human brain.

- Because of the expected superiority of future MBs in terms of size and speed, such machines will be able to discern patterns in data which we humans cannot. The irreducible representations formed by MBs will be more sophisticated and general and subtle than what our brains are capable of.

- It will be possible to mass-produce such brains. Once an MB has been evolved to the desired degree of sophistication (for a particular type of application), it can be mass-copied (like we copy software at present).

- The networking possible among a large number of distributed motes will result in the effective development of a distributed super-MB.

- In view of the fact that only limited success has been achieved by techniques of classical AI, there has been a trend to move towards evolutionary or adaptive robotics. Adaptive robotics is about imbuing robots with creative problem-solving capabilities. This means that they must *learn* from real-life experiences, so that, as time passes, they get better and better at problem-solving. A beginning has been already made in this direction by the development of, for example, the *Darwin* series of robots.

- There is an evolutionary underpinning to the situation that we humans are so good at skills like walking, visual recognition, talking, and manipulation of objects. The evolutionary natural-selection processes gave our body and large brain a highly developed proficiency for these tasks, as these were essential for the battle for survival against competitors and predators. For achieving this, the human brain evolved into a *special* kind of computer, rather than a general-purpose or *universal* computer.

- Skills like number crunching are of recent origin, and therefore our brains are no match for computers. Mathematical skills were not necessary for the survival of our early ancestors. For machines, however, calculating is easier than reasoning, and reasoning is easier than perceiving and acting.

- The field of artificial life embodies a parallel approach to autonomous robotics. The two fields complement each other, and have several regions of overlap. Together, they represent the evolution of smart structures along Darwinian and Lamarckian lines.

- One of the necessary conditions for rapid progress in machine intelligence is that sophisticated but low-cost universal computers be available to researchers.
- The ratio of memory-size to speed (megabytes divided by MIPS) has remained remarkably constant in universal computers, over the entire history of the computer industry.
- Distributed supersensory systems can not only have swarm intelligence, they can also undergo evolution. Like in the rapid evolution of the human brain, both the gene pool and the meme pool are instrumental in this evolution of *distributed intelligence*. This ever-evolving superintelligence and knowledge sharing can be available to each agent of the network, leading to a snowballing effect.
- Speech recognition, machine vision, and the AI gadgetry used in smart cars are some examples wherein the conventional AI approach employing expert systems and symbolic logic, etc. has not been very successful, and where the availability of really intelligent machine-brains based on the neocortical model will make a big difference in the near future.
- Intelligent MBs could replicate like software. Modules of specialized learning, memory and intelligence could be swapped or implanted, with incredible consequences. When this happens, it would be the artificial equivalent of the evolution of the meme pool in Nature. Such a process would also give centre-stage to Lamarckism, with its attendant higher speed of artificial evolution.
- It appears inevitable that, at a certain stage in evolutionary history, our own creations, namely intelligent robots, will excel us in all aspects. In particular, they will take over from us the further evolution of machine intelligence.
- Humans may be able to coevolve with intelligent robots.
- Automated generation and distribution of information and knowledge will be a key aspect of economic activity.
- In due course, outer space will be colonized by our superintelligent robots. Perhaps we humans will also have a role in that.

References

Adleman, L. M. (11 November 1994). 'Molecular computation of solutions to combinatorial problems'. *Science,* 266: 1021.

Adleman, L. M. (August 1998). 'Computing with DNA'. *Scientific American,* 279: 34.

Appenzeller, J. and E. Joselevich (2005). 'Carbon nanotubes for data processing'. In Waser, R. (ed.), *Nanoelectronics and Information Technology: Advanced Electronic Materials and Novel Devices,* 2nd edition, p. 471. KGaA: Wiley-VCH Verlag.

Axelrod, R. (1984). *The Evolution of Cooperation.* New York: Basic Books.

Balch, T., F. Dellaert, A. Feldman, A. Guillory, C. L. Isabell, Z. Khan, S. C. Pratt, A. N. Stein and H. Wilde (July 2006). 'How multirobot systems research will accelerate our understanding of social animal behaviour'. *Proc. IEEE,* 94(7): 1445.

Ball, P. (4 February 2006). 'Walk this way'. *New Scientist,* 189: 40.

Boyd, J. (4 February 2006). 'Now hear this', *New Scientist,* 189: 44.

Brooks, R. A. (13 September 1991). 'New approaches to robotics'. *Science,* 253: 1227.

Chislenko, A. (1996). 'Networking in the mind age'. Blog on the website http://www.lucifer.com/~sasha/home.html

Cuesta, F. and A. Ollero (2005). *Intelligent Mobile Robot Navigation*. Berlin: Springer.

Dawkins, R. (1986). *The Blind Watchmaker: Why the Evidence of Evolution Reveals a Universe Without Design*. New York: Norton.

Dawkins, R. (1989). *The Selfish Gene*. Oxford: Oxford University Press.

Dawkins, R. (1998). *Unweaving the Rainbow*. Allen Lane: Penguin.

Dyson, F. J. (1985). *Origins of Life*. Cambridge: Cambridge University Press.

Dyson, G. B. (1997). *Darwin Among the Machines: The Evolution of Global Intelligence*. Cambridge: Perseus Books.

Ekert, A. (2006). 'Quanta, ciphers, and computers'. In G. Fraser (ed.), *The New Physics for the Twenty-First Century*. Cambridge, U. K.: Cambridge University Press, p. 268.

Fox, D. (5 November 2005). 'Brain Box'. *New Scientist*, p. 28.

Fukuyama, F. (2002). *Our Posthuman Future*. New York: Farrar, Straus and Giroux.

Goldhaber-Gordon, D., M. S. Montemerlo, J. C. Love, G. J. Opiteck, and J. C. Ellenbogen (1997). 'Overview of nanoelectronic devices'. *Proc. IEEE*, 85(4): 521.

Grabowski, R., L. E. Navarro-Serment and P. K. Khosla (November 2003). 'An army of small robots'. *Scientific American,* p. 63.

Hawkins, J. and S. Blakeslee (2004). *On Intelligence*. New York: Times Books (Henry Holt).

Hollis, R. (October 2006). 'Ballbots'. *Scientific American, India*, 1(17): 54.

Horgan, J. (July 1994). 'Can science explain consciousness?' *Scientific American,* p. 72.

Huang, G. T. (4 February 2006). 'Get a grip'. *New Scientist*, 189: 46.

Jelinek, F. (1997). *Statistical Methods for Speech Recognition*. Cambridge MA: MIT Press.

Joy, B. (April 2000). 'Why the future does not need us'. *Wired* magazine.

Katic, D. and M. Vukobratovic (2003). *Intelligent Control of Robotic Systems*. Dordrecht: Kluwer.

Kelly, K. (1994). *Out of Control: The New Biology of Machines, Social Systems, and the Economic World*. Cambridge: Perseus Books.

Kennedy, J. (2006). 'Swarm intelligence'. In Zomaya, A. Y. (ed.), *Handbook of Nature-Inspired and Innovative Computing: Integrating Classical Models with Emerging Technologies*. New York: Springer, p. 187.

Krichmar, J. L., D. A. Nitz, J. A. Gally and G. M. Edelman (8 February 2005), 'Characterizing functional hippocampal pathways in a brain-based device as it solves a spatial memory task'. *PNAS, USA*, 102: 2111.

Kurzweil, R. (1998). *The Age of Spiritual Machines: When Computers Exceed Human Intelligence*. New York: Viking Penguin.

Kurzweil, R. (2005). *The Singularity is Near: When Humans Transcend Biology*. New York: Viking Adult.

Lieber, C. M. (September 2001). 'The incredible shrinking circuit'. *Scientific American,* 285(3): 50.

Lipton, R. J. (28 April 1995). 'DNA solution of hard computational problems'. *Science* 268: 542.

Marks, P. (16 September 2006). 'Antisocial robots go to finishing school'. *New Scientist*, p. 28.

Minsky, M. (1986). *The Society of Mind*. New York: Simon and Schuster.

Moravec, H. (1988). *Mind Children: The Future of Robot and Human Intelligence*. Cambridge: Harvard University Press.

Moravec, H. (1998). 'When will computer hardware match the human brain?' *J. Evolution and Technology*, 1: 1.

Moravec, H. (1999a). *Robot: Mere Machine to Transcendent Mind*. Oxford: Oxford University Press.

Moravec, H. (December 1999b). 'Rise of the robots'. *Scientific American,* p. 124.

Nolfi, S. and D. Floreano (1998). 'Co-evolving predator and prey robots: Do "arms races" arise in artificial evolution?'. *Artif. Life*, 4: 311.

Nolfi, S. and D. Floreano (2000). *Evolutionary Robotics: The Biology, Intelligence, and Technology of Self-Organizing Machines*. Cambridge, Massachusetts: MIT Press.

Scassellati, B., C. Crick, K. Gold, E. Kim, F. Shic, and G. Sun (2006). 'Social development'. *IEEE Computational Intelligence Magazine*, 1: 41.

Schaub, B. (14 October 2006). 'My android twin'. *New Scientist*, 192: 42.

Shapiro, E. and Y. Benenson (May 2006). 'Bringing DNA computers to life'. *Scientific American,* 294: 32.

Sipper, M. (2002). *Machine Nature.* New Delhi: Tata McGraw-Hill.

Smith, L. S. (2006). 'Implementing neural models in silicon'. In Zomaya, A. Y. (ed.), *Handbook of Nature-Inspired and Innovative Computing: Integrating Classical Models with Emerging Technologies.* New York: Springer, p. 433.

Stix, G. (February 2005). 'Nanotubes in the clean room'. *Scientific American,* 292(2): 66.

Thompson, K. G. (2006). 'Autonomous mental development: Soaring beyond tradition'. *IEEE Computational Intelligence Magazine,* 1: 50.

Thurn, S., W. Burgard and D. Fox (2005). *Probabilistic Robotics.* Cambridge, Massachusetts: MIT Press.

Uchibe, E. and M. Asada (July 2006). 'Incremental coevolution with competitive and cooperative tasks in multirobot environment'. *Proc. IEEE,* 94(7): 1412.

Ulaby, F. T. (July 2006). 'The legacy of Moore's law'. *Proc. IEEE,* 94(7): 1251.

Webb, B. (December 1996). 'A cricket robot'. *Scientific American,* 275: 62.

Weng, J. (1998). In C. W. Chen and Y. Q. Zhang (eds.), *Learning in Computer Vision and Beyond: Development in Visual Communication and Image Processing.* New York: Marcel Dekker.

Weng, J. and W.-S. Hwang (2006). 'From neural networks to the brain: Autonomous mental development'. *IEEE Computational Intelligence Magazine,* 1: 15.

Weng, J., J. McClelland, A. Pentland, O. Sporns, I. Stockman, M. Sur and E. Thelen (2001). 'Autonomous mental development by robots and animals'. *Science,* 291: 599.

11

THE FUTURE OF SMART STRUCTURES

We are very near to the time when virtually no essential human function, phys-ical or mental, will lack an artificial counterpart. The embodiment of this convergence of cultural developments will be the intelligent robot, a machine that can think and act as a human, however inhuman it may be in physical or mental detail. Such machines could carry on our cultural evolution, including their own construction and increasingly rapid self- improvement, without us, and without the genes that built us. When that happens, our DNA will find itself out of job, having lost the evolutionary race to a new kind of competition.

— Hans Moravec (1988), *Mind Children*

Current research on smart structures is generally focused on smart sensor systems, and on actuator mechanisms in materials exhibiting nonlinear response (ferroics, soft matter, nanostructured materials). There is also an increasing amount of activity on artificial neural networks and other branches of computational intelligence. The smartness approach to engineering was originally introduced for large systems, and was driven by an ever-present push from the military and the aerospace industries. Civil engineering and consumer applications provided an additional impetus. All along, there has been a constant trend towards miniaturization of smart systems, embodied at present in the 'household' term MEMS. A recent book by Varadan, Vinoy and Gopalakrishnan (2006) provides a good account of the present situation.

In this concluding chapter, we first summarize the existing applications of smart structures and systems, and then take a look at their probable future. At present, most researchers seem to be reconciled to the view that, for the time being, we are nowhere near Mother Nature when it comes to building really smart ('brainy') structures and systems. Developments in the fields of nanostructured materials and soft matter, as also computer science and brain science, do make some room for optimism, but only some. Therefore, it appears prudent to make two separate kinds of projections about smart structures: short-term projections (STPs), and long-term projections (LTPs). STPs are about what is certainly possible to achieve. LTPs involve a certain degree of speculation about what *could* be achievable, and naturally, expert opinion is bound to be divided on such matters.

The key element in the development of really smart structures, both biological and artificial, is *evolution*. Before human beings, endowed with that marvellous and powerful mental organ called the neocortex, came on the scene, Darwinian evolution was proceeding at its own slow pace, with no specific end-goals. The gene pool and the meme pool of the humans has brought in the all-important new factor that we can tinker with and speed up the evolutionary processes, particularly for the machines (smart structures) we want to develop. Creating an artificial equivalent of the human brain is one such audacious wish.

There are two important features of artificial evolution, which make it different from biological evolution: We humans can set an end-goal to which we want our software and hardware to evolve; and Lamarckian evolution is no longer taboo. Moravec (1988, 1998, 1999a, b) has estimated that evolution engineered by humans can be 10 million times faster that what had been going on in natural phenomena before we decided to interfere.

11.1   Present applications of smart structures and systems

Smart structures and systems already find a wide range of applications (Varadan, Vinoy and Gopalakrishnan 2006) in machine tools, photolithography, process control, health monitoring, consumer electronics, aircraft, submarines, the automotive industry, civil engineering, biomedical systems, and the computer industry.

Piezoceramic transducers are used for vibration control in machine tools, as also during the manufacture of microelectronic circuits.

Shape memory materials find a variety of applications (cf. Chapter 9), including those for the online shape-optimization of aerodynamic surfaces, construction of artificial muscles, and active control of *in vivo* drug-delivery devices.

Fibre-optic sensors are used for health-monitoring of fibre-reinforced ceramics, metal-matrix composites, and other structural composites.

Shake-stabilization of hand-held video cameras is effected by using piezoceramic and MEMS accelerometers and rotation-rate sensors, and fibre-optic gyros.

In the aviation industry, vibration control and twist control of helicopter rotor blades, and adaptive control of aircraft control surfaces, is achieved by using piezoceramic stack actuators, PZT and MEMS accelerometers, and magnetostrictive mounts. Similarly, active control of noise is carried out by employing piezoceramic pick-ups and error sensors, PZT audio resonators and analogue voice coils, and digital signal processor chips. In smart submarines, piezoceramic actuators provide acoustic signature suppression of submarine hulls.

The automotive industry is a major user of smart devices. Increased comfort level involves the use of rain monitors, occupant identifiers, HVAC sensors, air pollution sensors, etc. Safety features include the use of piezoelectric yaw-axis rotation sensors for antiskid antilock braking, ceramic ultrasonic 'radar' for collision avoidance and parking assistance, MEMS accelerometers for airbag control, and electronic stability controls for four-wheel independent autobraking. A variety of materials also find use in chromogenic mirrors and windows.

Smart buildings use a variety of IR, visible-range, and fibre-optic sensors and communication systems for better safety, security, and energy saving. For example, smart windows can cut air-conditioning costs.

Smart structures find widespread uses in the computer industry. Piezo-accelerometers provide error-anticipating signals for correcting for head-motion-related read/write errors. Bimorph-type piezo-positioners and asperity-detector arms are used in high-density disc drives.

Autonomous robots are a special and very important class of smart systems. We have discussed some details of their present operations in Section 10.4.

11.2   Short-term projections

- The ongoing research on developing smart buildings, smart bridges, smart air-craft, smart automobiles, smart fluids, etc. will continue to make rapid progress. Better and lower-cost sensors and actuators will become available for such applications, particularly because of the high degree of integration possible in MEMS and NEMS.

- Medical applications of smart materials like Ni-Ti are already a big industry. More progress will come through miniaturization and use of high-energy-density ferroic materials.

- Expert systems and other concepts in the field of artificial intelligence have a niche of their own in smart-structure applications, and there is a lot more to come. There is plenty of scope for limited-application robots.

- The next-generation data storage devices in the computer industry will use smart read/write head micropositioners based on piezoceramic and MEMS accelerometers and rotation-rate sensors, as also piezoceramic and fibre-optic gyros.

- Tools of computational intelligence are getting refined and powerful at a rapid rate, and will continue to do so as the cost of 'supercomputing' power will drop drastically. This progress will have a direct bearing on the level of sophistication of the smart structures we develop.

- Nanoscience and nanotechnology are going to make a qualitative change to our lives. Drexler's assertions about building nanomachines (nanoassemblers and nanoreplicators) atom-by-atom or molecule-by-molecule have been contested by some eminent chemists. One of the reasons given by Smalley (2001) for the impossibility of Drexler's dream was that '... the manipulator fingers on the hypothetical self-replicating nanobot ... are ... too sticky: the atoms of the manipulator hands will adhere to the atom that is being moved'. It is conceivable that a way out of this difficulty will be found by analogy with ideas from colloidal science. Atoms of inert gases can be perhaps made to play the role played by surfactants in colloids. One could inject single atoms into a pool of inert gas atoms, so as to form a very weakly bonded coating of inert atoms around the injected atoms (at low-enough temperatures, and perhaps at high pressures). Use of these 'dressed' atoms, rather than their bare version, could help in taking care of Smalley's objection: The dressed atoms could be undressed on site, just before their incorporation into the nanostructure being assembled.

- CNT-reinforced nanocomposites offer a combination of low density and remarkable mechanical properties like extremely high strength, stiffness, and resilience. A number of applications based on them are likely to emerge in the near future. Moreover, there is more to CNTs than just exceptional mechanical properties. Even electrical devices based on them have been developed (cf. Service 2001). The fascinating thing about artificial composites is that, as discussed in the appendix on composite materials, there are a whole range of parameters which can be optimized at the design stage for producing a material of one's choice.
- As argued by Moravec (1988), increasingly sophisticated smart structures will evolve in a sequence similar to that which occurred in Nature, culminating in (and then going beyond) the neocortical-level intelligence of humans: 'That animals started with small nervous systems gives confidence that today's small computers can emulate the first steps toward humanlike performance'.
- There is going to be a further crumbling of barriers among different branches of science. In this book we have seen the ubiquity of complexity in a variety of real-life situations. The realization is dawning that self-organized criticality (SOC) is indeed a very pervasive feature of the universe. It includes the functioning of the human brain (which we want to create artificially), as also the very fabric of life and evolution (in which we are interested because we want our machines to evolve to become truly intelligent).
- The futures of smart structures and humans are going to be far more strongly intermingled than most of us seem to realize at present. There is already a need to change our approach to education and human resource development (HRD) (Culshaw 1999). This book is a modest effort at emphasizing the need for a *comprehensive* approach to HRD, so far as the future of smart-structures research is concerned.

11.3   Long-term projections

Our intelligence, as a tool, should allow us to follow the path to intelligence, as a goal, in bigger strides than those originally taken by the awesomely patient, but blind, processes of Darwinian evolution. The route is from the bottom up, and the first problems are those of perception and mobility, because it is on this sensorimotor bedrock that human intelligence developed.

– Hans Moravec (1988), *Mind Children*

As we saw in Chapters 7 and 10, Hawkins' model of human intelligence gives hope that truly intelligent machines may not, after all, be that distant into the future. But building massively parallel configurations, having complex and ever-changing connectivity patterns, is quite a challenge. One has to pin high hopes on nanotechnology for providing the needed breakthroughs.

When genuine artificial intelligence does come into being, there will be a justification for using the term *intelligent structures*, instead of the term 'smart structures' used in the title of this book.

We list here a few thoughts and possibilities for the future of smart structures

- Although Hawkins favours silicon, it is conceivable that other kinds of computers, exploiting the self-assembly possibilities in soft matter, will provide better, stable, cheaper and more compact alternatives.

- Smart structures of the future will overlap strongly with real artificial life (RAL). The living body and the living brain are worthy models to emulate (or, are they?). They are time-proven, and therefore realistic. And they are economical in terms of energy efficiency.

- Nature's factory for fabricating the living body involves the use of nanomachines like molecules of proteins or RNA, and a variety of organelles. The ribosome, for example, is an assembly of RNA and proteins that makes proteins by an assembly-like process. Similarly, topoisomerase is a nanomachine that unwinds double-stranded DNA. Nature produces these tiny machines by the processes of polymerization and molecular self-assembly. Human beings will do well to mimic Nature for creating certain types of artificial smart structures.

- The biological cell is the ultimate self-replicating nanomachine. Any machine needs an energy source, a supply of raw materials, and instructions for doing its job. The biological cell has a membrane which selectively allows matter and energy to come inside. Some of the molecules so received are used for fuelling the energy needs. Other molecules are modified and used for manufacture, maintenance, movement, and defence. The DNA has the information for fabrication and self-replication. mRNA is a temporary transcript of this information, which is conveyed to ribosomes for making specific proteins. The proteins, in turn, are responsible for building everything needed by the organism, and for providing locomotion when needed. Thus, *linking or polymerization* is used by the cell for manufacturing large linear molecules, and these molecules then spontaneously fold themselves into three-dimensional functional structures. The information about the folding is coded in the shapes and the sequence of the building blocks of the polymers. DNA, RNA, and the proteins are all manufactured by this processing strategy. There is *linear synthesis, followed by molecular self-assembly*. And catalysis plays a central role in many of the manufacturing processes in the cell. We are giving this description here to emphasize the suggestion that the best smart structures we can make are likely to be made by a similar approach.

- A *hybrid approach* holds great promise as well. Instead of trying to make machines the way Drexler advocated, one could start with existing biological molecules and modify them. It has not been possible to create life in the laboratory, but one can manipulate natural materials (like in gene splicing) to create new life.

- There can also be a hybrid approach for device fabrication. Biological nan-odevices and, say, MEMS cover different grounds in terms of size, energy, economics, materials, and approach to manufacturing. As argued by Schmidt and Montemagno (2004), a combined approach can yield better results. *Artificial bionanomachines* are likely to be parts of smart structures of the future.

- One more kind of hybrid approach involves integrating the best features of genetic algorithms, neural networks, fuzzy logic and expert systems through an interface program. GAs are best suited for optimization, NNs for forecasting, fuzzy logic for decision making, and expert systems for various kinds of diagnostics. There are going to be more and more attempts to benefit from the different types of capabilities, expertise, and intelligence, all interacting with one another and with the environment through a controlling interface program (Chatterjee 2005).

- Interesting possibilities exist for the creation and investigation of neuron–semiconductor interfaces. Computers and brains involve very different conduction processes. In computers it is electrons in a solid, and in the brain it is ions in a polar fluid. The vastly different mobilities of the charge carriers present a major challenge in the efforts towards integrating real neurons with IC chips (Fromherz 2005). This field of research is in its infancy, but exciting developments are in store in the near future.

- Smart sensor systems have their jobs cut out for them. What can be envisaged is that they would find use as subsystems or slaves in smart structures controlled by a master machine-brain with true intelligence. An even more exciting scenario is that advanced vivisystems will come into existence in which each member of the swarm is not as unintelligent as a bee or an ant or a neuron, but is, in fact, a truly intelligent machine-brain. It is not easy to predict what such *supersmart structures* will do!

- Machines, smart structures included, will become more and more biological, just as the biological is getting more and more engineered (Kelly 1994). A somewhat different viewpoint is that of Moravec (1999a). The biological regime, though very impressive in terms of smartness and intelligence, is made of fragile stuff (proteins), and its computational processes (run by 'squirting chemicals' in a fluid) are slow, typically in the millisecond range. A changeover from the carbon-based organic processes to, say, silicon-based inorganic processes for all things intelligent, will have its advantages.

- Evolution of machine intelligence is *inevitable*. At present we humans are instrumental in bringing this about. But soon (within half a century), things will start getting out of our control because the intelligent robots we develop will surpass us in just about everything. It is only a matter of time before we humans are rendered irrelevant, and our 'mind children' (the superintelligent robots) (Moravec 1988) evolve with ever-increasing speed, and colonize the universe, not just physically, but also as an omnipresent intelligence. Such a development will turn genetic evolution on its head.

- Moravec (1999a) points out that in recent centuries culture has overtaken genetics. The information held in our libraries is now thousands of times more than that in our genes. This ever-increasing cultural rather than biological information will be available to our mind children, and will form the basis of their rapid evolution.

- The inevitability of colonizing other planets, by humans and/or robots, has been expressed by a number of thinkers. Smart robots will have several advantages compared to humans for such ventures. As Moravec (1999a) said, 'even now, it is relatively cheap to send machines into the solar system since the sunlight-filled vacuum is as benign for mechanics, electronics, and optics as it is lethal for the wet chemistry of organic life'. Venturing into outer space may come either as a matter of intention by enterprising humans, or as an act of intelligent rogue robots who consider themselves superior to humans. A robot–human joint venture looks like a much better scenario. The desirability of cutting the umbilical chord that ties us to Mother Earth cannot be doubted. 'The garden of earthly delights will be reserved for the meek, and those who would eat of the tree of knowledge must be banished. What a banishment it will be! Beyond Earth, in all directions, lies limitless outer space, a worthy arena for vigorous growth in every physical and mental dimension' (Moravec 1999a).

- Kurzweil (1998, 2005) anticipates the emergence of 'the strongest of strong AI'. Humans will probably coevolve with these 'artificial' superintelligences, via neural implants that will enable the humans to upload their carbon-based neural circuitry into the hardware they themselves were instrumental in developing. They will then live forever as bits of data flowing through 'artificial' hardware. That will mark a complete blurring of the distinction between the living and the nonliving.

References

Chatterjee, P. (April 2005). 'Man, machine and intelligence'. *Electronics for You*, p. 27.

Culshaw, B. (1999). 'Future perspectives: Opportunities, risks, and requirements in adaptronics', in H. Janocha (ed.), *Adaptronics and Smart Structures: Basics, Materials, Design, and Applications.* Berlin: Springer-Verlag.

Fromherz, P. (2005). 'The neuron–semiconductor interface'. In Willner, I and E. Katz (eds.), *Bioelectronics: From Theory to Applications*. Wiley-VCH Verlag GmbH and Co., KGaA.

Kelly, K. (1994). *Out of Control: The New Biology of Machines, Social Systems, and the Economic World.* Cambridge: Perseus Books.

Kurzweil, R. (1998). *The Age of Spiritual Machines: When Computers Exceed Human Intelligence*. New York: Viking Penguin.

Kurzweil, R. (2005). *The Singularity is Near: When Humans Transcend Biology*. New York: Viking Adult.

Moravec, H. (1988). *Mind Children: The Future of Robot and Human Intelligence*. Cambridge: Harvard University Press.

Moravec, H. (1998). 'When will computer hardware match the human brain?' *J. Evolution and Technology*, 1: 1.

Moravec, H. (1999a). *Robot: Mere Machine to Transcendent Mind*. Oxford: Oxford University Press.

Moravec, H. (December 1999b). 'Rise of the robots'. *Scientific American,*: p. 124.

Schmidt, J. and C. Montemagno (2004). 'Biomolecular motors'. In Di Ventra, M., S. Evoy and J. R. Heflin (eds.), *Introduction to Nanoscale Science and Technology*. Dordrecht: Kluwer.

Service, R. F. (2001). 'Assembling nanocircuits from the bottom up'. *Science*, 293: 782.

Smalley, R. E. (September 2001). 'Of chemistry, love and nanobots'. *Scientific American,* 285: 68.

Varadan, V. K., K. J. Vinoy and S. Gopalakrishnan (2006). *Smart Material Systems and MEMS: Design and Development Methodologies*. Chichester, England: Wiley.

APPENDIX

A1. Artificial intelligence

While cybernetics scratched the underside of real intelligence,
artificial intelligence scratched the topside.
The interior bulk of the problem remains inviolate.
– Hans Moravec (1988)

The field of artificial intelligence (AI) took birth with the publication of the paper 'A Logical Calculus of the Ideas Immanent in Nervous Activity' by McCulloch and Pitts (1943). It was argued that the brain could be modelled as a network of logical operations (e.g. 'and', 'or', 'nand'). This was the first attempt to view the brain as an information-processing device. The McCulloch–Pitts model demonstrated that a network of very simple logic gates could perform very complex computations. This fact influenced substantially the general approach to the design of computers.

In classical AI, the basic approach is to try to make computers do things which, if done by humans, would be described as intelligent behaviour. It is intended to be the study of principles and methodologies for making 'intelligent' computers. It involves representation of knowledge, control or strategy of procedures, and searching through a problem space, all directed towards the goal of solving problems (Clearwater 1991). The problem with classical AI is that, by and large, it has faired poorly at delivering real-time performance in dynamically changing environments.

In a conventional computer, we have circuitry and software. The former is continuous, and the latter symbolic. Before the 1960s, two distinct schools of thought had evolved in the theory of 'intelligent' systems (Newell 1983): One went the continuous or cybernetics way (cf. Wiener 1965), and the other the symbolic or AI way.

Cyberneticians were mainly concerned with pattern recognition and *learning* algorithms. Their continuous approach predominantly involved parallel processing,

And people doing AI work focussed on the development of *expert systems* doing specific 'intelligent' jobs like theorem-proving, game-playing, or puzzle-solving. The AI symbolic approach was basically a serial-computation approach. Statistical models were used, and training by using expert information available *a priori* was carried out. The trained AI system, when confronted with a new piece of sensory input, proceeded to serially compute the most likely classification for the new data, and then took the data as per its training. In this whole exercise the software was not changed; only the parameters of the statistical model were adjusted in the light of new experience.

From the 1960s onwards, the field of AI has encompassed pattern recognition also. There has also been an increasing interest in the problems of neuroscience, so that the distinction between AI and cybernetics has been becoming more and more blurred.

Lately, there has been a resurgence of interest in applying the AI approach to what have been traditionally regarded as tough problems in materials science (Takeuchi *et al.* 2002). This is partly a fallout of the fact that, apart from experimental science and theoretical science, *computational science* has emerged as a distinctly different field of research. A large number of scientific problems are just too tough to be tackled in any other way than by modelling them on a computer. It was mathematically proved by Poincare, as early as in 1889, that there is no analytic solution for the dynamics of even the three-body problem, not to speak of the more intractable N-body interactions. Things have not improved much since then. Numerical solutions, using a computer, are often the only option available. Theoretical scientific complexity apart, quite often it is very expensive, if not impossible, to conduct certain experiments. Computer simulation can again help a great deal. It was only inevitable that, in due course, AI methods also made their contribution to this scenario. Some computational problems are so hard that only the AI approach can yield sensible results. The result, in materials science, is a marked increase in the efficiency with which new materials and processes are being discovered. *High throughput* is a current paradigm for checking and implementing new strategies for developing materials and processes. We give here a glimpse of this approach.

There are three broad categories of applications of AI techniques in materials science (Maguire *et al.* 2002): Intelligent process control; discovery of useful new materials and processes; and advanced computational research.

Intelligent process control

The idea here is to do real-time integration of a distributed network of sensors with sophisticated process models and materials-transformation models that capture the coupled effects of chemical reactivity and transport phenomena. Expert systems are built into the real-time process control so as to take immediate and on-line corrective action where and when needed, all along the process pathway, rather than merely adjusting some general process parameters. A detailed model about the process is built into the system. Real-time sensory data coming from a large number of carefully chosen and positioned sensors are continually processed by the computer for taking 'expert' decisions for achieving the desired end-product in the most efficient and cost-effective manner. Availability of detailed physico-chemical information about the thermodynamics of the system, along with models for interpreting the data, makes the computer steer the process intelligently along the most desirable trajectory in phase space.

Discovery of useful new materials

AI techniques are being used for data mining and rapid mapping of phase diagrams for designing new materials. Information about the properties desired for the new material is fed in, and the system makes a search by pattern-matching.

Advanced computational research

Here the computer itself becomes an experimental/theoretical method for investigating phenomena which cannot be tackled in any other way. N-body interactions are an example. Nanotechnology is another. There is so much in the nascent field of nanotechnology that is unexplored that an AI approach to the modelling of systems at the nanoscale, and training the AI assembly accordingly for problem-solving, can pay rich dividends. Once the system has been trained, the speed of simulation is independent of the complexity of the underlying interaction potentials (whether two-body or 10-body). And the algorithm can be used with any type of equations of motion.

The 'intelligence' in artificial intelligence is nowhere near the true intelligence of the human brain. Therefore the successes of AI have been of a rather limited nature, or at least not what one would expect from genuinely intelligent systems. The reason is that machine intelligence has so far been not modelled substantially on the neocortical model of the human brain (Hawkins and Blakeslee 2004).

The current use of statistical reasoning techniques is leading to a revival of interest in large-scale, comprehensive applications of AI. The latest example of this is the use of statistical techniques for achieving a modicum of success in the translation of languages, i.e. *machine translation* (MT) (cf. Stix 2006).

We should also mention here a recent attempt to overcome some of the shortcomings of the conventional AI approach by taking recourse to statistical reasoning. An artificial brain called 'Cyc' has been put on the internet (see Mullins 2005, for some details). Developed by Doug Lenat for over two decades, it is supposed to develop *common sense*. And interaction with the world will make it more and more experienced, and therefore better able to behave like human beings. It is based on the hope that if we can build up a database of common-sense *context-sensitive* knowledge and expert systems, we can come closer to the dream of human-like intelligence (although still not in the sense emphasized by Hawkins and Blakeslee 2004). In Cyc, each new input of information is compared and correlated with all the existing facts in the database, so that context-sensitivity develops and the system becomes cleverer with growing experience. The knowledge exists in the memory in the form of logical clauses that assert the stored truths, using the rules of symbolic logic.

A2. Cell biology

All tissues in animals and plants are made up of cells, and all cells come from other cells.

A cell may be either a *prokaryote* or an *eukaryote*. The former is an organism that has neither a distinct nucleus and a membrane, nor other specialized organelles. Examples include bacteria and blue–green algae. We shall not discuss such cells further.

Unicellular organisms like yeast are eukaryotes. Such cells are separated from the environment by a semi-permeable *cell membrane*. Inside the membrane there is a nucleus and the *cytoplasm* surrounding it.

Multicellular organisms are all made up of eukaryote-type cells. In them the cells are highly specialized, and perform the function of the organ to which they belong.

The nucleus contains *nucleic acids*, among other things. With the exception of viruses, two types of nucleic acids are found in all cells: RNA (ribonucleic acid) and DNA (deoxyribonucleic acid). Viruses have either RNA or DNA, but not both (but then viruses are not cells).

DNA contains the codes for manufacturing various proteins. Production of a protein in the cell nucleus involves *transcription* of a stretch of DNA (this stretch is called a *gene*) into a portable form, namely the *messenger RNA* (or mRNA). This messenger then travels to the cytoplasm of the cell, where the information is conveyed to a 'particle' called the *ribosome*. This is where the encoded instructions are used for the synthesis of the protein. The code is read, and the corresponding amino acid is brought into the ribosome. Each amino acid comes connected to a specific *transfer RNA* (tRNA) molecule; i.e. each tRNA carries a specific amino acid. There is a three-letter recognition site on the tRNA that is complementary to, and pairs with, the three-letter code sequence for that amino acid on the mRNA.

The one-way *flow of information* from DNA to RNA to protein is the basis of all life on earth. This is the *central dogma* of molecular biology.

DNA has a double-helix structure. Each of the two backbone helices consists of a chain of phosphate and deoxyribose sugar molecules, to which are attached the bases adenine (A), thymine (T), cytosine (C), and guanine (G), in a certain sequence. It is a sequenced polymer. A phosphate molecule and the attached sugar molecule and base molecule constitute a *nucleotide base*. The sequence in which they exist decides the genetic code of the organism. A strand of DNA is a *polynucleotide*, or an *oligonucleotide*.

The two helices in the double-helix structure of DNA are loosely bonded to each other, all along their length, through hydrogen bonds between complementary base pairs: Almost always, A bonds to T, and C bonds to G.

Just as DNA can be viewed as a sequence of nucleotide bases, a protein involves a sequence of amino acids. Only 20 amino acids are used for synthesizing all the proteins in the human body. Three letters (out of the four, namely the bases A, T, C, G) are needed to code the synthesis of any particular protein. The term *codon* is used for the three consecutive letters on an mRNA. The possible number of codons is 64. The linking of most of the amino-acid-triplets for synthesizing a protein can be coded by more than one codon. Three of the 64 codons signal the 'full stop' for the synthesis of a protein.

There are $\sim 60 - 100$ trillion cells in the human body. In this multicellular organism (as also in any other multicellular organism), almost every cell (red blood 'cells' are an exception) has the same DNA, with exactly the same order of the nucleotide bases.

The nucleus contains 95% of the DNA, and is the control centre of the cell. The DNA inside the nucleus is complexed with proteins to form a structure called *chromatin*.

The fertilized mother cell (the *zygote*) divides into two cells. Each of these again divides into two cells, and so on. Before this cell division (*mitosis*) begins, the chromatin condenses into elongated structures called *chromosomes*.

A gene is a functional unit on a chromosome, which directs the synthesis of a particular protein. As stated above, the gene is transcribed into mRNA, which is then translated into the protein.

Humans have 23 pairs of chromosomes. Each pair has two nonidentical copies of chromosomes, derived one from each parent.

During cell division, the double-stranded DNA splits into the two component strands, each of which acts as a template for the construction of the complementary strand. At every stage, the two daughter cells are of identical genetic composition (they have identical *genomes*). In each of the 60 trillion cells in the human body, the genome consists of around three billion nucleotides.

At appropriate stages, *cell differentiation* starts occurring into various distinct types of cells, like muscle cells, liver cells, neurons, etc. The term *stem cells* is used for the primal undifferentiated cells. Because of their ability to differentiate into other cells, stem cells are used by the body to act as a repair system, replenishing other cells when needed.

How does cell differentiation occur, and with such high precision? Kauffman (1993) started investigating this problem in 1963. It was known at that time that a cell contains a number of *regulatory genes*, which can turn one another on and off like switches. This implied that there are *genetic circuits* within the cell, and the genome is some kind of a *biochemical computer*. There must be an algorithm which determines the messages sent by genes to one another, and deciding the switching on or off of appropriate genes (i.e. their active and inactive states). This computing behaviour, in turn, determines how one cell can become different from another.

Thus, at any given instant, there is a *pattern* of on and off genes, and the pattern changes over time with changing conditions. As Kauffman realized, this was a case of parallel computing, and the genome has several, stable, *self-consistent patterns* of activation (each responsible for a specific cell differentiation). How could such a high degree of order, namely such specific and complicated genetic configurations, arise through the trial and error of gradual evolution? The whole thing had to be there *together*, and not partially, to be functional at all.

Kauffman concluded that the order must have appeared right in the beginning, rather than having to evolve by trial and error. But how?

He could answer the question successfully through the *cellular-automata* approach. In his model, each cell of the automaton represented a gene. It was known at that time that each regulatory gene was connected to only a few other genes (this number is now known to be between 2 and 10). In Kauffman's model, each cell of the automaton, i.e. each node of the network, was taken as receiving two inputs, i.e. each gene was modelled as connected to two other genes. This sparsely (but not too sparsely) connected network was a good choice. If the connectivity is too little, the system would quickly settle to an uninteresting stable state. And if a network is too densely connected, it becomes hypersensitive and chaotic, not moving towards any stable order or pattern.

Kauffman defined some *local rules* for his 100-node cellular automaton, and started the process on a computer with a random initial configuration of the nodes of the network. It was found that the system tended to order towards a small number of patterns. Most of the nodes just froze into an off or on state, and a few oscillated through ~10 configurations, just like the regulatory genes of the real genome.

Spurred by this initial success, larger and larger networks were investigated on the computer. It could be established that, in agreement with data from the real world, the total number of differentiated cell types in an organism scales roughly as the square root of the number of genes it has.

Thus Kauffman's work established that complex genetic circuits could come into being by spontaneous self-organization, without the need for slow evolution by trial and error. It also established that *genetic regulatory networks are no different from neural networks*.

Such networks are examples of systems with *nonlinear dynamics*. The not-too-sparsely connected network of interacting genetic 'agents' is a nonlinear system, for which the stable pattern of cycles corresponds to a basin or attractor (cf. appendices on nonlinear dynamics and on chaos).

The term *ontogeny* is used for the development of a multicellular being from one single cell, namely the zygote. As we have seen above, ontogeny involves cellular division and cellular differentiation. *Embryogenesis* is another term used for the ontogeny of animals, especially human beings.

A3.   Chaos and its relevance to smart structures

> *For the want of a nail, the shoe was lost;*
> *for the want of a shoe the horse was lost;*
> *and for the want of a horse the rider was lost,*
> *being overtaken and slain by the enemy,*
> *all for the want of care about a horseshoe nail.*
> – Benjamin Franklin

Nonlinear dynamics is discussed in Appendix A9. Here we focus on a possible consequence of nonlinearity, namely chaos.

Chaos theory deals with unstable conditions where even small changes can cascade into unpredictably large effects (Abarbanel 2006). As meteorologist Edward Lorenz said, in the context of the complexity of phenomena that determine weather, the flap of a butterfly's wings in Brazil might set off a tornado in Texas, for example.

Chaos is a complex phenomenon that seems random but actually has an underlying order (Langreth 1992). The irregular-looking time evolution of a chaotic system is characterized by a strong dependence on initial conditions, but is, nevertheless, deterministic. The apparent irregularity stems from the nonlinearities in the equations of motion of the system, which magnify the small but inevitable errors in fixing the initial conditions to such an extent that long-time behaviour becomes seemingly erratic and practically unpredictable.

In a *dissipative* or non-Hamiltonian system, energy is dissipated by friction, etc., and any required movement can be maintained only by using external driving forces. Dissipative systems are characterized by the presence of *attractors*. These are bounded regions in phase space (e.g. fixed points, limit cycles, etc.) to which the trajectory of the dissipative system gets 'attracted' during the time evolution of the system. The attraction occurs due to the dissipative nature of the system, which results in a gradual shrinking of the phase-space volume accessible to the system.

Strange attractors are very unlike the simple attractors mentioned above. They provide a basis for classifying dissipative chaotic systems (Gilmore 2005). What is strange about them is the sensitive dependence of the system on initial conditions. They typically arise when the flow in phase space contracts the volume elements in some directions but stretches them along others. Thus, although there is an overall contraction of volume in phase space (characteristic of a dissipative system), distances between points on the attractor do not necessarily shrink in all directions. Likewise, points which are close initially may become exponentially separated in due course of time.

A chaotic attractor typically has embedded within it an infinite number of *unstable periodic orbits*: The periodic orbits are unstable even to small perturbations. Any such perturbation or displacement of the periodic orbit (e.g. due to noise) grows exponentially rapidly in time, taking the system away from that orbit. This is the reason why one does not normally observe periodic orbits in a free-running chaotic system.

A *chaotic attractor* is a geometric object that is neither point-like nor space-filling. It typically has *fractal* (or nonintegral) dimensions, which is another reason why it is called strange.

A breakthrough in chaos theory occurred when Ott, Grebogi and Yorke (OGY) (1990) showed that it should be possible to convert (or stabilize, or synchronize) a chaotic attractor to one of its possible periodic motions by applying small, time-dependent, feedback-determined perturbations to an appropriate system parameter. Pecora and Carroll (1990, 1991) also made a seminal, independent contribution to this field at about the same time.

OGY also pointed out the tremendous application potential of this idea. Any of a number of different periodic orbits can be stabilized, so one can choose the one best suited for optimizing or maximizing the performance of the system. What is more, if the need arises (because of changing environmental conditions), one can quickly and easily switch from one chosen orbit to another by changing the applied time-dependent perturbation. This can be done without having to alter the gross system configuration. *The relevance of this to smart-structure applications is obvious: Such structures, by definition, are those which can alter their response functions suitably to achieve an objective even under changing environmental conditions.*

As discussed by OGY, the availability of this flexibility is in contrast to what happens for attractors which are not chaotic but, say, periodic. Small changes in the system parameters will then change only the orbit slightly, and one is stuck with whatever performance the system is capable of giving. There is no scope for radical improvement without changing the gross configuration of the system, something not practical for real-life dynamic systems (e.g. an aircraft in flight).

The work of OGY has shown that, in a chaotic system, multi-use or multi-exigency situations can be accommodated by switching the temporal programming of the small perturbations for stabilizing the most appropriate periodic orbit in phase space. Such multipurpose flexibility appears to be essential for the survival of higher life forms. OGY speculated that chaos may be a necessary ingredient in the regulation of such life forms by the brain. It follows that *the design of really smart structures will have to factor this in.*

We discuss *asymptotic stability* in the appendix on nonlinear systems. As pointed out there, asymptotic stability implies irreversibility, so one is dealing with a dissipative system rather than a conservative system. Such a system can approach a unique attractor reproducibly because it can eliminate the effects of perturbations, wiping out all memories of them. Asymptotic stability, arising from the dissipative nature of a system, is a very beneficial effect in living systems. By contrast, a conservative system will keep a memory of the perturbations. A conservative system cannot enjoy asymptotic stability.

The heart of a living being is an example of an asymptotically stable dissipative system. The chaotic feature in the functioning of the heart may be actually responsible for preventing its different parts from getting out of synchronization. In the light of OGY's work on chaotic systems, one can conclude that the heart-beat rate is controlled and varied by the brain by applying appropriate time-dependent perturbative impulses. Garfinkel *et al.* (1992) induced cardiac arrhythmia in a rabbit ventricle by using the drug ouabain, and then succeeded in stabilizing it by an approach based on chaos theory. Electrical stimuli were administered to the heart at irregular time intervals determined by the nature of the chaotic attractor. The result was a conversion of the arrhythmia to a periodic beating of the heart.

The idea that it is possible to steer a chaotic system into optimum-performance configurations by supplying small kicks that keep sending the system back into the chosen unstable periodic orbit has wide-ranging applicability to chemical, biological, electronic, mechanical and optical systems. Its physical realization was first reported by Ditto, Rauseo and Spano (1990). They applied it to a parametrically driven ribbon of a magnetoelastic material. Application of magnetic field to such a material modifies

its stiffness and length. The thin ribbon was clamped at the base and mounted vertically. A field as small as 0.1 to 2.5 Oe, applied along the length of the vertical ribbon,
changed (decreased) the Young's modulus by an order of magnitude, causing the initially stiff and straight ribbon to undergo gravitational buckling, making it sway like
an inverse pendulum. A combination of dc and ac magnetic fields was applied, and
the frequency of the ac field was ~1 Hz. A suitable ratio of the dc and ac fields made
the buckling and unbuckling of the ribbon chaotic, rather than periodic. The swaying
of the ribbon was recorded as a function of time by measuring the curvature near its
base. The measurements yielded a time series of voltages, V(t).

From this time series a certain desired mode of oscillation was selected for stabilization. The experimenters waited till the chaotic oscillations came close to this
mode. A second set of magnetic perturbations, determined from the knowledge of
$V(t)$, was applied. Use of these feedback perturbations resulted in a stable *periodic*
orbit. Chaos had been controlled.

It is worthwhile to recapitulate here what can be achieved in the control of chaotic
systems, and how they score over linear, conservative systems. The latter can do only
one thing well (Ditto and Pecora 1993). By contrast, nonlinear systems and devices
can handle several tasks. If we measure the trajectory of a chaotic system, we cannot
predict where it would be on the attractor at some time in the distant future. And yet
the chaotic attractor itself remains the same in time. What is more, since the chaotic
orbits are ergodic, one can be certain that they would eventually wander close to the
desired periodic orbit. Since this proximity is assured, one can capture the orbits by
a small control. This feature is crucial for exercising control.

Deliberate building of chaos into a system can provide sensitivity and flexibility
for controlling it. *Instability can be a virtue if the system involved is chaotic*. There
are typically an infinite number of unstable periodic orbits coexisting in a chaotic
system, offering a wide choice for selecting (*synchronizing*) an orbit (Aziz-Alaoui
2005), and for switching from one orbit to another, all with the underlying objective
of optimizing performance, as well as easily altering the orbit for a smart tackling of
changing environmental conditions.

A4.   Composites

Composites are made of two or more components or phases, which are strongly
bonded together in accordance with some desired connectivity pattern. They are carefully patterned inhomogeneous solids, designed to perform specific functions. Some
authors emphasize the presence of interfaces in composites, and define a composite
as a macroscopic combination of two or more distinct materials with recognizable
interfaces among them (Miracle and Donaldson 2001).

Composites can be configured in an infinite number of ways, and that offers
immense scope for design for achieving or enhancing certain desirable properties, as
well as for suppressing undesirable ones. Sometimes, *new* properties, not possessed
by any of the constituents separately, can also emerge.

Composites can be either *structural*, or *nonstructural* (often called *functional*). Both are important from the point of view of applications in smart structures. Structural composites were introduced as early as in the late nineteenth century, and this subject is therefore already quite highly developed (see Beetz 1992; Hansen 1995; Chung 2001). Functional composites, on the other hand, are of relatively more recent origin.

A typical structural composite has a matrix, a reinforcement, and a filler. The matrix material binds the other materials in the composite, giving it its bulk shape and form. The reinforcing component, usually in the form of filaments, fibres, flakes or particulates, determines to a large extent the structural properties of the composite. The filler meets the designed structural, functional, and other requirements.

Apart from the relative sizes and concentrations of the various phases constituting a composite, a factor of major importance is their *connectivities*. Connectivity has been defined as the number of dimensions in which a component of a composite is self-connected (Newnham, Skinner and Cross 1978; Newnham and Trolier-McKinstry 1990a, b). There are 10 possible connectivity classes for a diphasic composite in three dimensions, when no distinction is made between, say, the classes 1–3 and 3–1. When such a distinction is made, six additional connectivity classes arise. The 16 connectivity classes are: 0–0, 1–0, 0–1, 2–0, 0–2, 3–0, 0–3, 1–1, 2–1, 1–2, 3–1, 1–3, 2–2, 3–2, 2–3, and 3–3.

We consider an example to explain the meaning of these symbols. It will also illustrate the difference between, say, 1–3 connectivity and 3–1 connectivity. Imagine a composite in which poled rods or fibres of the piezoelectric ceramic PZT are embedded in a polymer matrix. Here the polymer matrix is the major phase, and it is connected to itself (self-connected) in all three directions or dimensions. The PZT phase is self-connected only in one dimension, namely along its length, so this is a 1–3 connectivity composite.

Contrast this with a situation in which one takes a rectangular block of poled PZT ceramic, drills parallel holes in it along one direction, and fills the holes with a polymer. This is a 3–1 composite, where a convention has been followed that the connectivity of the 'active' phase should be written first; PZT is the active phase (Pilgrim, Newnham and Rohlfing 1987).

For a review of several commercial applications of composite piezoelectric sensors and actuators, see Newnham *et al.* (1995).

Of particular interest for smart structures are the 2–2 composites, commonly known as laminated composites. Fibre-reinforced polymer-matrix laminated composites are particularly well suited for embedding sensors (Hansen 1995).

The properties of a composite may be categorized as *sum* properties, *combination* properties, and *product* properties.

The density of a composite is an example of a *sum* property; it is the weighted arithmetic mean of the densities of the constituent phases. Other examples of sum properties are electrical resistivity, thermal resistivity, dielectric permittivity, thermal expansion, and elastic compliance (van Suchtelen 1972; Hale 1976). The value of a sum property can depend strongly on the connectivity pattern of the composite (cf. Wadhawan 2000).

An example of a *combination* property of a composite is provided by the speed of acoustic waves in a biphasic composite (i.e. a composite made from two phases). The speed depends on two properties, namely Young's modulus and density, and the mixing rule for the Young's moduli of the two phases is not the same as that for density. Further complications are caused by the fact that the mixing rules are different for transverse and longitudinal acoustic waves (Newnham 1986).

Product properties of composites can be particularly fascinating (van Suchtelen 1972). Consider a biphasic composite with an X-Y effect in Phase 1, and a Y-Z effect in Phase 2. Application of a force X invokes a response Y in Phase 1, and then Y acts as a force on Phase 2 to invoke a response Z. The net result is an X-Z effect, which is a product property, not present in Phase 1 or Phase 2 individually.

For example, Phase 1 may be magnetostrictive, and Phase 2 piezoelectric. Suppose we apply a magnetic field **H**. Phase 2, being nonmagnetic, is not influenced by it directly. The magnetic field produces magnetostrictive strain in Phase 1 (the X-Y effect). Assuming that the two phases are coupled adequately, the strain will act on Phase 2, and produce a dipole moment (through the inverse piezoelectric effect) (the Y-Z effect). The net result (the X-Z effect) is that magnetic field induces an electric dipole moment in the composite; this is called the magnetoelectric effect. Note that neither Phase 1, nor Phase 2, may be magnetoelectric, but the composite is.

We discuss the symmetry of composite systems in a separate appendix. The emergence of the magnetoelectric effect as a product property is a consequence of the fact that when the symmetries of the two constituent phases are superimposed, the net symmetry is lower than either of the component symmetries. A lower symmetry means a lower set of restrictions on the existence of a property in a material. The symmetries of the two component phases prevent the occurrence of the magnetoelectric effect in them. But the lower symmetry of the composite allows this effect to occur.

Transitions in composites

Phase transitions, including field-induced phase transitions, can occur in any of the constituent phases of a composite. In addition, connectivity transitions are also possible. A number of examples have been discussed by Pilgrim, Newnham and Rohlfing (1987). The connectivity pattern can be altered continuously by changing the volume fractions of the component phases, or by changing their relative size scales. At a critical value of these parameters, the composite acquires a new connectivity, with a drastic change in macroscopic properties (Newnham and Trolier-McKinstry 1990a).

Nanocomposites

In a nanocomposite, at least one of the phases has at least one of the dimensions below 100 nm. Many of them are biphasic. There are three main types of them (Cammarata 2004): nanolayered, nanofilamentary, and nanoparticulate. All of them have a very high ratio of interface area to volume. This can result in totally new and *size-tuneable* properties. Take the example of a nanolayered semiconductor, made from alternating layers of epitaxially matched GaAs and $GaAl_xAs_{1-x}$. For layer thicknesses below the electronic mean free path in the bulk form of the two materials, quantum confinement effects arise, drastically affecting the electronic and photonic properties. What is more, the properties can be tuned by altering the thicknesses of the two layers.

A5.   Crystallographic symmetry

Symmetry considerations form an integral part of the philosophy of physics. Noether's theorem gives an indication of why this is so. According to this theorem (cf. Lederman and Hill 2005): *For every continuous symmetry of the laws of physics, there must exist a conservation law; for every conservation law, there must exist a continuous symmetry.*

Symmetry of physical systems is described in the language of *group theory*. In this appendix we introduce the definition of a group, and give a very brief description of crystallography in terms of symmetry groups.

Atoms in a crystal have nuclei and charge clouds of electrons around them. The chemical bonding among the atoms results in a certain spatial distribution of the electron cloud, which we can describe in terms of a density function, $\rho(x, y, z)$.

Certain coordinate transformations (translations, rotations, reflections, inversion) may map the density function onto itself; i.e. leave it invariant. The set of all such *symmetry transformations* for a crystal forms a 'group', called the *symmetry group of the crystal.*

What is a group? A group is a *set* with some specific properties. A set is a collection of objects (or 'members', or 'elements') which have one or more common characteristics. The characteristics used for defining a set should be sufficient to identify its members. A collection of integers is an example of a set, as also a collection of cats.

A group is a set for which a rule for combining ('multiplying') any two members of the set has been specified, and which has four essential features which we shall illustrate here by considering the example of the symmetry group of a crystal. (For this example, the set comprises of symmetry transformations of the crystal.)

Suppose a rotation θ_1 about an appropriate axis is a symmetry operation, and a rotation θ_2 about the same or different axis is another symmetry operation. Since each of them leaves the crystal invariant, their successive operation (or 'product', denoted by $\theta_1 \theta_2$) will also leave the crystal invariant, so the product is also a symmetry operation, and therefore a member of the set of symmetry operations. This is true for all possible products. We say that the set has the property of *closure*.

It also has the property of *associativity*. What this means is that if θ_3 is another, arbitrarily chosen, symmetry operation, then $(\theta_1 \theta_2) \theta_3$ has the same effect as $\theta_1 (\theta_2 \theta_3)$.

The set includes an *identity* operation, which simply means that not performing any coordinate transformation is also a symmetry operation.

Lastly, for every symmetry operation (or 'element' of the set), the *inverse* element is also a symmetry operation. For example, if a rotation θ_1 is a symmetry operation, so is the negative rotation $-\theta_1$.

A set of elements, for which a law of composition or multiplication of any two elements has been defined, is a group if it has the properties of closure and associativity, and if identity and inverse elements are also members of the set. A simple example of a group is the set of all integers, with addition as the law of composition.

A crystal has the distinctive feature that an atom or a group of atoms or molecules can be identified as a *building block* or *unit cell*, using which the whole crystal can be generated by repeating it along three appropriately identified vectors, say, $\mathbf{a}_1$, $\mathbf{a}_2$, $\mathbf{a}_3$. These vectors are called the *lattice vectors* because the set of all points

$$\mathbf{r} = n_1\mathbf{a}_1 + n_2\mathbf{a}_2 + n_3\mathbf{a}_3 \qquad (A1)$$

for all integral values of n_1, n_2, n_3 generates a lattice of *equivalent points*. This also means that every lattice, and thence any crystal based on that lattice, has *translational symmetry*. The lattice and the crystal are invariant under lattice translations defined by eqn A1 for various values of the integral coefficients.

Apart from the translational symmetry, a lattice may also have *rotational or directional symmetry*. The rotational symmetry of a crystal lattice defines the *crystal system* it belongs to.

All crystals in three-dimensional space belong to one or the other of only seven crystal systems: triclinic, monoclinic, orthorhombic, trigonal, tetragonal, hexagonal, and cubic.

For each of these crystal systems, one can identify a unit cell which has the distinct shape compatible with the rotational symmetry of that crystal system. For example, the unit cell is a cube for any crystal belonging to the cubic crystal system. Similarly, the unit cell has the shape of a square prism for any crystal belonging to the tetragonal crystal system.

The rotational symmetry of a crystal is described by a particular type of group, called the *point group*. It is a set of all crystallographic symmetry operations which leave at least one point unmoved. Since only directional symmetry is involved, there are no translations to be considered, and all the operations of reflection, rotation, or inversion can be applied about a fixed plane, line or point.

There are only 32 distinct *crystallographic point groups*, seven of which describe the directional symmetry of the seven crystal systems.

Each crystallographic point group is a set of mutually compatible rotations, reflections or inversion operations. The fact that the crystal also has the all-important translational symmetry puts severe restrictions on what can qualify as a crystallographic rotational symmetry (cf. Wadhawan 2000 for details). For example, a rotation of $2\pi/5$ cannot be a symmetry operation for a crystal. In fact, the only permissible rotations are 2π, $2\pi/2$, $2\pi/3$, $2\pi/4$, and $2\pi/6$. These correspond to one-fold, two-fold, three-fold, four-fold, and six-fold axes of symmetry, respectively. The corresponding symmetry operations of the point group are denoted by 1, 2, 3, 4, and 6.

Some crystal structures possess *inversion symmetry*, denoted by the symbol i or $\bar{1}$. Some others may possess a composite symmetry comprising a combination of inversion and any of the four permitted rotational symmetries. For example, whereas the operation i takes a point (x, y, z) to $(-x, -y, -z)$, and the operation denoted by the symmetry element 2_z takes (x, y, z) to $(-x, -y, z)$, a composite operation which is a combination of these two, and is denoted by $\bar{2}$, takes (x, y, z) to $(x, y, -z)$. (Incidentally, $\bar{2}$ happens to have the same effect as a reflection (m_z) across a plane normal to the z-axis.)

Thus the elements of the 32 crystallographic point groups consist of symmetry operations 1, 2, 3, 4, 6, $\bar{1}$ $(= i)$, $\bar{2}$ $(= m)$, $\bar{3}$, $\bar{4}$, and $\bar{6}$, and their mutually compatible combinations.

There are only 14 distinct types of crystal lattices in three dimensions. These are called *Bravais lattices*. Since there are only seven crystal systems, it follows that a crystal system can accommodate more than one Bravais lattices. For example, there are three Bravais lattices belonging to the cubic crystal system: simple cubic (sc), body-centred cubic (bcc), and face-centred cubic (fcc). The centring mentioned here refers to the fact that if one insists on choosing a cube-shaped unit cell to reflect the full directional symmetry of the crystal system and the crystal lattice, then the cell would have lattice points, not only at the corners, but also at the body-centre $(^1/_2\ ^1/_2\ ^1/_2)$ (in the case of the bcc lattice), or the face centres $(^1/_2\ ^1/_2\ 0)$, $(^1/_2\ 0\ ^1/_2)$, $(0\ ^1/_2\ ^1/_2)$ (in the case of the fcc lattice).

The full atomic-level (or *microscopic*) symmetry of a crystal is described by its *space group*. A crystallographic space group is a group, the elements of which are all the symmetry operations (lattice translations, rotations, reflections, etc.) that map the crystal structure onto itself. The only translational symmetry a crystal can have is that described by one of the 14 *Bravais groups* (i.e. the groups describing the symmetry of the Bravais lattices). Therefore, to specify the space-group symmetry of a crystal we have to identify its Bravais lattice, as well as the symmetry operations involving rotation, reflection and inversion, and their mutually consistent combinations. There are 230 crystallographic space groups in all.

All crystals having the same point-group symmetry are said to belong to the same *crystal class*. Thus all crystals can be divided into 32 crystal classes.

Eleven of the 32 crystallographic point groups have the inversion operation as a symmetry operation. They are called the 11 *Laue groups*.

The remaining 21 point groups are *noncentrosymmetric*. For one of them, all components of the piezoelectric tensor are identically equal to zero. The remaining 20 allow the occurrence of the piezoelectric effect.

Out of these 20 *piezoelectric classes*, 10 are *polar classes*. For them the point-group symmetry is such that there is at least one direction (axis) which is not transformed into any other direction by the symmetry operations comprising the group. This means that it is a direction for which symmetry does not result in a cancellation of any electric dipole moment that may exist because of the charge distribution in the unit cell. For example, suppose the only rotational symmetry a crystal has is a two-fold axis, say along the z-direction. A spontaneously occurring dipole moment (P_z) along this direction will not get *al*tered or cancelled by any other rotational symmetry, because none exists. Thus the 10 polar classes of crystals are characterized by the occurrence of *spontaneous polarization*. They are the 10 *pyroelectric classes* because they exhibit the pyroelectric effect: The spontaneous polarization changes with temperature, resulting in an occurrence of additional charge separation on crystal faces perpendicular to this *polar direction or axis*.

Since the 10 polar classes are a subset of the 20 piezoelectric classes, all pyroelectric crystals are necessarily piezoelectric also.

A crystalline material may exist, not as a single crystal, but as a *polycrystal*, i.e. as an aggregate of small crystals (*crystallites*). Since such an assembly of crystallites does not have the periodicity of an underlying lattice, there is no restriction on the allowed rotational symmetry. Of special interest is the occurrence of axes of ∞-fold symmetry. In fact, if the crystallites are oriented randomly, any direction in the polycrystal is an ∞-fold axis.

Point groups involving at least one ∞-fold axis are called *limit groups* or *Curie groups*. Limit groups are relevant for dealing with polycrystalline specimens of ferroic materials (Wadhawan 2000). For example, if we are dealing with a polycrystalline ferroelectric, we can *pole* it by applying a strong enough electric field at a temperature a little above the temperature of the ferroelectric phase transition, and cooling it to room temperature under the action of the field. A substantial degree of domain reorientation occurs under the action of the field, so that the spontaneous polarization in different crystallites ('grains') and domains tends to align preferentially along the direction of the applied field. Whereas there was no net polarization before poling, the specimen acquires a preferred direction, along which there is a net macroscopic spontaneous polarization.

What is the point-group symmetry of the poled specimen? It is the same as that of the superimposed electric field, and is denoted by the symbol ∞m: There is an axis of ∞-fold symmetry along the direction of the poling electric field; in addition, there is also mirror symmetry (denoted by m) across all planes passing through this polar axis. One can visualize this as the symmetry of a cone or a single-headed arrow.

We consider some examples of the description of space-group symmetries of crystals mentioned in this book.

$BaTiO_3$ has the so-called perovskite structure. Its cubic phase (existing above 130°C) has the symmetry $Pm\bar{3}m$. Here P means that the underlying Bravais lattice is *primitive*; i.e. only one lattice point is associated with every unit cell. The rest of the symbol is actually a brief version of $m_{[100]}\bar{3}_{[111]}m_{[110]}$. The first symbol tells us that there is a mirror plane of symmetry, the normal of which points along the [100] direction, or the x-direction.

The international convention is such that, if there is a symbol 3 in the second place of the point-group symbol, it means that one is dealing with the cubic crystal system, with a unit cell that has the shape of a cube. Only a cubic unit cell can have a $\bar{3}$ axis of symmetry along its body-diagonal, i.e. along the [111] direction.

The third symbol, $m_{[110]}$, denotes the presence of a mirror plane of symmetry normal to the [110] direction, or the xy-direction.

On cooling, the cubic phase of $BaTiO_3$ enters the tetragonal phase at 130°C, which has the symmetry $P4mm$. Here P, of course, means a primitive unit cell. In the rest of the symbol, a 4 in the first position implies, by convention, that one is dealing with the tetragonal crystal system. In fact, the full point-group symbol is $4_{[001]}m_{[100]}m_{[110]}$. The four-fold axis of symmetry is along the [001] direction, or z-direction. The second symbol indicates the presence of a mirror plane of symmetry normal to the [100] direction, and the third symbol represents a mirror plane normal to the [110] direction. The point-group $4mm$ is one of the 10 polar groups, allowing the tetragonal phase of $BaTiO_3$ to exhibit pyroelectricity and ferroelectricity. The shape of the unit cell is that of a square prism.

On further cooling, $BaTiO_3$ passes to a phase of symmetry $Amm2$, and thereafter to a phase of symmetry $R3c$. Both have polar point-group symmetries. The point-group symmetry of the $Amm2$ phase is $mm2$. It belongs to the orthorhombic crystal system. By convention, $mm2$ stands for $m_{[100]}m_{[010]}2_{[001]}$. That is, the first mirror plane is normal to the x-axis, and the second is normal to the y-axis. The two-fold axis, which is also the polar axis for this point group, is along the z-axis. The symbol A in $Amm2$ means that the underlying Bravais lattice is A-face (or yz-face) centred, and is therefore not a primitive lattice. The unit cell has the shape of a rectangular brick, and its face normal to the x-axis has a lattice point at its centre. Thus, there are two lattice points per unit cell; one at a corner of the unit cell, and the other at the centre of the A-face.

In $R3c$, a 3 in the first place (rather than the second place, as in a cubic point group) indicates that we are dealing with the trigonal or rhombohedral crystal system. The Bravais lattice is primitive, but we write $R3c$, rather than $P3c$ (by convention). The underlying point group is $3m$. There is a three-fold axis along [001], and a c-glide normal to [100]. The notional mirror plane associated with the c-glide operation is parallel to the three-fold axis, and its normal is along [100]. The glide operation is a composite symmetry operation. In the present case, it amounts to reflecting a point across the plane, and then translating this point by $\mathbf{c}/2$.

We consider two more space-group symbols, used in Chapter 9. These are Cm and $P4/mmm$.

In Cm, the occurrence of a lone m (with no other symbols for the point-group part) indicates that it is for a crystal belonging to the monoclinic system. And C tells us that the Bravais lattice is C-face centred.

The full form of $P4/mmm$ is $P(4_{[001]}/m_{[001]})m_{[100]}m_{[110]}$, which, by now, should be self-explanatory to the reader. The underlying point-group symmetry ($4/mmm$) is centrosymmetric, rather than polar.

A6. Electrets and ferroelectrets

An electret is a dielectric solid that has been 'electrized' or quasi-permanently polarized by the simultaneous application of heat and strong electric field (Pillai 1995). On cooling to room temperature from an optimum high temperature, the dielectric under the action of the strong electric field develops a fairly permanent charge separation, manifested by the appearance of charges of opposite signs on its two surfaces. Thus even a nonpolar material can be made polar, exhibiting pyroelectricity and piezoelectricity.

High temperature and electric field are not the only ways of introducing quasi-permanent charge separation in an insulating material. Other options include the use of a magnetic field in place of electric field (*magnetoelectrets*), and photons and other ionizing radiation (*photoelectrets*, *radioelectrets*, etc.). Mechanical stress has also been used, instead of electric or magnetic fields.

If the dipole moment of an electret can be made to switch sign reversibly (almost like a ferroelectric), we speak of a *ferroelectret* (Bauer, Gerhard-Multhaupt and Sessler 2004).

Initially the materials used for making electrets were waxes, wax mixtures, and other organic substances. It was realized in due course that the use of suitable polymers can result in higher dipole moments, as well as a better permanency of the dipole moments. The polymers used also have superior thermomechanical properties, and can be readily processed into thin or thick films of requisite shapes and sizes.

Polymers can be either polar (like polyvinylidene fluoride (PVDF)), or nonpolar (like polyethylene (PE), or polytetrafluoroethylene (PTFE; better known as 'Teflon')). Electrets have been made from both types, and involve a number of polarization mechanisms (Pillai 1995).

Electrets find a wide range of device applications, including those in sensors, actuators, and robotics (cf. Nalwa 1995).

A7.   Glass transition

Any noncrystalline solid is a glass. A glass is a disordered material that lacks the periodicity of a crystal, but behaves mechanically like a solid. Because of this non-crystallinity, a whole range of relaxation modes and their temperature variation can exist in a glass. In fact, an empirical definition of glass, due to Vogel (1921) and Fulcher (1925), states that a glass is one for which the temperature dependence of relaxation time is described by the equation

$$\tau = \tau_0 e^{T_0/(T-T_f)} \tag{A2}$$

where τ_0, T_0, and T_f are 'best-fit' parameters. This is the well-known *Vogel–Fulcher equation* (cf. Tagantsev 1994; Angell 1995). The parameters τ_0 and T_0 depend on the temperature range of measurement.

A thermodynamic definition of glass can be given in terms of two experimental criteria, namely the existence of a *glass transition*, and the existence of a *residual entropy* at $T = 0$ K (cf. Donth 2001; Debenedetti and Stillinger 2001). A conventional or canonical glass is usually obtained by a rapid cooling or *quenching* of a melt (to prevent crystallization). It is thus a state of frozen disorder, and can therefore be expected to have a nonzero configurational entropy at $T = 0$K.

One can associate a glass transition temperature T_g with a glass-forming material. For $T \gg T_g$, it is a liquid. As the temperature is decreased, the density increases. As the density approaches the value for the solid state, its rate of increase with decreasing temperature becomes smaller. T_g is the temperature such that this rate is high above it, and low below it.

Other properties also change significantly around T_g. In particular, there is a large increase of viscosity below T_g, and the specific heat suddenly drops to a lower value on cooling to T_g.

Spin glasses and orientational glasses (including relaxor ferroelectrics) have features in common with canonical glasses. But there is also an important point of difference. Their glass transition can take place even at low cooling rates. Although they are characterized by a quenched disorder, sudden cooling or quenching is not necessary for effecting it.

The term *glassy behaviour* is used in the context of systems that exhibit noncrystallinity, nonergodicity, hysteresis, long-term memory, history-dependence of behaviour, and multiple relaxation rates. Multiferroics usually display a variety of glassy properties.

A8.   Nonextensive thermostatistics

Entropy is an all-important concept in thermodynamics, invoked for understanding how and why one form of energy changes (or does not change) to another. As introduced by Clausius in 1865, the term entropy was a measure of the maximum energy available for doing useful work. It is also a measure of order and disorder, as expressed later by the famous Boltzmann equation:

$$S = k_B \ln W \tag{A3}$$

The entropy S is thus a product of the Boltzmann constant, k_B, and the logarithm of the number of (equally probable) microstates of the system under consideration. So defined, entropy is an *extensive* state parameter; i.e. its value is proportional to the size of the system. Also, for two independent systems, such an entropy for the combined system is simply the sum of the entropies of the two individual systems.

This equation for entropy, though a workhorse of physics and thermodynamics for over a century, has had its share of failures, and has therefore been generalized by Tsallis (1988, 1995a, b, 1997). Tsallis began by highlighting the three premises on which Boltzmann thermodynamics is based (cf. Tirnakli, Buyukkilic and Demirhan 1999):

- The effective microscopic interactions are short-range (in relation to the linear size of the system).
- The time range of the microscopic memory of the system is short compared to the observation time (i.e. one is dealing with 'Marcovian processes').
- The system evolves, in some relevant sense, in a Euclidean-like space-time. A contrary example is that of (multi)fractal space-time.

Such a system has the property of thermodynamic extensivity (or *additivity*). Boltzmann thermodynamics fails whenever any of these conditions in violated. There is a plethora of situations in which this happens. By 'failure' of the Boltzmann formalism is meant the *divergence* of standard sums and integrals appearing in the expressions for quantities like partition function, internal energy, entropy (Tsallis 1995b). As a result, one is left with no 'well-behaved' prescriptions for calculating, for example, specific heat, susceptibility, diffusivity, etc. Contrary to the predictions of Boltzmann thermodynamics, these quantities are always measured to be finite, rather than infinite.

Tsallis (1988) remedied this very serious situation by generalizing Boltzmann thermodynamics by introducing two postulates. The first postulate generalizes the definition of entropy, and the second postulate generalizes the definition of internal energy (which is another extensive state parameter in Boltzmann thermodynamics).

In Tsallis thermostatistics, an *entropic index q* is introduced, with $q = 1$ coming as a special case corresponding to conventional thermostatistics. The generalized entropy is postulated as defined by

$$S_q = k_B \frac{1 - \sum_{i=1}^{W} p_i^q}{q - 1} \tag{A4}$$

with

$$\sum_i p_i = 1 \tag{A5}$$

Here q is a real number, and $\{p_i\}$ are the probabilities of the W microscopic states.

The entropy so defined is nonnegative but *nonextensive*. Its limiting value for $q = 1$ is the standard (extensive) entropy, interpreted in the 1870s by Gibbs in terms of statistical mechanics:

$$S_1 = -k_B \sum_i^W p_i \ln p_i \tag{A6}$$

This equation reduces to the Boltzmann equation for the equiprobability case, i.e. when $p_i = 1/W$.

Tsallis entropy has the *pseudo-additivity* property. If A and B are two *independent* systems, i.e. if $p_{ij}^{A+B} = p_i^A p_j^B$, then

$$\frac{S_q(A + B)}{k_B} = \frac{S_q(A)}{k_B} + \frac{S_q(B)}{k_B} + (1 - q)\frac{S_q(A)}{k_B}\frac{S_q(B)}{k_B} \tag{A7}$$

Thus $(1 - q)$ is a measure of the *nonextensivity* of the system. Moreover, the entropy is greater than the sum for $q < 1$, and less than the sum for $q > 1$. The system is said to be *extensive* for $q = 1$, *superextensive* for $q < 1$, and *subextensive* for $q > 1$.

The second postulate introduced by Tsallis is the following generalized equation for internal energy:

$$U_q \equiv \sum_{i=1}^{W} p_i^q \varepsilon_i \tag{A8}$$

Here $\{\varepsilon_i\}$ is the energy spectrum of the microscopic states.

In this formalism, the canonical ensemble equilibrium distribution is obtained by first defining the generalized partition function as

$$Z_q \equiv \sum_{i=1}^{W} [1 - (1-q)\beta\varepsilon_i]^{1/(1-q)} \tag{A9}$$

where

$$\beta = 1/(k_B T) \tag{A10}$$

One then optimizes S_q under suitable constraints (namely $\sum_i p_i = 1$ and $U_q \equiv \sum_{i=1}^{W} p_i^q \varepsilon_i$) as follows:

$$p_i = \frac{1}{Z_q} = [1 - (1-q)\beta\varepsilon_i]^{1/(1-q)} \tag{A11}$$

This expression is the generalization of the standard expression for the Boltzmann weight, namely $e^{-\beta\varepsilon_i}$.

The Boltzmann factor is therefore no longer an exponential always. It can be a power law. Power laws are strongly linked to fractal behaviour, and are encountered in a large variety of natural phenomena. In nonextensive systems, the correlations among individual constituents do not decay exponentially with distance, but rather obey a power-law dependence (cf. Section 5.5.2 on self-organized criticality, where power-law dependence is the central theme).

A visit to the website http//:tsallis.cat.cbpf.br/biblio.htm gives some idea of the huge number of very basic scientific problems which have yielded to Tsallis thermostatistics. Here is a small sampling:

- Spin glasses and the replica trick.
- Theory of perceptions, notably the theory of human visual perception.
- The travelling-salesman problem.
- The ubiquity of Levy distributions in Nature.
- Non-Gaussian behaviour of the heartbeat.
- Stellar polytropes.
- Two-dimensional turbulence.
- Peculiar velocity distribution of galaxy clusters.
- Nanostructured materials.

- Earthquakes, flocking patterns of birds, clouds, mountains, coastlines, and other self-organizing systems that exhibit fractal behaviour.
- Time-dependent behaviour of DNA and other macromolecules.

Tsallis has argued that his postulated expression of entropic nonextensivity 'appears in a simple and efficient manner to characterize what is currently referred to as complexity, or at least some types of complexity'. The basic idea is that any small number raised to a power less than unity becomes larger. For example, $0.4^{0.3} = 0.76$. Thus, if an event is somewhat rare (say $p = 0.4$), the fact that $q = 0.3$ makes the effective probability larger (0.76). Tsallis gives the example of a tornado to illustrate how low-probability events can grow in weight for nonextensive systems. Unlike the air molecules in normal conditions, the movements of air molecules in a tornado are highly *correlated*. Trillions and trillions of molecules are turning around in a correlated manner in a tornado. A vortex is a very rare (low-probability) occurrence, but when it is there, it controls everything because it is a nonextensive system.

A9. Nonlinear dynamical systems

Over the last few decades, the somewhat arbitrary compartmentalization of science into various disciplines has been getting more and more porous. The science of nonlinear systems is one major reason for this changing perspective.

Nonlinear phenomena in large systems tend to be very complex. The ready availability of huge computing power has made all the difference when it comes to investigating them. Nonlinear phenomena are now more tractable than ever before, and (to distort the original meaning of Phil Anderson's famous remark), more is indeed different, in the sense that more computational power has led to a *qualitative* change in nonlinear science; it has made it all-pervasive. The basic unity of all science has become more visible. The diverse range of the contents of this book is an example of that.

A nonlinear system is characterized by the breakdown of the principle of linear superposition: the sum of two solutions of an equation is not necessarily a solution. The output is not proportional to the input; the proportionality factor is not independent of the input. This makes field-tuneability of properties a wide-ranging reality, a situation of direct relevance to the subject of smart structures.

The book *Exploring Complexity* by Nicolis and Prigogine (1989) is an excellent and profound introduction to the subject of nonlinear systems (also see Prigogine 1998). We recapitulate here a few basic ideas from that book, just to introduce the reader to *the vocabulary of complexity*. Some of the terms we use here are explained in the Glossary.

The time-evolution of a system, described by a set of state parameters $\{\chi_i\}$, can be influenced by the variations of some *control parameters* λ:

$$\partial \chi_i / \partial t = F_i(\{\chi_i\}, \lambda) \qquad (A12)$$

Whatever the form of F_i, in the absence of constraints these equations must reproduce the state of equilibrium:

$$F_i(\{\chi_{i,eq}\}, \lambda_{eq}) = 0. \tag{A13}$$

For a *nonequilibrium* steady state (equilibrium is also a steady state), this generalizes to

$$F_i(\{\chi_{i,s}\}, \lambda_s) = 0. \tag{A14}$$

For a *linear* system, if χ is the unique state-variable, eqn A12 can take the form

$$d\chi/dt = \lambda - k\chi, \tag{A15}$$

where k is some parameter of the system. This yields a stationary-state solution:

$$\lambda - k\chi_s = 0. \tag{A16}$$

A plot of χ against λ is a straight line, as one would expect for a linear-response system. For a nonlinear system, eqn A16 will not hold, and the plot will not be a straight line, making the system amenable to all kinds of complex behaviour.

Having arrived at a stationary state, the system stays there if there is no perturbation. For a *conservative system*, χ_s is a state of mechanical equilibrium. For a *dissipative system*, it can be a state of stationary nonequilibrium or a state of equilibrium.

In any real system, there are always perturbations, either internal (e.g. thermal fluctuations), or external (because the system is always communicating with the environment), so that the so-called stationary state really gets modified to

$$\chi(t) = \chi_s + x(t), \tag{A17}$$

where x denotes the perturbation. χ_s now serves as a *reference state* for the system.

We now consider all possible ways in which the system may respond to the deviations imposed by $x(t)$. There are four possible scenarios.

- *Case 1.* The simplest possibility is that, after some transient jitter or adjustment, the system comes close to the reference state χ_s. This is a case of *point stability*. An example is that of a pendulum, which when displaced slightly from its vertical equilibrium position, tends to settle towards that position again. This final state is described as an *attractor*; it 'attracts' the system towards itself.

  If we are interested, not in the response of one stationary state, but of a whole trajectory of them, we deal with of *orbital stability*, rather than point stability.

- *Case 2.* If $\chi(t)$ approaches χ_s asymptotically as time progresses, χ_s is said to be an *asymptotically stable* state. Like in Case 1, the argument can be extended to *asymptotic orbital stability*. Asymptotic stability implies irreversibility, so we are dealing with a dissipative system here. Such a system can approach a unique attractor reproducibly because it can eliminate the effects of perturbations, *wiping out all memories of them*. Asymptotic stability is a very beneficial effect of irreversibility in Nature (Nicolis and Prigogine 1989; Prigogine 1998).

By contrast, conservative systems keep a memory of the perturbations. A conservative system cannot enjoy asymptotic stability.

- *Case 3.* In this category, perturbations have a strong destabilizing effect, so that, as time passes, $\chi(t)$ does not remain near χ_s. We speak of *point instability* and *orbital instability* of χ_s. Such a situation is possible for both conservative and dissipative systems.

- *Case 4.* It can happen that a system is stable against small initial perturbations, but unstable against large initial perturbations. The χ_s is then said to be *locally stable* and *globally unstable*. If, on the other hand, there is stability against any initial value of the perturbation, we have *global stability*. In such a case, χ_s is said to be a *global attractor*. An example is that of thermodynamic equilibrium in an isolated system.

Bifurcation

Let us focus on dissipative systems; conservative systems cannot have asymptotic stability.

Consider a variable x, controlled by a parameter λ through the following rate equation:

$$dx/dt = f(x, \lambda) = -x^3 + \lambda x \qquad (A18)$$

The *fixed points* (steady states) are given by

$$-x_s^3 + \lambda X_s = 0 \qquad (A19)$$

This equation has three solutions: x_0 and $x_{\pm}$. Apart from the trivial solution $x_0 = 0$, the other two solutions are given by

$$-x_s^2 + \lambda = 0 \qquad (A20)$$

Only positive λ gives meaningful solutions of this equation:

$$x_{\pm} = \sqrt{\lambda} \qquad (A21)$$

The solution $x_0 = 0$ is independent of λ, and the other two solutions correspond to two distinct branches of the plot of x_s against λ. Thus, for $\lambda \geq 0$ the horizontal curve bifurcates into two branches. This is known as *pitchfork bifurcation* (cf. Stewart 1982).

The fixed point x_0 is globally asymptotically stable for $\lambda < 0$, and unstable for $\lambda > 0$. And x_+ and x_- are asymptotically but not globally stable.

At the bifurcation point $\lambda = 0$, the solutions $x_{\pm}$ cannot be expanded as a power series of this control parameter; this point is a *singularity*.

Thus, multiple, simultaneously stable, solutions can exist for nonlinear systems described by some very simple mathematical models. The example discussed here demonstrates the ability of the system to bifurcate or to switch to perform regulatory tasks.

We have so far discussed only a one-dimensional phase space, in which the phase trajectories can be only straight half-lines, converging to or diverging from the fixed points. Much more flexibility of behaviour becomes available in two-dimensional phase space. Features such as *periodic attractors* and *limit cycles* become possible in such a phase space.

The complexity of possible nonlinear behaviour increases enormously as we go to still higher-dimensional phase spaces. Particularly interesting is the existence of *strange attractors*. The property of asymptotic stability of dissipative systems allows the possibility of attracting *chaos*: chaos becomes the rule, rather than the exception, in such systems, as trajectories emanating from certain parts of phase space are inevitably drawn towards the strange attractor (Crutchfield, Farmer and Packard 1986; Grebogi, Ott and Yorke 1987; Gleick 1987; Kaye 1993; Ott and Spano 1995). We discuss chaos in a separate appendix.

A10.   Symmetry of composite systems

Suppose we have two systems described by symmetry groups G_1 and G_2. The systems could, for example, be the two phases constituting a biphasic composite material. Or we could have a crystal of symmetry G_1 on which a field of symmetry G_2 has been applied. What is the net symmetry of the composite system in each of these examples?

Common sense tells us that if some symmetry operation is present in both G_1 and G_2, then it would be present in the composite system also. And if a symmetry operation is present only in G_1 or only in G_2, but not in both, then it would not be present as a symmetry of the composite system. In other words, only the *common* symmetry elements can survive when G_1 and G_2 are superimposed.

This fact is embodied in what is called the *Curie principle of superposition of symmetries*, which states that if two or more symmetries $(G_1, G_2, G_3, \dots)$ are superimposed, then the symmetry group (G_d) of the composite system has only those elements that are common to all the superimposed groups. Mathematically this is expressed by writing G_d as the *intersection group* of $G_1, G_2, G_3, \dots$:

$$G_d = G_1 \cap G_2 \cap G_3 \dots \tag{A22}$$

It is clear from this that G_d cannot be a higher symmetry than the component symmetries G_1, G_2, G_3, etc.; it can at the most be equal to any of them:

$$G_d \subseteq G_i, \quad i = 1, 2, 3, \dots \tag{A23}$$

In the appendix on composites, we discuss an example of how this lowering of symmetry, when two phases are superimposed in a composite material, makes possible the occurrence of the magnetoelectric effect.

Equation A23, which is a corollary of the Curie principle, embodies a very important theorem of crystal physics, called the *Neumann theorem*. Before we state the
theorem, let us assume that G_1, G_2, G_3, ... denote the symmetry groups for the
various *macroscopic* physical properties of a crystal. Since all these properties
occur in the same crystal, G_d can be identified with the point-group symmetry
of the crystal. The Neumann theorem simply states that the symmetry G_i possessed by any macroscopic physical property of the crystal cannot be lower than
the point-group symmetry of the crystal; it must be at least equal to it, if not higher
(eqn A23).

The subscript d in eqn A22 stands for *dissymmetrization* or symmetry-lowering.
In general, the symmetry is indeed lowered when we superimpose two or more
symmetries.

But there can be exceptions when we superimpose 'equal' objects in certain special
ways. The different domain types in a specimen of a ferroic material are an example
of equal objects. Each domain type has the same crystal structure as any other domain
type; only their mutual positions or orientations are different.

We consider here a simpler (geometrical) example of equal objects to illustrate
how the symmetry of a composite object formed from them can be *higher* than G_d.

Consider a rhombus (Fig. A1a), with one of its diagonals horizontal (parallel to the
x-axis); the other diagonal will naturally be vertical (parallel to the y-axis). Either of
them divides the rhombus into two equal parts, each part being an isosceles triangle.
Let us choose the vertical diagonal for this purpose.

We can view the rhombus as a composite object, formed by combining the two
equal triangles along the vertical diagonal. Each triangle has the same symmetry,
namely a mirror plane (or line) m_y perpendicular to the y-axis. Therefore, in the

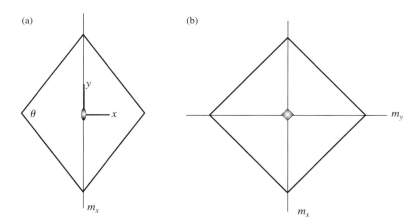

Fig. A1   Formation of a composite object (rhombus) from two equal isosceles triangles, having an apex
angle θ (a). The rhombus becomes a square when $\theta = 90°$ (b), even when there is no change in the
symmetries of the two isosceles triangles from which the square is constituted.

notation of eqn A22,

$$G_1 = G_2 = (1, m_y) \tag{A24}$$

Then

$$G_d = G_1 \cap G_2 = (1, m_y) \cap (1, m_y) = (1, m_y) \tag{A25}$$

But the actual symmetry (say G_s) of the composite object (the rhombus) is higher than this: There is present an additional mirror symmetry (m_x).

The symmetry group G_s can be obtained as an *extended group* from G_d:

$$G_s = G_d \cup M \tag{A26}$$

This generalization of the original Curie principle was suggested by Shubnikov, and eqn A26 is a statement of the *Curie–Shubnikov principle* of superposition of symmetries (cf. Wadhawan 2000).

M in eqn A26 is a *symmetrizer*. If the objects superimposed are unequal (imagine two unequal isosceles triangles superimposed as described above), then m_x is not a symmetry operation of the composite, and M is just an identity operation. But for the example of the rhombus discussed here,

$$M = (m_x, i) \tag{A27}$$

This can be rewritten as

$$M = (1, m_y)m_x = G_d m_x \tag{A28}$$

Substituting eqn A27 into eqn A26 we get

$$G_s = (1, m_y) \cup (m_x, i) = m_x m_y \tag{A29}$$

which describes correctly the symmetry of the rhombus.

Latent symmetry

As first discussed by Wadhawan (2000), a very interesting and important situation develops as we vary the apex angle, say θ, of the two equal isosceles triangles combined to construct the composite object, namely the rhombus. For $\theta = 90°$ the rhombus becomes a square (Fig. A1b), and then even eqn A29, which is a generalization of the original Curie principle, fails to give correctly the symmetry group for the square. We had introduced only one symmetrizer ($M = G_d m_x$) to explain the symmetry of the rhombus. More symmetrizers must be introduced to form a still larger extended group G_s which can describe the symmetry of the square (which has a four-fold axis of symmetry among its symmetry elements). The general definition of the symmetrizer is therefore as follows:

$$M = G_d g_2 \cup G_d g_3 \cup G_d g_4 \cup \ldots \tag{A30}$$

There is, however, another way of looking at what has happened here (Wadhawan 2000). Suppose we start with two equal *right-angled* isosceles triangles. For them,

$G_1 = G_2 = (1, m_y)$, as before; the fact that $\theta = 90°$ makes no difference to G_1 and G_2. There is no four-fold symmetry axis either in G_1 or in G_2, or in the recipe used for forming the composite (the square) from the two component triangles. Yet the four-fold axis does arise when the composite is formed. Wadhawan (2000) called the four-fold axis an example of *latent symmetry*. It is as if the four-fold axis lies dormant (i.e. is not manifest) in the symmetry of the two identical right-angled triangles, and manifests itself only when the composite is formed.

A formal group-theoretical treatment of this concept has been given by Litvin and Wadhawan (2001, 2002). Consider an object A of symmetry H. We can construct a composite object S from A by applying on it a set of transformations ('isometries') $\{g_1 = 1, g_2, \ldots g_m\}$:

$$S = \{A, g_2 A, g_3 A, \ldots g_m A\} \qquad (A31)$$

There is not much loss of generality if we assume that the isometries involved constitute a group, say G:

$$G = \{1, g_2, g_3, \ldots g_m\} \qquad (A32)$$

Latent symmetry, by definition (Litvin and Wadhawan 2002), is any symmetry of the composite S that is not a product of the operations or isometries of G and H.

A *partition theorem* has been proved in Litvin and Wadhawan (2002), which provides a sufficient condition that an isometry is a symmetry of a composite constructed from a component A by a set of isometries which constitute a group G.

A11.   Tensor properties

In this appendix we describe some basic notions about tensors, and introduce the tensor properties relevant to the subject matter of this book.

The properties of a material are specified as relationships between measurable quantities. For example, we can measure the mass m and the volume V of a specimen of a material, and it is the density ρ which connects these two measurables:

$$m = \rho V \qquad (A33)$$

Both m and V in this equation are scalars; they can be specified completely in terms of single numbers (in appropriate units), and therefore ρ also can be described completely by a single number. It is a *scalar property* of the material.

Tensors of rank 1

Let us consider a material (a single crystal, to be specific) which exhibits what is called the *pyroelectric effect*. It can occur in crystalline materials belonging to any of the 10 polar classes. The point-group symmetries of these classes of crystals are such that there is a direction in them (called the *polar axis*), along which there occurs a nonzero electric polarization or dipole moment, even when no electric field has been applied. Thus such a polarization is *spontaneous*, rather than *induced* by an external field.

The spontaneous polarization, naturally, varies with temperature. Let us say that a change ΔT in temperature results in a change $\mathbf{P}(P_1, P_2, P_3)$ in the polarization; the components P_1, P_2, P_3 of $\mathbf{P}$ are with reference to a Cartesian system of axes. We can express this relationship between ΔT and P_i as follows:

$$P_i = p_i \Delta T \tag{A34}$$

Here the proportionality constant p_i is a component of what is called the *pyroelectric tensor*.

(p_i) is a vector. All its three components must be specified for defining it completely, unlike the case of density, which requires only one number for a complete specification.

Both (P_i) and (p_i) are vectors. Another example of a vector is the position vector of a point in space: It is the straight line from the origin to the point in question. Such a line is defined by the three coordinates of the point: (x_1, x_2, x_3) or (x_i).

Suppose we make a transformation of the coordinate axes by rotating the reference frame in a general way. Then the coordinates (x_i) will change to, say, (x_i'). It should be possible to calculate the new coordinates in terms of the old ones:

$$x_i' = \sum_j a_{ij} x_j, \quad i = 1, 2, 3 \tag{A35}$$

One usually follows the convention that if some index is repeated on any of the sides of an equation (as j is on the RHS in the above equation), a summation over that index is implicitly there, and need not be written explicitly. So we can rewrite eqn A35 as

$$x_i' = a_{ij} x_j, \quad i = 1, 2, 3 \tag{A36}$$

Since (p_i) is a vector, just like the position vector (x_i), its components will change under a coordinate transformation according to an equation similar to eqn A36:

$$p_i' = a_{ij} p_j \tag{A37}$$

It should be noted that, since there is a summation over the index j, it does not matter what symbol we use for this index; the following equation says the same thing as the above equation:

$$p_i' = a_{im} p_m \tag{A38}$$

We say that (p_i) is a *tensor of rank* 1; this is because its components transform (under a general rotation of coordinate axes) like the components of a single position vector (x_i).

Tensors of rank 2

We next consider an example of a tensor property of rank 2. We discuss the response of an insulating or dielectric material to a small electric field (E_i). The applied field results in an electric displacement (D_j). There is no reason to presume that the vector (D_j) should be parallel to the vector (E_i). In general,

$$D_1 = \varepsilon_{11}E_1 + \varepsilon_{12}E_2 + \varepsilon_{13}E_3 \tag{A39}$$

Here the proportionality constants are measures of the dielectric permittivity response of the material. Similar equations can be written for D_2 and D_3. The three equations, representing the overall dielectric response of the material, can be written compactly as follows:

$$D_i = \varepsilon_{ij}E_j, \quad i = 1, 2, 3 \tag{A40}$$

The proportionality constants in this equation are components of the *dielectric permittivity tensor*. Since $i = 1, 2, 3$ and $j = 1, 2, 3$, there are nine such components in all, compared to just three for a vector, or tensor of rank 1.

What is the rank of this permittivity tensor? To answer this question, we have to see how the components of this tensor behave under a coordinate transformation. We can write equations similar to eqn A40 for the new frame of reference:

$$D'_i = \varepsilon'_{ij}E'_j, \quad i = 1, 2, 3 \tag{A41}$$

Since (E_i) and (D_i) are vectors,

$$E'_i = a_{ij}E_j \tag{A42}$$

$$D'_i = a_{ij}D_j \tag{A43}$$

Substituting from eqn A40 into eqn A43,

$$D'_i = a_{ij}\varepsilon_{jk}E_k \tag{A44}$$

Equation A42 can be inverted to get

$$E_k = a_{lk}E'_l \tag{A45}$$

Substituting this into eqn A44 we get

$$D'_i = a_{ij}a_{lk}\varepsilon_{jk}E'_l \tag{A46}$$

Comparing eqns A46 and A41,

$$\varepsilon'_{il} = a_{ij}a_{lk}\varepsilon_{jk} \tag{A47}$$

This result tells us that, when a coordinate transformation like rotation, reflection or inversion is carried out, the permittivity tensor transforms as a product of two position vectors or two coordinates. It is therefore called *a tensor of rank* 2, or a *second rank tensor*.

Any physical quantity which transforms according to eqn A47 is a tensor of rank 2. Its components are represented by *two* indices, unlike only one index needed for specifying the components (v_i) of a vector, or tensor of rank 1.

This can be generalized. A tensor of rank n, by definition, transforms as a product of n vectors or n coordinates, and therefore requires n indices for specifying its components.

Another example of a tensor of rank 2 is the magnetic permeability tensor (μ_{ij}):

$$B_i = \mu_{ij} H_j \tag{A48}$$

The dielectric permeability and magnetic permeability tensors are examples of *matter tensors*.

An example of a second-rank *field tensor*, is the stress tensor. Stress is defined as force (f_i) per unit area (A_i):

$$f_i = \sigma_{ij} A_j \tag{A49}$$

Being a field tensor, the stress tensor is independent of the crystal on which the stress is applied. By contrast, the strain tensor (e_{ij}), which is also a second rank tensor, is a *matter tensor*, and is influenced by the symmetry of the crystal (see below):

$$\Delta l_i = e_{ij} l_j \tag{A50}$$

Tensors of rank 3

The piezoelectric tensor (d_{ijk}) is the most familiar example of a tensor property of rank 3. It transforms as a product of three vectors:

$$d'_{ijk} = a_{il} a_{jm} a_{kn} d_{lmn} \tag{A51}$$

In a piezoelectric material, there is a coupling of electric polarization and mechanical strain. In the so-called *direct piezoelectric effect* (which can occur in a crystal belonging to any of the 20 piezoelectric crystal classes), application of stress produces electric polarization:

$$P_i = d_{ijk} \sigma_{jk} \tag{A52}$$

The *inverse* piezoelectric effect pertains to the development of strain when an electric field is applied to the piezoelectric crystal:

$$e_{ij} = d_{ijk} E_k \tag{A53}$$

In a smart material based on the piezoelectric effect, both the direct and the inverse piezoelectric effect can be involved: The direct effect is used for sensing, and this is followed by feedback through the inverse effect, which provides actuation through the strain developed.

As can be seen from eqns A49 and A50, the stress tensor and the strain tensor are *symmetric* tensors: $\sigma_{ij} = \sigma_{ji}$, and $e_{ij} = e_{ji}$. This symmetry is carried over to the piezoelectric tensor (eqn A52): $d_{ijk} = d_{ikj}$. Because of this 'intrinsic' symmetry, the piezoelectric tensor has only 18 independent components, instead of 27. It is therefore customary to denote it as $(d_{i\mu})$, with $i = 1, 2, 3$ and $\mu = 1, 2, \ldots, 6$. For example, for a stress applied along the x_3-axis, the polarization component P_3 along the same direction is determined by the d_{33} coefficient: $P_3 = d_{33}\sigma_3$.

The d-tensor discussed so far is the *piezoelectric charge tensor*: it determines the charge separation produced by the applied stress. It is often relevant to deal with what is called the *piezoelectric voltage tensor* (g). It determines the open-circuit voltage generated by the applied stress. The two tensors are related via the permittivity tensor:

$$g = d/\varepsilon \qquad (A54)$$

Hydrophones, used by the Navy in large numbers, have to sense very weak hydrostatic pressure (p). For such a situation, $\sigma_{11} = \sigma_{22} = \sigma_{33} = -p$. Equation A52 then yields

$$P_3 = d_{31}(-p) + d_{32}(-p) + d_{33}(-p) \qquad (A55)$$

Equating d_{31} with d_{32}, and introducing the symbol d_h to represent the hydrostatic piezoelectric charge coefficient, we can write eqn A55 as

$$P_3 = (d_{33} + 2d_{31})(-p) \equiv d_h(-p) \qquad (A56)$$

The hydrophone usually feeds into a very high-impedance load. Therefore, the voltage generated by the hydrophone can be written as

$$g_h = (d_{33} + 2d_{31})/\varepsilon_{33} \qquad (A57)$$

Tensors of rank 4

The piezoelectric effect described above is a *linear* effect: Reversal of the electric field reverses the mechanical deformation (cf. eqn A53). There is also a *quadratic* effect, determined by the fourth-rank *electrostriction tensor* (M_{ijkl}):

$$e_{ij} = M_{ijkl}E_k E_l \qquad (A58)$$

In this case, reversal of the sign of the electric field has no effect on the (electrostrictive) strain.

Another tensor of rank 4 is the *magnetostriction tensor*:

$$e_{ij} = Q_{ijkl}H_k H_l \qquad (A59)$$

Lastly, we consider the more familiar example of a fourth-rank tensor, namely the *elastic compliance tensor*, which determines the strain produced by the application of a small stress (Hooke's law):

$$e_{ij} = s_{ijkl}\sigma_{kl} \tag{A60}$$

The inverse of this tensor is the *elastic stiffness tensor*:

$$\sigma_{ij} = c_{ijkl}e_{kl} \tag{A61}$$

Effect of crystal symmetry on tensor properties

Matter tensors have to conform to the symmetry of the matter they describe. For crystals, the relevant symmetry is the point-group symmetry. It is not necessary to consider the full space-group symmetry for this purpose because the directional symmetry of macroscopic tensor properties is not influenced by crystallographic translational operations.

A macroscopic tensor property of a crystal cannot have less directional symmetry than the point-group symmetry of the crystal (Neumann theorem). For example, if the crystal has inversion symmetry, i.e. if its point group is one of the 11 Laue groups, then only those tensor properties can be nonzero which have inversion symmetry among the elements of their symmetry group. Let us take a look at how the properties of various ranks described above behave under an inversion operation.

We see from eqn A33 that, since both m and V remain the same under an inversion operation, so does the density ρ. This is true for all scalar or zero-rank tensor properties.

Next, we refer to eqn A34. Under an inversion operation, ΔT remains the same, but P_i changes to $-P_i$ for all i. Therefore, (p_i) changes to $(-p_i)$. Since the inversion operation is a symmetry operation, we must have $-(p_i) = (p_i)$ for each i. This is possible only if all the components of the pyroelectric tensor, a tensor of rank 1, are identically equal to zero.

Equation A40 describes a tensor of rank 2. Under inversion, since both (E_i) and (D_i) change signs, the permittivity tensor, a tensor of rank 2, remains invariant.

Lastly, we consider a tensor property of rank 3 (eqn A52). Like permittivity, the stress tensor in eqn A52 is also a tensor of rank 2. Therefore it remains invariant under an inversion operation. By contrast, the polarization tensor in the same equation, being a tensor of rank 1 (like the pyroelectric tensor) changes sign under inversion. These two facts can be consistent only if the piezoelectric tensor in eqn A52 is equal to its own negative, which means that all its components must be identically equal to zero.

We can now generalize. All tensor properties of odd rank are absent in crystals having inversion symmetry, and all even-rank tensor properties are permitted by this symmetry to exist in such crystals.

Likewise, we can investigate the effect of other directional or point-group symmetry elements present in a crystal on the tensor properties.

References

Abarbanel, H. D. I. (2006). 'Physics of chaotic systems'. In G. Fraser (ed.), *The New Physics for the Twenty-First Century*. Cambridge, U. K.: Cambridge University Press, p. 311.

Angell, C. A. (31 March 1995). 'Formation of glasses from liquids and biopolymers'. *Science*, 267: 1924.

Aziz-Alaoui, M. A. (2005). 'Synchronization of chaos', in J.-P. Francoise, G. L. Naber and T. S. Tsun (eds.), *Encyclopaedia of Mathematical Physics*, Vol. 5, p. 213. Amsterdam: Elsevier.

Bauer, S., R. Gerhard-Multhaupt and G. M. Sessler (February 2004). 'Ferroelectrets: Soft electroactive foams for transducers'. *Physics Today*, p. 37.

Beetz, C. P. (1992). 'Composite materials', in G. L. Trigg (ed.), *Encyclopedia of Applied Physics*, Vol. 4. New York: VCH Publishers.

Cammarata, R. C. (2004). 'Nanocomposites'. In Di Ventra, M., S. Evoy and J. R. Heflin (eds.), *Introduction to Nanoscale Science and Technology*. Dordrecht: Kluwer.

Chung, D. D. L. (2001). *Applied Materials Science*. London: CRC Press.

Clearwater, S. H. (1991). 'Artificial intelligence'. In G. L. Trigg (ed.), *Encyclopedia of Applied Physics*, Vol. 2, p. 1. New York: VCH Publishers.

Crutchfield, J. P., J. D. Farmer and N. H. Packard (1986). 'Chaos'. *Scientific American,* 254(12): 46.

Debenedetti, P. G. and F. H. Stillinger (8 March 2001). 'Supercooled liquids and the glass transition'. *Nature*, 410: 259.

Ditto, W. L. and L. M. Pecora (Aug. 1993). 'Mastering chaos'. *Scientific American*, 269: 62.

Ditto, W. L., S. N. Rauseo and M. L. Spano (1990). 'Experimental control of chaos'. *Phys. Rev. Lett.* 65: 3211.

Donth, E. (2001). *The Glass Transition*. Berlin: Springer.

Fulcher, G. S. (1925). 'Analysis of recent measurements of the viscosity of glasses'. *J. Amer. Ceram. Soc.* 8: 339.

Garfinkel, A., M. L. Spano, W. L. Ditto and J. N. Weiss (28 Aug. 1992). 'Controlling cardiac chaos'. *Science*, 257: 1230.

Gilmore, R. (2005). 'Chaos and attractors', in J.-P. Francoise, G. L. Naber and T. S. Tsun (eds.), *Encyclopaedia of Mathematical Physics*, Vol. 1, p. 477. Amsterdam: Elsevier.

Gleick, J. (1987). *Chaos: Making a New Science*. New York: Viking Penguin.

Grebogi, C., E. Ott and J. A. Yorke (1987). 'Chaos, strange attractors, and fractal basin boundaries in nonlinear dynamics'. *Science* 238: 632.

Hale, D. K. (1976). 'The physical properties of composite materials'. *J. Mater. Sci.*, 11: 2105.

Hansen, J. S. (1995). 'Introduction to advanced composite materials', in Udd, E. (ed.), *Fibre Optic Smart Structures*. New York: Wiley.

Hawkins, J. and S. Blakeslee (2004). *On Intelligence*. New York: Times Books (Henry Holt).

Kauffman, S. A. (1993). *The Origins of Order*. Oxford: Oxford University Press.

Kaye, B. (1993). *Chaos and Complexity: Discovering the Surprising Patterns of Science and Technology*. Weinheim: VCH.

Langreth, R. (1992). 'Engineering dogma gives way to chaos'. *Science*, 252: 776.

Lederman, L. M. and C. T. Hill (2005). *Symmetry and the Beautiful Universe*. New York: Prometheus Books.

Litvin, D. B. and V. K. Wadhawan (2001). 'Latent symmetry and its group-theoretical determination'. *Acta Cryst.* A57: 435.

Litvin, D. B. and V. K. Wadhawan (2002). 'Latent symmetry'. *Acta Cryst.* A58: 75.

Maguire, J. F., M. Benedict, L. V. Woodcock and S. R. LeClair (2002). 'Artificial intelligence in materials science: Application to molecular and particulate simulations'. In Takeuchi, I., J. M. Newsam, L. T. Willie, H. Koinuma and E. J. Amis (eds.), *Combinatorial and Artificial Intelligence Methods in Materials Science*. MRS Symposium Proceedings, Vol. 700. Warrendale, Pennsylvania: Materials Research Society.

McCulloch, W. and W. Pitts (1943). 'A logical calculus of the ideas immanent in nervous activity'. *Bull. Math. Biophys.* 5: 115.

Miracle, D. B. and L. Donaldson (2001). 'Introduction to composites', in D. B. Miracle and L. Donaldson (eds.), *Composites*, Vol. 21 of *ASM Handbook*. Materials Park, Ohio 44073-0002, USA: The Materials Information Society.

Moravec, H. (1988). *Mind Children: The Future of Robot and Human Intelligence*. Cambridge: Harvard University Press.

Mullins, J. (23 April 2005). 'Whatever happened to machines that think?'. *New Scientist*, p. 32.

Nalwa, H. S. (ed.) (1995). *Ferroelectric Polymers: Chemistry, Physics, and Applications*. New York: Marcel Dekker.

Newell, A. (1983). 'Intellectual issues in the history of artificial intelligence'. In Machlup, F. and U. Mansfield (eds.), *The Study of Information: Interdisciplinary Messages*. New York: Wiley.

Newnham, R. E. (1986). 'Composite electroceramics'. *Ann. Rev. Mater. Sci.* 16: 47.

Newnham, R. E., D. P. Skinner and L. E. Cross (1978). 'Connectivity and piezoelectric–pyroelectric composites'. *Mat. Res. Bull.*, 13: 525.

Newnham, R. E. and S. E. Trolier-McKinstry (1991a). 'Crystals and composites' *J. Appl. Cryst.*, 23: 447.

Newnham, R. E. and S. E. Trolier-McKinstry (1990b), "Structure–property relationships in ferroic nanocomposites". *Ceramic Transactions*, 8: 235.

Newnham, R. E., J. F. Fernandez, K. A. Murkowski, J. T. Fielding, A. Dogan and J. Wallis (1995). 'Composite piezoelectric sensors and actuators'. In George, E. P., S. Takahashi, S. Trolier-McKinstry, K. Uchino and M. Wun-Fogle (eds.), *Materials for Smart Systems*. MRS Symposium Proceedings, Vol. 360. Pittsburgh, Pennsylvania: Materials Research Society.

Nicolis, G. and I. Prigogine (1989). *Exploring Complexity: An Introduction*. New York: W. H. Freeman.

Ott, E., C. Grebogi and J. A. Yorke (1990). 'Controlling chaos'. *Phys. Rev. Lett.* 64: 1196.

Ott, E. and M. Spano (1995). 'Controlling chaos'. *Physics Today*, 48(5): 34.

Pecora, L. and T. Carroll (1990). 'Synchronisation in chaotic systems'. *Phys. Rev. Lett.*, 64: 821.

Pecora, L. and T. Carroll (1991). 'Driving systems with chaotic signals'. *Phys. Rev. A*, 44: 2374.

Pilgrim, S. M., R. E. Newnham and L. L. Rohlfing (1987). 'An extension of the composite nomenclature scheme'. *Mat. Res. Bull.*, 22: 677.

Pillai, P. K. C. (1995). 'Polymer electrets'. In Nalwa, H. S. (ed.) (1995), *Ferroelectric Polymers: Chemistry, Physics, and Applications*. New York: Marcel Dekker.

Prigogine, I. (1998). *The End of Certainty: Time, Chaos, and the New Laws of Nature*. New York: Free Press.

Stewart, I. N. (1982). 'Catastrophe theory in physics'. *Rep. Prog. Phys.* 45: 185.

Stix, G. (March 2006). 'The elusive goal of machine translation'. *Scientific American,* 294: 70.

Tagantsev, A. K. (1994). 'Vogel–Fulcher relationship for the dielectric permittivity of relaxor ferroelectrics'. *Phys. Rev. Lett.* 72: 1100.

Takeuchi, I., J. M. Newsam, L. T. Willie, H. Koinuma and E. J. Amis (eds.) (2002). *Combinatorial and Artificial Intelligence Methods in Materials Science*. MRS Symposium Proceedings, Vol. 700. Warrendale, Pennsylvania: Materials Research Society.

Tirnakli, U., F. Buyukkilic and D. Demirhan (1999). 'A new formalism for nonextensive physical systems: Tsallis thermostatistics', *Tr. J. Phys.*, 23: 21.

Tsallis, C. (1988). 'Possible generalizations of Boltzmann–Gibbs statistics'. *J. Stat. Phys.* 52: 479.

Tsallis, C. (1995a). 'Some comments on Boltzmann–Gibbs statistical mechanics'. *Chaos, Solitons and Fractals*, 6: 539.

Tsallis, C. (1995b). 'Non-extensive thermostatistics: brief review and comments'. *Physica A*, 221: 277.

Tsallis, C. (July 1997). 'Levy distributions'. *Physics World*: p. 42.

Van Suchtelen (1972). 'Product properties: A new application of composite materials'. *Philips Research Reports*, 27: 28.

Vogel, H. (1921). 'Das temperatur-abhangigkeitsgesetz der viskositat von flussigkeiten'. *Phys. Zeit.* 22: 645.

Wadhawan, V. K. (2000). *Introduction to Ferroic Materials*. Amsterdam: Gordon and Breach.

Wiener, N. (1965), 2nd edition. *Cybernetics: Or Control and Communication in Animal and the Machine*. Cambridge MA: MIT Press.

FURTHER READING

Moravec, H. (1988). *Mind Children: The Future of Robot and Human Intelligence*. Harvard: Harvard University Press.

Nicolis, G. and I. Prigogine (1989). *Exploring Complexity: An Introduction*. New York: W. H. Freeman.

Drexler, K. E. (1992). *Nanosystems: Molecular Machinery, Manufacturing and Computation*. New York: Wiley.

Kelly, K. (1994). *Out of Control: The New Biology of Machines, Social Systems, and the Economic World*. Cambridge: Perseus Books.

Brignell, J. and N. White (1994). *Intelligent Sensor Systems*. Bristol: Institute of Physics Publishing.

Lehn, J.-M. (1995). *Supramolecular Chemistry: Concepts and Perspectives*. New York: VCH.

Banks, H. T., R. C. Smith and Y. Wang (1996). *Smart Material Structures: Modeling, Estimation, and Control*. New York: Wiley.

Bak, P. (1996). *How Nature Works: The Science of Self-Organized Criticality*. New York: Springer.

Dyson, G. B. (1997). *Darwin Among the Machines: The Evolution of Global Intelligence*. Cambridge: Perseus Books.

Otsuka, K. and C. M. Wayman (eds.) (1998). *Shape Memory Materials*. Cambridge: Cambridge University Press.

Holland, J. H. (1998). *Emergence: From Chaos to Order*. Cambridge, Massachusetts: Perseus Books.

Prigogine, I. (1998). *The End of Certainty: Time, Chaos, and the New Laws of Nature*. New York: Free Press.

Kurzweil, R. (1998). *The Age of Spiritual Machines: When Computers Exceed Human Intelligence*. New York: Viking Penguin.

Moravec, H. (1999). *Robot: Mere Machine to Transcendent Mind*. Oxford: Oxford University Press.

Janocha, H. (1999) (ed.). *Adaptronics and Smart Structures: Basics, Materials, Design, and Applications*. Berlin: Springer.

Heudin, J.-C. (1999) (ed.). *Virtual Worlds: Synthetic Universes, Digital Life, and Complexity*. Reading, Massachusetts: Perseus Books.

Wadhawan, V. K. (2000). *Introduction to Ferroic Materials*. Amsterdam: Gordon and Breach.

Kauffman, S. A. (2000). *Investigations*. Oxford: Oxford University Press.

Whitesides, G. M. (Sept. 2001). 'The once and future nanomachines'. *Scientific American,* 285: 70.

Srinivasan, A. V. and D. M. McFarland (2001). *Smart Structures: Analysis and Design*. Cambridge: Cambridge University Press.

Varadan, V. K., X. Jiang and V. V. Varadan (2001). *Microstreolithogrphy and Other Fabrication Techniques for 3D MEMS*. New York: Wiley.

Wolfram, S. (2002). *A New Kind of Science*. Wolfram Media Inc. (for more details about the book, cf. www.wolframscience.com).

Sipper, M. (2002). *Machine Nature*. New Delhi: Tata McGraw-Hill.

Lehn, J.-M. (16 April 2002). 'Toward complex matter: Supramolecular chemistry and self-organization'. *PNA, USA*, 99(8): 4763.

Kleman, M. and O. D. Lavrentovich (2003). *Soft Matter Physics: An Introduction*. New York: Springer-Verlag.

Mainzer, K. (2004). *Thinking in Complexity: The Computational Dynamics of Matter, Mind, and Mankind* (4th edition). Berlin: Springer-Verlag.

Hawkins, J. and S. Blakeslee (2004). *On Intelligence*. New York: Times Books.

Di Ventra, M., S. Evoy and J. R. Heflin (eds.) (2004). *Introduction to Nanoscale Science and Technology*. Dordrecht: Kluwer.

Waser, R. (ed.) (2005). *Nanoelectronics and Information Technology: Advanced Electronic Materials and Novel Devices*, 2nd edition. KGaA: Wiley-VCH Verlag

Atkinson, W. I. (2005). *Nanocosm: Nanotechnology and the Big Changes Coming from the Inconceivably Small*, 2nd edition. New York: AMACOM: American Management Association.

Kelsall, R., I. Hamley and Geoghegan (eds.) (2005). *Nanoscale Science and Technology*. London: Wiley.

Kurzweil, R. (2005). *The Singularity is Near: When Humans Transcend Biology*. New York: Viking Adult.

Konar, Amit (2005). *Computational Intelligence: Principles, Techniques and Applications*. Berlin: Springer.

Singh, J. (2005). *Smart Electronic Materials: Fundamentals and Applications*. Cambridge: Cambridge University Press.

Raymond, K. W. (2006). *General, Organic, and Biological Chemistry: An Integrated Approach*. New Jersey: Wiley.

Zomaya, A. Y. (ed.) (2006). *Handbook of Nature-Inspired and Innovative Computing: Integrating Classical Models with Emerging Technologies*. New York: Springer.

Varadan, V. K., K. J. Vinoy and S. Gopalakrishnan (2006). *Smart Material Systems and MEMS: Design and Development Methodologies*. Chichester: Wiley.

Alon, U. (2007). *An Introduction to Systems Biology: Design Principles of Biological Circuits*. London: Chapman and Hall/CRC.

GLOSSARY

(Most of the definitions are taken from standard works. The book *Lexicon of Complexity* by F. T. Arecchi and A. Farini (1996) (published by Istituto Nazionale di Ottica, Florence) is a particularly valuable source for some carefully worded definitions.)

Action potential. The all-or-none electrical pulse that travels down an axon of a neuronal cell, normally from the cell body to the many synapses at the far end of the axon. Such pulses allow the transmission of information within nerves.

Actuator. A device that converts other forms of energy into mechanical energy or movement. A motion-generating device.

Adaptive structure. A structure that is able to cope successfully with its environment. Adaptation involves two components: assimilation and accommodation.

Algorithm. A formal computational rule.

Amino acid. A molecule with one or more carboxyl groups and one or more amino groups. Amino acids are building blocks for peptides and proteins.

Amphiphile. A molecule with the property that a part of it has affinity towards water, and another part that is repelled by water. Soap is an example. It consists of a hydrophilic (water-loving) head which is usually ionic, and a hydrophobic (water-hating) tail which is typically a hydrocarbon chain.

Amygdala. An almond-shaped structure in the brain that acts as an anatomical link between parts of the brain which perceive and the parts which govern emotions.

Analogue. A device using smoothly varying physical quantities to represent reality.

Android. A robot with human appearance.

Anticodon. A sequence of three bases in tRNA that is complementary to a codon on mRNA.

Artificial intelligence (AI). A field of enquiry aiming to reproduce the *behavioural* aspect of human intelligence in machines by means of algorithms. It involves representation of knowledge, control or strategy of procedures, and searching through a problem space, all directed towards the goal of solving problems.

Artificial life (AL). The study of man-made systems that exhibit behaviours characteristic of natural living systems, such as self-organization, reproduction, development, and evolution.

Artificial reality. Modelling and designing of worlds that do not exist, and to display them in three dimensions.

Assembler. Drexler (1992): Any programmable nanomechanical system able to perform a wide range of mechanosynthetic operations.

Associative memory. The ability of a neural net to go from one internal representation to another, or to infer a complex representation from a portion of it. Such a memory is *content addressable*.

(The) 'Astonishing' hypothesis. (Crick 1994): The hypothesis that a person's mental activities are *entirely* due to the behaviour of nerve cells, glial cells, and the atoms, ions, and molecules that make them up and influence them.

Atomic force microscope (AFM). Also called the scanning force microscope (SFM). A scanning probe microscope in which the property sensed as a function of position on the specimen is the force of interaction between the probing tip and the surface of the specimen. This very high-power microscope can image insulating or semiconducting surfaces directly, in ambient atmosphere, or even in a liquid.

Attractor. Final state of stability of a dissipative system. A geometric form in phase space that characterizes the long-term behaviour of a dissipative system. It is what the behaviour of the system settles down to, or is attracted to. For example, for a damped harmonic oscillator, the attractor is the point where the oscillator finally comes to rest. In this example, the attractor is a set of dimension zero. Similarly, an attractor that is a closed curve is a set of dimension unity. There can be *point* attractors, *periodic or quasi-periodic* attractors, and *chaotic* attractors. If a system on an attractor is given a small perturbation, it would tend to return toward it.

Augmented reality. Adding virtual information or objects which do not belong to the original scene; e.g. virtual objects included in real-time on to a live video.

Automata. Devices that convert information from one form to another according to a definite procedure.

Avatar. Incarnation.

Axon. The long signal-emitting extension of a neuron. A neuron usually has only one such output cable, although it often branches extensively.

Baldwin effect. A mechanism whereby individual learning can change the course of evolution. A few members of a population with better learning capabilities may have a higher probability of survival (say from predators), resulting in higher chances of their survival and procreation. As the proportion of such members grows, the population can support a more diverse gene pool, leading to quicker evolution, suited to meet that challenge.

Basin of attraction. The set of all initial conditions in the phase space of a dissipative system for which the phase trajectory moves asymptotically towards a particular attractor.

Bifurcation. A qualitative change in the topology of an attractor-basin portrait, and the attendant qualitative change in the dynamics, brought about by the quasi-static variation of the control parameter.

Bit. Contracted form of 'binary digit' (0 or 1).

Brownian motion. Motion of a particle in a fluid caused by thermal fluctuations.

Bush robot. Envisaged by Moravec, a bush robot is a fractal-branching ultra-dextrous robot of the future. It has 'a branched hierarchy of articulated limbs, starting from a macroscopically large trunk through successively smaller and more numerous branches, ultimately to microscopic twigs and nanoscale fingers'. Such a robot, with millions or billions of fingers, will be too complex to be controlled by humans. It will have to be an autonomous robot, backed by adequate computing power. In fact, the computational units will have to be small enough to be distributed throughout the bush, almost down to the individual fingers. Such a creature will be able to manipulate matter at nanometre length-scales.

Capillary interactions. A liquid is surrounded by a gas or a solid, or another liquid. In all cases, there are one or more interfaces, and the chemical bonding on one side of an interface is different from that on the other. Because of this, there is a surface energy or an interface energy per unit area of the interface. Capillary interactions are the forces that arise from the tendency of the system to minimize the interfacial area, so as to minimize the contribution of these interfaces to the overall free energy. For example, a drop of water or mercury in air takes a spherical shape, because such a shape has the least surface area.

Catalyst. A catalyst is a material which is not consumed in a reaction, but which alters (usually speeds up) the net rate of the reaction by taking part is some of its steps and getting released in the end. A *homogeneous* catalyst is dispersed in, and exists as, the same phase as the reactants (usually a gas mixture or a liquid solution). *Heterogeneous* catalysts are usually solids, which share an interface with the usually fluid reactants. Naturally, a large area of the interface can make a catalyst more efficient in speeding up the reaction. For this reason, nanostructured materials, including nanoporous materials, can serve as efficient catalysts for several types of chemical reactions.

Catastrophe. A sudden, discontinuous change in the state of a system. A system liable to undergo catastrophic change is described by a family of functions incorporating control parameters such that the number of equilibria changes as the control parameters are changed.

Cellular automata (CA). A class of computer algorithms, first introduced by John von Neumann in the late 1940s. They represent discrete dynamical systems, the evolution of which is dictated by local rules. They are usually realized on a lattice of cells (in a computer program), with a finite number of discrete states associated with each cell, and with a local rule specifying how the state of each cell is to be updated in discrete time steps. This approach involves approximations to partial differential equations, in which time, space, and also the signal, all take integral values only.

Ceramic. A compound of a metal and a nonmetal. An inorganic nonmetallic solid, usually consisting of small crystals ('grains') of a material which have been fused together by suitable heat-treatment such that the strength of chemical bonding between the grains is comparable to the strength of bonding within a grain. The majority of ceramics are oxide materials. This gives them a very high degree of permanence of existence because they are practically immune to corrosion.

Cerebellum. The lowermost part at the back of the brain, involved in the coordination of complex muscular movements.

Chaos. It is possible for certain simple deterministic systems, with only a few elements, to generate random behaviour. This randomness, which is fundamental, and which cannot go away even if more information is gathered, is called chaos.

Chaotic attractor. An attractor for a system with chaotic dynamics. Such a system exhibits exponentially sensitive dependence on initial conditions.

Chaotic system. That nonlinear deterministic system the behaviour of which depends sensitively on the initial conditions, although not every nonlinear system is chaotic. In the chaotic regime, the behaviour of a deterministic system appears random, although the randomness has an underlying geometrical form. In other words, chaos is deterministic; it is generated by fixed rules which do not themselves involve any elements of chance. Bak (1996) has emphasized that chaos theory cannot explain complexity.

Chromosome. A strand of DNA. A set of genes stringed together.

Church's thesis. According to this, 'what is human computable is universal-computer computable'. Thus it equates the information-processing capabilities of a human being with the intellectual capacity of a universal Turing machine.

Cocktail-party effect. The ability of the human brain to isolate and identify one voice in a crowd.

Coherence. Fixed phase relationship.

Colloidal dispersion. A heterogeneous system in which particles of dimensions less than 10 microns (either solids or droplets of liquid) are dispersed in a liquid medium.

Combinatorial chemistry. The science of using a combinatorial process of experiments for generating sets of molecular assemblies (compounds) from reversible connections between sets of building blocks. It aims at high-throughput synthesis and screening for identifying new useful compounds.

Complex system. Bak (1996): 'A system with large variability'. This variability exists in a self-similar way on a wide range of length-scales or other scales. Such a system usually consists of a large number of simple elements or agents, which interact with one another and the environment. According to Bak (1996), complexity is a consequence of criticality.

Computational complexity. The number of steps an algorithm has to take to get an answer.

Connectionism. The viewpoint in the theory of artificial neural networks that the complex connectivity of neurons required for problem-solving can be encoded in the local synapses and synaptic weights, and that it may not be generally necessary to move around the information over long paths.

Conservative system. A system (e.g. a frictionless pendulum) in which the dynamics is invariant with time reversal. A Hamiltonian system.

Content addressable memory. A memory very unlike the memory in a conventional computer. The latter uses symbolic logic, and the memory is stored in files with definite addresses, so the address must be specified to retrieve such memory. By contrast, in a connectionist system, a pattern of connections itself is the address of that pattern. Such a memory is content-addressable because the address is in that particular set of connections of the network. Such memory is flexible, fault-tolerant, and robust.

Control system. A goal-seeking system with non-goal-seeking components. The human brain is an example of this.

Copolymer. A polymer in which the fundamental monomer units are not all the same, but are two or more similar molecules.

Corpus callosum. A transverse band of nerve fibres that joins the two hemispheres of the brain and makes them function as a single unit.

Cortex. Also called the cerebral cortex, or neocortex. A pair of large folded sheets of nervous tissue, one on either side of the top of the head of a primate.

Critical exponents. Parameters characterizing the singular behaviour of thermodynamic functions in the vicinity of a critical point.

Critical phenomenon. Divergence of some measured quantity (like relaxation time or correlation length) in the vicinity of a 'critical point' heralding the occurrence of a phase transition.

Crystal. A solid in which atoms are arranged periodically. The periodicity of atomic structure may exist in one, two, or all the three dimensions. Diamond is a crystal, but glass is not.

Cybernetics. A term coined by Norbert Wiener in a book published in 1948. Comparative study of control and communication systems in the animate and the inanimate world, with special reference to feedback mechanisms and to the coupling of various variables. The study of self-organizing or learning machines.

Cyborg. Cybernetic organism. A term coined by NASA for a person whose physiological functioning is not 100% biological but is aided by, or depends on, mechanical or electronic devices.

Cytomimetic. Mimicking the biological cell (as large vesicles or liposomes do in many ways).

Dendrites. The large number of tree-like signal-receiving tentacles which each neuron has.

Deterministic. Relating to a dynamical system the equations of motion of which do not contain any random or stochastic parameters. It has been presumed for long that the time evolution of such a system is uniquely determined by the initial conditions. The fact, however, is that even deterministic laws have fuzziness. There are limits to the precision of computation, as well as of physics. These arise from factors such as: the Heisenberg uncertainty principle; the finite speed of light; the inverse relationship between information and entropy; and the finiteness of the time elapsed since the origin of the universe by the big bang (Davies 2005).

Deterministic algorithm. An algorithm the operation of which is known in advance.

Diamondoid structures. Drexler (1992): Structures that resemble diamond in a broad sense: strong, stiff structures containing dense, three-dimensional networks of covalent bonds, formed chiefly from first and second row atoms with a valence of 3 or more. Many of the most useful diamondoid structures (in nanosystems) are likely to be rich in tetrahedrally coordinated carbon.

Differentiation. Change of a cell to a cell of different type.

Digital organisms. Computer programs that can self-replicate, mutate, and adapt by natural selection.

Dissipative system. A dynamical system that displays irreversibility; e.g. a system with friction. Friction is a form of dissipation. If the phase-space trajectory of a dissipative system is a closed loop, this loop (i.e. the phase-space volume) shrinks with time. Because of this feature of 'negative divergence' in phase space, dissipative systems are typically characterized by the presence of attractors.

Distributed intelligence. In a living system the knowledge distributed over its constituents organizes itself into a whole which is greater than the sum of its parts. This is distributed intelligence; it far exceeds the intelligence of any of the individual constituents.

DNA. Deoxyribonucleic acid. The long molecule that has the genetic information encoded in it as a sequence of four different molecules called nucleotides (adenine (A), thymine (T), guanine (G), and cytosine (C)). There is a double backbone of phosphate and sugar molecules, each carrying a sequence of the bases A, T, G, C. This backbone is coiled into a double helix (like a twisted ladder). In this double-helix structure, base molecule A bonds specifically to T (via a weak hydrogen bond), and G bonds to C. This sequence of base pairs defines the primary structure of DNA.

DNA computing. The manipulation of DNA to solve mathematical problems (cf. Adleman 1998). The technique has the potential of high speed and massive parallelism.

Dopant. A trace impurity deliberately included in the atomic structure of a material to change its properties in a desired way.

Dynamical combinatorial chemistry (DCC). An evolutionary approach for bottom-up nanotechnology. It involves the use of a dynamical combinatorial library

(DCL) of intermediate components which, on the introduction of templates, may produce the desired molecular assembly through molecular recognition.

Dynamical system. A system described by a set of equations from which one can, in principle, predict the future, given the past. It is therefore a deterministic system.

Emergent properties. Properties emerging from the self-organization of the mutually interacting components of a complex system. These properties are not possessed by any individual component of the system, but are a manifestation of collective behaviour. In so-called *strong emergence*, additional laws or 'organizing principles' emerge at different levels of complexity, which must be invoked for explaining the emergent properties. This line of reasoning goes against the grain of reductionism.

Emulsion. A cloudy colloidal system of micron-sized droplets of one immiscible liquid dispersed in another (like oil in water). It is generally formed by vigorous stirring, and is thermodynamically unstable; the sizes of the droplets tend to grow with time. The presence of a surfactant can increase the stability.

Enzymes. Biological catalysts. They speed up the breaking up of large biomolecules into smaller fragments, or the addition of functional groups to biomolecules, or the occurrence of oxidation–reduction reactions. In almost all cases, enzymes are fairly large protein molecules.

Epitaxy. Ordered growth on a crystal plane to form a thin layer of crystalline material with controlled orientation and impurity content.

Ergodic system. A dynamical system in which, given reasonable amount of time, the phase-space trajectory comes arbitrarily close to any point in phase space. For such a system, time averages over dynamical variables are the same as phase-space averages. This can be possible only if the phase-space trajectory covers a finite region densely. Chaotic motion is always ergodic, but the converse need not be true.

Eukaryotic organism. A biological organism the cells of which have nuclei.

Evolution, Darwinian. Selective (i.e. more likely) reproduction of the fittest. Random crossover and mutation, plus nonrandom cumulative selection.

Evolution, Lamarckian. The principle of use and disuse, and the principle of inheritance of acquired characteristics.

Evolutionary robotics. An artificial-evolution approach to robotics. The control system of the robot is represented by an artificial chromosome, subject to the laws of genetics and natural selection.

Expert system. A computer program that is able to perform a task or answer a question requiring human expertise (e.g. medical, legal). Such systems have two kinds of knowledge: Readily available textbook knowledge; and domain-specific heuristics that can be used for problem-solving using that textbook knowledge. Expert systems employ symbolic reasoning and are used in classical AI configurations.

Exponential-time algorithm. An algorithm for which a bound on time complexity cannot be expressed as it can be for a polynomial-time algorithm.

Feedback. A mechanism by which some proportion or function of the output signal of a system is fed back to the input. It is a very important mechanism by which information can stabilize, propagate and evolve in a dynamic system. There is often a *set point* or equilibrium point or desired goal. If the output is, say, greater than the set point, we have a positive *error signal*. On sensing this, if the system decreases the output (i.e. takes *opposite* action) so as to reduce the error signal, we speak of *negative feedback*. This makes the system stabilize around the equilibrium point. If, in a real-world situation, the equilibrium points keep shifting, the system *evolves* with time so as to keep up with the new equilibrium point all the time. *Positive feedback* systems have just the opposite characteristics: 'Them that has, gets'.

Feedback inhibition. A process whereby the product of a metabolic pathway inhibits one or more of the molecules responsible for the synthesis of the product. The effect of negative feedback.

Feedforward loop (FFL). A very common network motif in biological circuits. It can perform tasks such as sign-sensitive delay, sign-sensitive acceleration, and pulse generation. It is essentially a three-node network or graph (X, Y, Z), in which there are directed edges from X to Y, X to Z, and Y to Z.

Fitness. In the context of evolutionary processes, fitness of a member of the population means its relative chances or capability to procreate.

Fixed point of a map. Consider a map $x_{i+1} = F(x_i)$. Its fixed point is the point $x_{i+1} = x_i$.

Flop. The fundamental operation of a logic gate in an electronic circuit.

Fluctuations. A term used in the statistical mechanics of open systems. It refers to the perturbing action of the environment.

Fluctuation–dissipation theorem. The environment not only causes *fluctuations* in an open system, it also introduces *damping*. The theorem relates the fluctuations and the damping.

Foams. Foams or cellular materials consist of an interconnected network of solid material forming a space-filling structure of cells. Such structures occur in Nature in cork, sponge, coral, and bone. Artificial foams can be made from polymers, metals, ceramics, and glass. Use of composites offers further flexibility of design.

Fractal. A self-similar structure; i.e. a structure which is similar to itself (or mathematically identical) at practically all levels of magnification or demagnification. Thus there is no characteristic or intrinsic length-scale or time-scale for a fractal system; the system is scale-invariant. A scale-invariant system must be self-similar, and *vice versa*. Bak (1996) has made the suggestion that the morphology of self-organized-criticality (SOC) processes is fractal.

Fractal dimension. A measure of the ruggedness and space-filling ability of a structure.

Free energy. A measure of the capacity of a system to do work; a reduction in free energy should yield an equivalent quantity of work. In its simplest form, it is defined as the internal energy of a system *minus* the product of its temperature and entropy. According to the second law of thermodynamics, all processes occur so as to minimize the free energy. This can be done either by minimizing the internal energy, or by maximizing the entropy or randomness, or by a combination of these two options.

Frustration. A situation in which there are many ground-state (or equally likely) configurations (as in a spin glass).

Functional materials. Materials distinct from structural materials. The later have mainly a structural or edifice-supporting role. The physical and chemical properties of functional materials (crystals, ceramics, composites, polymers, soft matter) are significantly sensitive functions of temperature, pressure, electric field, magnetic field, mechanical stress, optical wavelength, adsorbed gas molecules, pH value, or any of a host of other environmental parameters. This field dependence can be exploited to achieve a specific purpose. Smart materials are a subset of functional materials.

Functionalism. A theory of the mind according to which it should be possible to extract minds from brains and put them into computers, and once this is done, consciousness can be interpreted as nothing more than the running of a very complex computer program. It does not matter what the hardware happens to be: squishy neurons or hard silicon.

Fuzzy logic. Logic involving a certain amount of inference and intuition. Such logic is used by intelligent systems (like human beings), but not by the older, conventional AI systems. In it, propositions are not required to be either true or false, but may be true or false to different degrees (unlike classical binary logic). An example: If A is (HEAVY 0.8) and B is (HEAVY 0.6), then A is 'MORE HEAVY' than B. By contrast, in classical binary logic, A and B may be members of the same class (both in HEAVY class or both in NOT-HEAVY class) or different classes.

Fuzzy sets. Sets in which the transition from membership to nonmembership is gradual, rather than abrupt. In such sets an element may partly belong to one set and partly to another.

Gel. A material consisting of two components: a liquid and a network of long polymer molecules that hold the liquid in place and so give the gel some solidity.

Gene. A unit of heredity. A specific sequence of DNA or RNA molecules which acts like a code, either by itself or in conjunction with other genes, for the manifestation of a particular characteristic of an organism. A gene specifies the structure of a particular protein. Genes are exchanged during crossover (sexual reproduction). A gene may take on one of several values (or 'alleles'). The gene is transcribed into mRNA; the latter is then translated into the protein. Genes are conserved across species.

Gene expression. The set of interactions among biomolecules involved in transcribing the DNA in a gene and 'translating' the resulting mRNA to make a protein.

Genetic algorithm (GA). An algorithm for controlling the evolution of a program. GAs are based on the Darwinian notion of greater chance of survival of the fittest, leading to evolution by cumulative natural selection. Typically, a randomly generated 'population' of individuals in the form of binary strings is first stored in a computer. An algorithm for the evolution of the population is specified. The artificial or computational equivalents of natural processes such as: variation ('crossover') by mutation and sexual reproduction; inheritance; fitness; competition; and greater chance of survival of the fittest, are modelled in the GA. GAs can result in very efficient and intelligent searches for solutions of a large diversity of computational problems, even in huge phase spaces. Where applicable, they can provide optimum solutions with minimal computational effort. The two most important characteristics of a GA are: use of a fixed-length binary representation; and heavy use of crossover.

Genetic code. The correlation or mapping between the 64 codons and the 20 amino-acid residues (for humans).

Genetic programming (GP). It combines biological metaphors from Darwinian evolution with computer-science approaches of machine learning, enabling automatic generation of computer programs which can adapt themselves for open-ended tasks. The website http://www.aimlearning.com provides additional information, as well as free GP software.

Genome. The total information encoded in all the DNA of an organism. The genome is organized into one or more chromosomes. Humans have a genome that is $\sim 3 \times 10^9$ base pairs in size.

Genotype. Genetic constitution of an organism; the genetic blueprint encoded in its strings of DNA chains. Genotypes correspond to the 'search space', and phenotypes to the 'solution space'.

Glial cell. A cell in the nervous system that is not a nerve cell but performs some supporting function.

Goal-seeking system. A system with a feedback mechanism to achieve or maintain a predetermined or desired state or goal.

Godel's theorem. A theorem on undecidability in mathematics. It states that in all but the simplest mathematical systems there may be propositions that cannot be proved or disproved by any finite mathematical or logical process. The proof of a given proposition may call for an infinitely large number of logical steps (cf. Chaitin 2006, for an update).

Graph. A set of points (vertices) and lines (edges), such that each line can connect exactly two points. A point may be shared by more than one lines. A point may also be connected to itself by a loop.

GTYPE. Generalized genotype, referring to the self-description of a creature in real or virtual world. A largely unordered set of low-level rules for use in artificial-life (AL) computations.

Hamiltonian system. A conservative, autonomous mechanical system with no energy dissipation. The phase trajectory of such a system exhibits 'zero divergence'.

Hebbian rule. A model according to which if neuron A repeatedly participates in firing neuron B, then the efficiency of A for firing B goes up.

Heuristic techniques. Techniques that constrain the search for the solution of a problem. A heuristic is a piece of knowledge that is used for reducing the amount of search, but is not guaranteed to work.

Hippocampus. A part of the brain, so called because its shape resembles a sea horse. Probably involved in the temporary storage or coding of long-term episodic memory.

Homeostasis. The ability of higher animals to maintain an internal constancy (e.g. a constant body temperature).

Host–guest chemistry. A smaller (guest) molecule encased by a larger (host) molecule. A situation of relevance to nanoscale self-assembly. In this branch of chemistry one studies the interactions between two molecules with the aim of either mimicking or blocking an effect caused by interactions among other molecules.

Hydrogel. Water-swollen cross-linked polymeric structure.

Hydrolysis. Splitting of a chemical group by the agency of a water molecule.

Hydrophobic effect. The net effect resulting from the attraction among nonpolar groups when they exist in an aqueous solution.

Hydrophone. A device for detecting sound underwater. It is usually a piezoelectric transducer, which converts the input mechanical vibrations into electrical signals.

Hypothalamus. A small region in the brain, the size of a pea. It secretes hormones and is involved with the control of hunger, thirst, sex, etc.

Hysteresis. The behaviour of a nonlinear system such that an attractor path is not (immediately) reinstated on the reversal of the control sweep.

Infomorphs. Advanced informational entities. A term introduced by Platt (1991). Distributed infomorphs have no permanent bodies, but possess highly sophisticated information-handling capabilities.

Intelligent. Everybody 'knows' what this word means, and that is a problem. The fact is that an intelligent person (or system) should be able to use a memory-based model of the environment to make continuous predictions of future events, and it is this ability to make predictions about the future that is the crux of intelligence. By this definition, only human beings (and some mammals with a cortex) are intelligent, and practically nothing else. Not yet. Hopefully, in the near future this 'real' intelligence will be built by us in machines or structure or systems. When that happens, 'smart structures' will become 'intelligent structures or systems', but only then. Till then we must make a distinction between intelligence as in 'artificial intelligence' (AI), and *real* intelligence as in the human cortical brain.

Invariant. A quantity describing dynamical behaviour such that its numerical value does not depend on the choice of phase-space coordinates.

Learning. Automatic programming. Incorporation of past experience into the neural-network pattern.

Learning, competitive. An essentially nonassociative statistical scheme that employs simple, unsupervised learning rules, so that useful hidden units develop.

Learning, forced. Learning employing an intermediate continuum of learning rules based on the manipulation of the content of the input stimulus stream.

Learning rule. The recipe for altering the synaptical weights in a neural net whenever the net learns something in response to new inputs.

Learning, supervised. A strategy for training of neural nets in which the desired output pattern is also given along with a series of inputs. Such nets are used as associative memories.

Learning, unsupervised. In this approach to the training of neural nets, the correct class to which a given input pattern belongs is not specified. Spontaneous learning.

Linear (or serial) processing. Computation done as a sequence of steps.

Liouville theorem. 'The phase space divergence of a Hamiltonian system is zero'.

Liposomes. Spherical structures comprising a closed lipid shell enclosing an aqueous interior. Also called vesicles.

Lithography. A series of processes for making components by masking, illumination, and etching.

Logic gate. Hardware device for changing the state (flop) in a circuit according to signal input.

Long-range order. Existence of statistical correlations among arbitrarily remote parts of a system.

Mammals. Animals with backbones which possess hair and suckle their young.

Map. An equation of the form $x_{t+1} = F(x_t)$, where the 'time' t is discrete and integer-valued. An example of a map is: $x_{n+1} = \lambda x_n(1 - x_n)$.

Meme. A term introduced by Dawkins to describe the cultural equivalent of the term 'gene'. In his neo-Darwinian theory of evolution of humans from one of the apes, there is a coevolution of the gene pool and the meme pool. A meme can be an idea, a behavioural pattern, a novel and appealing piece of music, or a fundamental approach to mathematics, etc., which seems to jump from one brain to another, as if taking a life of its own. Culture can be regarded as a cooperative cartel of memes. The big strides made by humans in science, art, philosophy, etc. are attributed to the coevolution of memes and genes.

Machine-phase matter. A term introduced by Drexler to describe nanomatter that is distinct from the other, familiar forms of matter (solid, liquid, gas). It would comprise of nanomachinery, i.e. its volume would be filled with active molecular machines and nanomachines.

Magnetoresistance. A property whereby the electrical resistance of a material changes when the magnetic field applied to it is changed.

Mechanosynthesis. Drexler (1992): Chemical synthesis controlled by mechanical systems operating with atomic-scale precision, enabling direct positional selection of reaction sites.

Memory, short-term (creation of). Creation of short-term memory in the brain amounts to a stimulation of the relevant synapse, which is enough to temporarily strengthen or sensitize it to subsequent signals.

Memory, long-term (creation of). (Fields 2005): Creation of long-term memory in the brain begins with the creation of short-term memory, which involves stimulation of a relevant synapse. Strong or repeated stimulation temporarily strengthens the synapse, and, in addition, sends a signal to the nucleus of the neuron to make the memory permanent. The signal was once believed to be *carried* by some unknown molecule. It was supposed that this molecule travels and gets into the nucleus, where it activates a protein called CREB. This protein, in turn, activates some genes selectively, causing them to be transcribed into messenger RNA, which then leaves the nucleus. The instructions carried by the mRNA are translated into the production of synapse-strengthening proteins that diffuse throughout the cell. Only a synapse already temporarily strengthened by the original stimulus is affected by these proteins, and the memory gets imprinted permanently. Recent research has shown that it is unnecessary to postulate that the synapse-to-nucleus signalling is by some hypothetical molecule. Strong stimulation is enough to depolarize the cell membrane, causing the cell to fire action potentials of its own, which, in turn, cause voltage-sensitive calcium channels to open. The calcium influx into the nucleus leads to the activation of enzymes, which activate CREB. The rest of the mechanism is as described above.

Micelles. Thermodynamically stable aggregates of amphiphilic molecules. Micelles can take a variety of shapes, and are examples of self-assembly of molecules, determined mainly by the need to minimize the free energy coming from the hydrophobic interaction.

Modularity. Existence of nearly independent subsystems in a system.

Molecular assembler. See 'assembler'.

Molecular machine. Drexler (1992): A mechanical device that performs a useful function using components of nanometre scale and defined molecular structure.

Molecular manufacture. Drexler (1992): The production of complex structures via nonbiological mechanosynthesis (and subsequent assembly operations).

Moore's law. Gordon Moore predicted that the number of circuit elements that could be placed on a silicon chip would double every $\sim$18 months. A more detailed statement of the law is as follows: Every three years, (i) device size would reduce by 33% ; (ii) chip size would increase by 50% ; and (iii) the number of circuit elements on a chip would quadruple. The estimated numbers are not invariant with time. Currently, the surface area of an IC chip is getting halved every 12 months.

Motes. Small, low-cost, smart, wireless sensor systems, with only moderate computing abilities, which can communicate with one another over short distances, and form autonomous networks for monitoring a variety of phenomena.

Motor cortex. The parts of the cerebral cortex concerned mainly with the planning and execution of movements.

mRNA. Messenger RNA. A form of RNA that carries information (codons) transcribed from DNA for the primary structure of the peptide it can make. The mRNA is read by ribosomes for assembling a protein according to the mRNA sequence. mRNA is similar to DNA, except that thymine (T) is replaced by uracil (U).

Mutation. A permanent change in the primary structure of DNA.

Nanobot. Nanosized robot.

Nanocomposite. A multiphase material where one or more of the phases have at least one dimension of order 100 nm or less.

Nanoelectronics. The emerging science and technology of using nanometre-scale electronic components (transistors, diodes, relays, logic gates) made from organic molecules, nanotubes and semiconductor nanowires. In present-day *micro*electronics, the basic building block is normally the semiconductor-based transistor, which acts as an amplifier of signals, and also as a switch for turning electric current on or off. In nanoelectronics, organic molecules or nanotubes may be used for this purpose. In microelectronic circuitry, metallic wires link the transistors for performing the various logical or arithmetic operations. In nanoelectronics the connecting wires will have to be nanotubes or other materials only a nanometre or two in thickness.

Nanomachine. A system in the nanometre regime, with moving parts.

Nanostructure science and technology. (M. C. Rocco, in Stix 2001): This emerging field is defined in a restrictive sense to cover materials and systems which have at least one dimension of $\sim$1–100 nm, are designed through processes that exhibit fundamental control over the physical and chemical attributes of molecular-scale structures, and which can be combined to form larger structures.

Nanotube. The two well-known forms of elemental carbon are diamond and graphite. Graphite comprises of two-dimensional sheets of carbon atoms, arranged as repetitive hexagonal loops, with each arm of a hexagon shared by two adjoining loops. Processes are possible by which these hexagonal layers role over and fuse at the edges to form tubes of carbon. These are called nanotubes because their diameter is $\sim$1.4 nm. In nanoelectronics, such tubes can function both as connection wires and as tiny transistors.

Natural selection. Any natural limitation that favours the survival and reproduction of individuals in a population who possess a specific hereditary character over those who do not possess it.

Near-field scanning optical microscope (NSOM). One of the many kinds of scanning-probe microscopes (SPMs) (some other examples being the STM and the AFM). The difference in the various types is the property scanned from point to point on the surface being investigated. The NSOM uses visible light for this topographical imaging, so it can map transmittance and fluorescence, etc.

Network motifs. Patterns of interconnections occurring in complex networks in numbers significantly larger than those in randomized networks.

Neural network. An artificial neural network is a computational configuration designed to learn by emulating the neural network of the human brain in a rudimentary manner. Although fairly successful, the existing models have so far failed to develop long-term memory.

Neuron. Nerve cell.

Noncovalent interactions. Interactions among atoms, molecules, or ions, not involving a sharing of valence electrons.

Nondeterministic algorithm. A guess-and-check algorithm. If the guessed solution is not correct, guess and check another solution.

NP-class problem. NP stands for 'nondeterministic polynomial'. NP-class problems are those for which a nondeterministic algorithm exists that can verify the solution of the problem in time that increases in a *polynomial* manner (rather than, say, exponential manner) with the increasing size of the problem. Roughly speaking, a problem is of complexity NP if one can guess a solution of the problem and then verify its correctness in polynomial time. For an NP problem, a proposed solution can be *verified* 'quickly'. By contrast, a P problem is that which can be *solved* 'quickly' even without being given a solution. It is widely believed that $P \neq NP$.

NP-complete. A class of problems such that, for a particular NP-complete problem, every instance of every problem in NP can be converted to an instance of this particular problem, and this conversion can be effected in polynomial time. No effective computer algorithms have been developed for *solving* such problems. The only known approach to solving them requires an amount of computational time proportional to an exponential function of the size of the problem, or to a polynomial function on a nondeterministic computer that, in effect, guesses the correct answer. A fast solution of any NP-complete problem (if such a solution exists) can be translated into a fast solution of any other.

NRAM. Nanotube random-access memory. Such memories are expected to be radiation-resistant, apart from offering a host of other anticipated advantages.

Nucleic acid. A chain-like molecule, found mostly in the nucleus of the biological cell. There are two types: DNA and RNA. The sequence of 'nucleotides' along the length of the chain determines the hereditary characteristics, passed on by parents to offspring.

Oligonucleotide. A molecule made up from two to ten nucleotide residues.

Oligopeptide. A molecule made up from two to ten amino acid residues.

Open system. A system interacting with an environment, such that its evolution with time is not governed by deterministic equations of motion.

Order parameter. A thermodynamic quantity identified by Landau that determines the critical behaviour of a system which undergoes a phase transition.

Organelle. Any organized or specialized structure that forms part of a biological cell.

P-class problem. For a problem (say of size n) belonging to this class of complexity, there exists a deterministic algorithm that solves it in polynomial time n^j, where j is an integer. See 'NP-class problem' for more information.

Parallel processing. Computation in which several steps are performed simultaneously.

PDP models. Parallel-distributed-processing models. The other terms used for these are: neural nets; connectionist models;neuromorphic systems. They comprise of a large number of *processing elements* (PEs), each interacting with others via excitatory and inhibitory connections. The fact that these are large systems, and that the interconnectedness is primarily local, results in fault-tolerance. Learning or training amounts to the adjustment of the degree of interconnectedness or *weights* associated with the connections.

Peptide bond. The amide bond that connects amino acids comprising a peptide or protein.

Perceptron. A neural network that undergoes supervised learning. The supervision is meant to achieve some kind of specialization. For example, a *photo-perceptron* has optical signals as stimuli. It was originally designed to recognize letters and other patterns placed in front of its 'eyes'.

Phase of a material. A piece of matter consisting of atoms or molecules organized in a specific geometric structure.

Phase space. A space spanned by the positions and momenta of the particles in a system. The state of the system at a particular time is represented by a point with specific coordinates in this space (the *representative point*). This point moves with time, and traces a *phase-space trajectory*, which may be a closed curve, or an open curve.

Phase transition. A system, e.g. a crystalline material, tends to adopt the most stable configuration or phase (positions of atoms, etc.) for a given temperature, pressure, or any other control parameter. For a different set of control parameters, a different phase may be more stable. The system is then said to have made a transition from one phase to the other.

Phenotype. The characteristics manifested by an organism; the structure created by it from the instructions in its genotype. The organism itself. Genotypes correspond to the 'search space', and phenotypes to the 'solution space'.

Physical fundamentalism. The dogma that only physical science deserves the title of true knowledge, and that although other belief systems may have utility for the groups that practise them, ultimately they are just made-up stories.

Pixels. Picture elements.

Planar technology. The technology for making IC chips, invented in 1957 by J. Hoerni. It is a package of steps for layering, and for etching metals and chemicals in silicon wafers, using templates called photomasks. This landmark approach made possible the production of IC chips cleanly and with a very high degree of reproducibility. It also enabled the creation of a huge variety of circuit types (at low cost) just by changing the pattern on the photomask. It introduced modular design in chip technology, because engineers could now draw from libraries of simple circuits already designed by others, and assemble extremely complex and sophisticated electronic circuits at low cost.

Poling. A polycrystalline ceramic is a mass of grains, fused together by heat treatment. Each grain is a tiny crystal ('crystallite'), and these crystallites are oriented randomly. If the material is a ferroic material (or rather, if the material is in a ferroic phase), each crystallite or grain of the ceramic can have domains, with domain walls separating the domains (provided the grains are not so small that surface effects prevent the occurrence of domain walls, and then the whole grain is just one single domain). Note that these domain boundaries are within each grain, and are distinct from grain boundaries, which separate grains from their neighbouring grains. To be specific, let us assume that the material is in a ferroelectric phase. Then the domains are ferroelectric domains. Since the grains are oriented randomly, the net dipole moment of the specimen is zero, or nearly zero, in spite of the fact that, locally, each unit cell of the polycrystal has a nonzero spontaneous polarization. But we can take advantage of the fact that the domains of a ferroelectric can be switched to other equivalent orientations by applying a sufficiently large electric field. Thus we can make one domain grow at the cost of another by applying this field (typically a few kV per mm). This process is called poling. After poling, the specimen acquires a nonzero electric dipole moment. Similarly, if we are dealing with a ferroelastic material, we can use uniaxial stress to achieve mechanical poling. Ditto for magnetic poling. The general fact is that any kind of poling becomes possible only if there are more than one orientation states (or domain states) available to the material, so that the poling field can do the switching from one state to another. This is possible only when a material is in a ferroic phase. Therefore, *only ferroic materials can be poled*.

Polymer. A large molecule formed by the chemical bonding of a large number of identical molecular units or *monomers*.

Polynomial-time algorithm. An algorithm for which the time complexity function is $O(p(n))$, where p is some polynomial function of input size n.

Population. A group of organisms occupying a specific area.

Post-humanism. A conscious attempt at altering the route and destination of human evolution. The aim is to use genetics and technology for having better bodies and brains.

Primary structure. The sequence of amino-acid residues (in proteins), or DNA/RNA residues (in polynucleotides).

Primates. Animals of the highest order of mammals, which include humans, apes, and monkeys.

Programming complexity. In a neural network (NN), data and instructions are not separable (unlike in a von Neumann computer). Programming complexity for an NN is the number of operations that must be performed by the optimum NN pattern for the solution of a problem.

Prokaryotic organisms. Organisms whose cells do not have nuclei.

Protein-folding prediction. A protein molecule consists of a sequence of amino acids. It has a primary, a secondary, and a tertiary structure. If we know the linear sequence of amino acids in a protein, we know its primary structure. This linear sequence of amino acids organizes itself into several secondary motifs like α-helices, β-sheets, connected by short loops. Further, the secondary structural motifs self-organize and fold into one or more compact three-dimensional globular domains; this is the tertiary structure. Prediction of the three-dimensional folded conformation of a naturally occurring protein molecule, given only its amino-acid sequence (and no knowledge of the folding of a similar sequence) has turned out to be a highly nontrivial, almost intractable, task. Drexler (1981, 1992) has argued that *de novo* protein-folding *design* should be easier than protein-folding *prediction*.

PTYPE. Generalization of the notion of phenotype in biology. It refers to the set of structures and behaviours that emerges from the implementation of the GTYPE in a virtual environment.

PZT. Abbreviation for lead zirconate titanate. A well-known piezoelectric material, commonly used in polycrystalline form.

Recombinant DNA. A molecule having DNA from two or more sources.

Reductionism. The philosophy that complex phenomena can be explained, at least in principle, in terms of their simplest elements. The basic premise is that every-thing can be *ultimately* explained in terms of the bottom-level laws of physics. This assumes the availability of *unlimited* time and computing power, and is an unrealistic assumption.

Released reality. Computer simulations in which real-world constraints (like the inability to reverse time, or to escape from the law of gravitation) are released or abandoned.

Replication. Duplication. In particular, use of DNA as a template for making new DNA molecules.

Residue. The part of a molecule that remains when the molecule has been incorporated into a larger molecule.

'Resist' material. Usually a polymer, employed in lithography. On exposure to photons or energetic particles like electrons or ions, the irradiated areas of the resist undergo structural or chemical modifications (e.g. cross-linking of chains in the polymer), such that their solubility to a solvent becomes different from that of unexposed areas. This is exploited for imprinting a pattern on a substrate.

Resonance (mechanical). Every solid object has a 'natural frequency' at which it has a propensity to oscillate. This frequency is determined by the shape and size of the solid and by its rigidity. If external vibrations impinging on the solid happen to be of the same frequency, the object 'resonates', which means that its oscillations are amplified, rather than reduced.

Ribosome. A complex comprising rRNA and proteins, where protein synthesis occurs.

RNA polymerase (RNAp). An enzyme (a complex of several proteins) that transcribes DNA for the formation of RNA.

Robust. A robust system or phenomenon is that the fundamental characteristics of which are not very sensitive to small external influences or perturbations.

rRNA. Ribosomal RNA. A type of RNA that combines with proteins to form ribosomes.

Scanning probe microscopy (SPM). A whole class of techniques in which a sharp probe tip is scanned across the specimen, and the interaction between the two is analysed to obtain nanometre-scale information about the surface topography of the specimen. A variety of probes can be used.

Scanning tunnelling microscope (STM). The first SPM, invented in 1986 by Binnig and Rohrer. In it, information about the topography of a conducting surface is obtained by applying a voltage to the probing tip and measuring the tunnelling current as a function of position. This current is a measure of the vertical distance between the tip and the surface, and therefore provides information about the topology at a nanometre scale. It also provides information about the electronic structure of the surface.

Search. Exploration of a parameter space or hypothesis space or problem space, according to well-defined rules.

Search space. The parameter space that must be explored by a problem-solving system for arriving at an optimum solution.

Secondary structure. In proteins, it is the folding of the chain of amino-acid residues. It involves the formation of hydrogen bonds between amide N–H groups and amide C=O groups from different parts of the polypeptide chain. In polynucleotides, it is the helix formed by the interactions between two strands of DNA.

Self-assembly. Autonomous organization of components into patterns or structures without human intervention. Spontaneous assembly of smaller subunits into higher ordered aggregates. The components bounce around in a fluid medium, get stuck and unstuck, until stable, least-free-energy, aggregates get formed.

Self-organization. 'Spontaneous but information-directed generation of organized functional structures in equilibrium conditions' (Lehn 2002).

Self-organized criticality (SOC). The fact (discovered by Per Bak) that large dynamical systems tend to drive themselves to a robust critical state with no characteristic spatial or temporal scales.

Sensory transcription networks. Transcription networks in the cell that respond to internal or external signals (like the presence of nutrients or stress) and lead to changes in gene expression.

Smart sensor. A sensor (often a microsensor) with integrated microelectronic circuitry.

Soft computing. Computing with allowance for the possibility of error and randomness. Computing involving a combined use of fuzzy logic, neural networks, and evolutionary (or genetic) computing. It adds learning capabilities to fuzzy systems. The methodologies used in soft computing are tolerant of imprecision, uncertainty, and partial truth. Where applicable, such an approach serves to achieve better performance, higher autonomy, greater tractability, lower cost of solution of problems, and a better rapport with reality. There is a hybridization of fuzzy logic with neural networks and/or evolutionary computation. The term 'soft computing' was coined by Zadeh, who described it as a 'consortium of methodologies which, either singly or in combination, serve to provide effective tools for the development of intelligent systems'.

Soliton. Solitary wave; a spatially localized wave travelling with a constant speed and shape. Generally a solution of some particular nonlinear wave equations.

Soma. The scientific word for cell body. Here is an example of a more general usage of this term: 'The ant colony is a superorganism differentiated into "germ plasm" (queens and males) and "soma" (workers)'.

Spin glass. (Mydosh 1993): A random, mixed-interacting, magnetic system characterized by a random, yet cooperative, freezing of spins at a well-defined temperature T_f below which a highly irreversible, metastable frozen state occurs without the usual long-range spatial magnetic order.

Spintronics. An electron not only has a charge, but also a magnetic spin which can take two alternative values. In conventional electronics only the charge is used in the design of circuits, and the spin is ignored. Spintronics aims at exploiting the spin also in electron and hole transport. It gives rise to potential new applications in electronics and quantum computing.

Stochastic. Nondeterministic but statistically well-characterized. In a stochastic system, the evolution in time is not uniquely determined, but, nonetheless, definite probabilities can be assigned to the different possibilities.

Stochastic algorithm. A probabilistic algorithm.

Strange attractor. An attractor with fractal (nonintegral) dimensions.

Superelasticity. The extremely compliant and yet *reversible* behaviour exhibited by a material in a certain temperature range around a martensitic phase transition. Usually such a material also exhibits the shape-memory effect. The effective compliance becomes very large (or the stiffness becomes very small) because the mechanical energy given to the material in the form of applied stress is used for transforming the material to the martensitic phase (with the resultant large change of macroscopic strain), rather than for producing only the normally expected small amount of induced strain in the original phase.

Superelectrostriction. A term introduced in this book by analogy with 'super-elasticity'. Typical electrostrictive response, in the form of strain proportional to the square of a small applied electric field, is rather small. But if the electric field causes domain-wall movement and/or a phase transition (e.g. from the rhombohedral phase to the tetragonal phase in the case of the relaxor ferroelectric PZN-PT(95.5/4.5)), an anomalously large strain can result. This anomalously large electrostrictive response is superelectrostriction.

Superpiezoelectricity. A term introduced in this book by analogy with 'supere-lasticity'. Typical piezoelectric response, in the form of strain produced by a small applied electric field, is rather small. But if the applied electric field causes domain-wall movement and/or a phase transition (e.g. from an antiferroelectric phase to a ferroelectric phase in the case of Pb(Zr, Ti, Sn)O_3), an anomalously large strain can result. This anomalously large piezoelectric response is superpiezo-electricity.

Supramolecular polymers. Monomeric units held together by specific directional secondary interactions.

Supramolecular structures. Large molecules formed by the mainly noncovalent grouping or bonding of smaller molecules.

Surfactants. Surface-active agents. Usually, amphiphilic compounds. They get read-ily adsorbed at oil–water or air–water interfaces, and reduce the surface tension there. Their amphiphilic nature makes them preferentially segregate to surfaces, where they can be active, like in detergents.

Swarm intelligence. A problem-solving capability emerging from the interactions of simple information-processing units (e.g. bees in a beehive).

Symbiogenesis. An adjunct to Darwinism, ascribing the complexity of living organ-isms to a succession of symbiotic associations between simpler living forms. A case of cooperation, rather than competition, between species. The forging of coalitions leading to higher levels of complexity.

Symbolic reasoning. A mathematical approach in which ideas and concepts are represented by symbols like words, phrases and sentences, which are then processed according to the logic fed into the computing machine.

Symmetric synaptic connection. A situation in which the synaptic strength and sign of the synapse from neuron i to neuron j is the same as that of another synapse from neuron j to neuron i.

Synapse. The region of connection between two neurons. A neuron in the human cortex makes five to ten thousand synapses with other neurons.

Synaptic weight. A parameter (with both a magnitude and a sign) that characterizes the nature (i.e. excitatory or inhibitory) and extent of influence one neuron has on another with which it has formed a synapse.

Semantic consistency and completeness. A formal system is semantically consistent, under a particular external interpretation, if and only if it proves only true statements. It is semantically complete if all true statements can be proved.

Syntactic consistency and completeness. A formal system is syntactically, or internally, consistent if and only if the system never proves both a statement and its negation. It is syntactically complete if one or the other is always proved.

Templated self-assembly. Spontaneous aggregation of components into a structure induced by a secondary ordering entity (Choi *et al.* 2000).

Thalamus. An important region of the forebrain. It is the gateway to the cortex, since all the senses (excepting smell) must relay through it to get to the cortex.

Thermal fluctuations. Statistical deviations about the mean value of the thermal energy of a system.

Transcription. The act of using a part of the primary structure of DNA (gene) as a template for producing RNA.

Transcription factor. The transcription rate of genes is regulated by specific proteins called transcription factors, which are usually in two states, namely active and inactive. These proteins rapidly switch between the two states, typically on microsecond time-scales. When active, the protein binds specific sites on DNA to influence the transcription rate.

Transcription network. A graph or network in which the nodes are genes, and the edges are transcriptional interactions. For example, if there are two nodes X and Y, and there is an edge directed from X to Y, it means that the protein encoded by gene X is a transcription factor that regulates gene Y.

Transducer. A device or medium that converts mechanical energy into electrical energy, or *vice versa*.

Transition elements. Those elements in the periodic table in which the filling of preceding subshells (3d, 4d, 5d, or 5f) is completed *after* the partial filling of subsequent subshells. This happens because states with the minimum possible energy are filled first, and then states with higher and higher energies. The first transition element is Sc (Z = 21), for which the 4s shell already has two electrons and the 3d shell has just got its first electron (electronic configuration $1s^2 2s^2 2p^6 3d^1 4s^2$). The next transition element, Ti, has two electrons in the 3d shell ($3d^2$), and so on. The transition elements can be conveniently grouped in terms of the partially filled penultimate subshell: 3d[Sc(21), Ti(22), V(23), Cr(24), Mn(25), Fe(26), Co(27), Ni(28), Cu(29), Zn(30)]; 4d[Y(39), Zr(40), Nb(41), Mo(42), Tc(43), Ru(44), Rh(45), Pd(46), Ag(47), Cd(48)]; 5d[*Lanthanides*: La(57), Hf(72), Ta(73), W(74), Re(75), Os(76), Ir(77), Pt(78), Au(79), Hg(80)]; 6d[*Actinides*: Ac(89), Rf(104), Db(105), Sg(106), Bh(107), Hs(108), Mt(109), Uun(110), Uuu(111), Uub(112)]; 4f[Ce(58), Pr(59), Nd(60), Pm(61), Sm(62), Eu(63), Gd(64), Tb(65), Dy(66), Ho(67), Er(68), Tm(69), Yb(70), Lu(71)]; 5f[Th(90), Pa(91), U(92), Np(93), Pu(94), Am(95), Cm(96), Bk(97), Cf(98), Es(99), Fm(100), Md(101), No(102), Lr(103)].

Translation. Biosynthesis of a protein by the use of the primary structure of mRNA as a template.

Travelling-salesman problem: What is the shortest distance a salesman must travel to visit a given number, N, of cities, visiting each city only once, and return to the city of origin? This NP-complete problem becomes highly nontrivial when N is large, and is the archetypal example of *combinatorial optimization*. For given locations of the N cities, one has to determine the least possible *cost function*, which in this case is the total length of the 'closed tour'.

tRNA. Transfer RNA. A type of RNA that transports amino-acid residues to the site of protein synthesis.

Turing machine. A universal computer, in the sense that it can simulate all other general-purpose computers.

Turing test. A test designed to check whether a computer can think and act 'intelligently' the way a human can.

Universal computer. A general-purpose programmable computer. Such mass-produced computers are neither too slow nor too fast for their memory capacity. One megabyte per MIPS (million instructions per second) is a good, optimum combination for a universal computer. *Special-purpose computers* deviate from this optimum ratio because they are specially designed (or evolved) for special tasks. The human brain is an example of a special-purpose computer evolved by Nature for ensuring the survival of our early ancestors, who had to develop strong abilities for recognition and navigation for this purpose. Ability for fast number-crunching was not among the skills needed for survival. That is why our brains are no match for the speed with which universal computers can perform this task.

Universal robots. Mass-produced, general-purpose, autonomous robots (yet to be developed). It is expected that such 'brainy' machines will act on our behalf in the real world as 'literal-minded' slaves (in the near future), just as universal computers are doing at present in the world of data.

VCL. Virtual combinatorial library. A potential library made up of all possible combinations in number and nature of the available components.

Vesicles. See 'liposomes'.

Virtual reality (VR). Modelling of an existing real environment, and visualizing it in 3D.

Virtual worlds (VW). Heudin (1999): The study of computer programs that implement digital worlds as wholes with their own 'physical' and 'biological' laws. Whereas VR has largely concerned itself with the design of 3D graphical spaces, and AL with the simulation of living organisms, VW is concerned with the simulation of worlds considered as wholes and with the synthesis of digital universes.

Virus. A small genome wrapped in a tough protein shell, and capable of replication only inside a plant, animal, or fungal cell.

Viscoelastic. Time-dependent response of a material to applied stress. Usually there is an elastic response in the beginning (as for a solid), after which the material starts to flow like a liquid.

Volatile memory. Memory that can 'evaporate'. Typically, information in a computer memory is lost when the power is switched off. Because of leakage processes, even when the power is on, the memory has to be 'refreshed' every 0.1 second or so.

Von Neumann machine. A computer in which the program resides in the same storage as the data used by that program. Practically all modern computers are of this type.

Weight, synaptical. The nature and strength of an inter-unit connection in a neural net. The 'nature' can be excitatory (positive weight) or inhibitory (negative weight). The 'strength' refers to the degree of influence that the unit from which the inter-connection begins (the *presynaptic neuron*) has on the incident unit (the *postsynaptic neuron*).

References

Arecchi, F. T. and A. Farini (1996). *Lexicon of Complexity*. Firenze (Florence): Istituto Nazionale di Ottica.

Bak, P. (1996). *How Nature Works: The Science of Self-Organized Criticality*. New York: Springer.

Binnig, G. and H. Rhorer (August 1985). 'The scanning tunnelling microscope'. *Scientific American,* 253(2): 40.

Chaitin, G. (March 2006). 'The limits of reason'. *Scientific American,* 294: 74.

Choi, I. S., M. Weck, B. Xu, N. L. Jeon and G. M. Whitesides (2000). 'Mesoscopic, templated self-assembly at the fluid–fluid interface'. *Langmuir*, 16: 2997.

Crick, F. (1994). *The Astonishing Hypothesis: The Scientific Search for the Soul*. London: Simon and Schuster.

Davies, P. (5 March 2005). 'The sum of the parts'. *New Scientist*, 185: 34.

Drexler, K. E. (1981). 'Molecular engineering: An approach to the development of general capabilities for molecular manipulation'. *Proc. Natl. Acad. Sci. USA*, 78: 5275.

Drexler, K. E. (1992). *Nanosystems: Molecular Machinery, Manufacturing and Computation*. New York: Wiley.

Fields, R. D. (February 2005). 'Making memories stick'. *Scientific American,* 292(2): 58.

Heudin, J.-C. (1999) (ed.). *Virtual Worlds: Synthetic Universes, Digital Life, and Complexity*. Reading, Massachusetts: Perseus Books.

Lehn, J.-M. (16 April 2002). 'Toward complex matter: Supramolecular chemistry and self-organization'. *PNA, USA*, 99(8): 4763.

Mydosh, J. A. (1993). *Spin Glasses: An Experimental Introduction*. London: Taylor and Francis.

Platt, C. (1991). *The Silicon Man*. Bantam Spectra Books.

Stix, G. (September 2001). 'Little big science'. *Scientific American,* 285: 26.

INDEX

transcription factors 60, 322
transcription networks (TNs) in a cell 60, 61, 322
transducer 322
transfer function 49
transition metal oxides (TMOs) 100
transitions in composites 275
translation (in a biological cell) 323
translational symmetry of a crystal 277
travelling-salesman problem (TSP) 42, 51, 183, 323
tree-structure approach to algorithms 53
triamantane 151
tRNA (transfer RNA) 268, 323
truly intelligent structures 19
travelling-salesman problem (TSP) 42, 51, 183, 323
tunnelling magnetoresistance (TMR) 98, 223
Turing Machine 5, 28, 53, 323
Turing test 5, 29, 236, 323
tweed patterns associated with ferroelastic phase transitions 88
two-way shape-memory effect (TWSME) 216, 221

undecidability theorem of computer science 54
undecidable behaviour (in computation science) 143
undifferentiated cells 58
unexpected bugs arising from the aggregation of bugless parts 41
unit cell of a crystal 277
universal computers 28, 53, 244, 246, 323
universality classes (in phase transitions) 80
universal law for vivisystems 40
universality classes for cellular automata 142
universal robots 323
unstable periodic orbits 271
unsupervised learning 48, 242
UV-moulding (UVM) 198–199

vants (virtual ants) 42
variants (or domain types) 81

variety (importance of) in evolution 53
varistor 11
VCL (virtual combinatorial library) 132, 323
very smart structures/materials 4, 18, 102, 141, 221
vesicles (or liposomes) 140–141, 312
vibration control in smart structures 104, 225, 258
vibronic character of wave function 99
virtual ants (vants) 42
virtual combinatorial library (VCL) 132, 323
virtual reality (VR) 323
virtual worlds (VWs) 323
virus 323
viruses (in PCs) 41, 202
viscoelastic material 125, 324
vivisystems 37, 200, 221, 238, 240, 262
Vogel–Fulcher equation 281
volatile memory 324
von Neumann machine 324
von Neumann's model for computers 44

wafer bonding 197
weather-brain 239
webby nonlinear causality 38
weight, synaptical 324
Why is life full of surprises? 54
Widrow–Hoff learning rule 50
wise structures 19, 239
world-view of a machine-brain 238, 251

X-ray lithography 162, 197

Y-Ba-Cu-O 107, 159

Z-DNA 167, 169
zero-field-cooling (ZFC) 96
zirconia (ZrO_2) 107, 160
ZnO 150
zygote 269–270

DISCARDED
CONCORDIA UNIV. LIBRARY

CONCORDIA UNIVERSITY LIBRARIES
MONTREAL